Conceptual Boundary Layer Meteorology

Conceptual Boundary Layer Meteorology

The Air Near Here

Edited by

April L. Hiscox
Associate Professor of Geography, University of South Carolina, Columbia, South Carolina, United States

ACADEMIC PRESS
An imprint of Elsevier

ELSEVIER

Academic Press is an imprint of Elsevier
125 London Wall, London EC2Y 5AS, United Kingdom
525 B Street, Suite 1650, San Diego, CA 92101, United States
50 Hampshire Street, 5th Floor, Cambridge, MA 02139, United States
The Boulevard, Langford Lane, Kidlington, Oxford OX5 1GB, United Kingdom

Notices
Knowledge and best practice in this field are constantly changing. As new research and experience
broaden our understanding, changes in research methods, professional practices, or medical treatment
may become necessary.

Practitioners and researchers must always rely on their own experience and knowledge in evaluating and
using any information, methods, compounds, or experiments described herein. In using such information
or methods they should be mindful of their own safety and the safety of others, including parties for whom
they have a professional responsibility.

To the fullest extent of the law, neither the Publisher nor the authors, contributors, or editors, assume any
liability for any injury and/or damage to persons or property as a matter of products liability, negligence or
otherwise, or from any use or operation of any methods, products, instructions, or ideas contained in the
material herein.

ISBN 978-0-12-817092-2

For information on all Academic Press publications
visit our website at https://www.elsevier.com/books-and-journals

Publisher: Candice Janco
Acquisitions Editor: Jennette McClain
Editorial Project Manager: Lena Sparks
Production Project Manager: Prem Kumar Kaliamoorthi
Cover Designer: Christian Bilbow

Typeset by STRAIVE, India

Working together
to grow libraries in
developing countries

www.elsevier.com • www.bookaid.org

Contents

Contributors

Wayne M. Angevine Cooperative Institute for Research in Environmental Sciences, University of Colorado, and NOAA Chemical Sciences Laboratory, Boulder, CO, United States

Robert S. Arthur Lawrence Livermore National Laboratory, Livermore, CA, United States

Marc Aubinet TERRA Teaching and Research Centre, University of Liege, Gembloux, Belgium

Kodi L. Berry NOAA National Severe Storms Laboratory, Norman, OK, United States

Gil Bohrer Department of Civil, Environmental and Geodetic Engineering, The Ohio State University, Columbus, OH, United States

Brian Butterworth Cooperative Institute for Research in Environmental Sciences, University of Colorado-Boulder; NOAA Physical Sciences Laboratory, Boulder, CO, United States

Ankur R. Desai University of Wisconsin-Madison, Madison, WI, United States

Peter K. Hall Avangrid Renewables, Portland, OR, United States

April L. Hiscox Department of Geography, University of South Carolina, Columbia, SC, United States

Eric Kutter School of Earth and Environmental Sciences, Queens College of the City University of New York, New York, United States

Alexandria G. McCombs Department of Geography and Geosciences, Salisbury University, Salisbury, MD, United States

Russell K. Monson Department of Ecology and Evolutionary Biology, University of Colorado, Boulder, CO, United States

Carmen J. Nappo CJN Research Meteorology, Oak Ridge, TN, United States

Sreenath Paleri University of Wisconsin-Madison, Madison, WI, United States

Brian Viner Atmospheric Technologies Group, Savannah River National Laboratory, Aiken, SC, United States

Theresia Yazbeck Department of Civil, Environmental and Geodetic Engineering, The Ohio State University, Columbus, OH, United States

Chuixiang Yi School of Earth and Environmental Sciences, Queens College of the City University of New York, New York, United States

Acknowledgments

If you are holding this book, I suspect you are a little like me. I did not set out to be a boundary layer meteorologist, a microclimatologist, a geographer, or any of the other titles that could apply to my professional life. I came to this field in a very circuitous way. At one point in my career, I thought I would be designing cell phone systems. But, through a series of luck and persistence I found the boundary layer. This beautiful yet mysterious part of the atmosphere captured my brain in a way nothing else had. It has continued to hold my interest for years. This is the book I needed when I first started working in the boundary layer. To me there is nothing more important than the air that we breathe. It is essential to being human, to breathe. So, to you the reader, I acknowledge your curiosity and desire to learn. I hope within these pages you find the spark to ignite your own explorations of the air near here.

No book is a singular endeavor and there are some specific people who have made what started as an idea come to be the book you are now holding.

I would like to thank Dr. Alexandria McCombs for her help in the initial stages of this book. She is responsible for some of the clever chapter titles, general structure, and choice of authors. While her life path kept her from being a co-editor, she is a co-creator.

I would like to thank all the authors of this book. They took on the challenge of explaining complex concepts in an accessible way and the results are better than I could have hoped. I would also like to specifically acknowledge Dr. Ankur Desai, Dr. Gil Bohrer, and Dr. Alison Steiner for many conversations at various conferences around the country regarding the struggles of teaching boundary layer meteorology to those without meteorology backgrounds. Those conversations reinforced this idea and ultimately inspired me to make it happen and see it through. I cannot go without thanking Dr. David R. Miller for taking me on as a student all those years ago and teaching me all about the boundary layer without the help of a book like this! To him I am forever grateful.

Finally, Dr. Mark Macauda, my life partner, constant cheerleader, and patient proofreader. We all owe him a debt of gratitude, for without him this book would never have been finished.

Working in the in-between: Defining the boundary layer

April Hiscox, PhD
Department of Geography, University of South Carolina, Columbia, SC, United States

1.1 What is this book?

It seems every book, chapter, or lecture about the atmospheric boundary layer starts nearly the same way: with a figure of the boundary layer, its various sublayers, and the general changes in its structure over a 24-h period. I, myself, have shown and drawn this same figure countless times in various classes and presentations. I even once drew it in the sand on the beach, trying to explain to a friend exactly what it is I study. It is a quick and straightforward way to describe the extreme complexity and variability of the boundary layer. As such, it would be the natural way to start this text, which is rooted in explaining complexity in an accessible way. However, this book is *also* meant to be different from the traditional approach, so you won't find that figure here.[1] Instead, the first six chapters of this book will lead you to be able to draw that figure yourself, by treating each of those sublayers and complex conditions as separate concepts. The remaining chapters will address more specific and practical problems in the boundary layer.

This book is not a replacement for any of the comprehensive boundary layer textbooks currently available (Lee, 2018; Foken, 2017; Stull, 2001; Garrett, 1992). I have dog-eared and well-worn copies of all of these on my own shelf and continue to refer to them often. Those texts, however, all have one thing in common: the authors assume their readers possess a background in calculus and physics, as well as an understanding of the whole atmospheric system. In other words, they are written for advanced students of atmospheric science or meteorology. Increasingly, those who need to understand the boundary layer do not have that comprehensive background. They are people working in government agencies (local, regional, or national). They are social scientists relating physical conditions to human behavior. They are individuals looking for new and creative solutions to climate problems. They are air quality regulators, citizen scientists, and environmental activists, who need an understanding of science to communicate it to the lay person. The exchanges between the land and air surface contribute to some of the greatest social and environmental issues (i.e., climate change, air pollution) facing society today. Thus, there is a great need for understanding the boundary layer outside of the strictly academic setting. Maybe even more importantly, there is a need for understanding the complexity of the air that surrounds us.

---

[1] If you really are a traditionalist and just want a visual of the boundary layer skip ahead to the end of Chapter 6.

Conceptual Boundary Layer Meteorology. https://doi.org/10.1016/B978-0-12-817092-2.00008-4

That is where this book comes in – it offers a conceptual view of what is going on and why it matters. The writing is meant to be less formal and more accessible. It is also meant to be welcoming. You, reader, are the target audience for this book. Whether you want to do graduate-level research in land–air interactions without an undergraduate degree in meteorology and need a place to start, or if you want to understand why the wind always seems to blow from one direction on a sunny day – you will find something here.

This chapter is the start. Here you will learn the language of the boundary layer. All scientific disciplines have their own technical language, and boundary layer meteorology is no different. However, the boundary layer, by its very nature, is a bridge between other disciplines, so it has a technical language that is rooted in meteorology but borrows from ecology, hydrology, biology, oceanography, and others. So, we must start with the common language that threads it all together.

1.2   Defining the boundary layer

The study and exploration of the boundary layer has come from a variety of places.[2] In this book you will find authors who define themselves as meteorologists, atmospheric scientists, hydrologists, engineers, and geographers. You will see they are working in both academic and professional settings with titles such as "Professor," "Scientist," "Researcher," "Consultant," or "Meteorologist." Maybe most interesting is that they have degrees in fields such as physics, civil engineering, geography, and even electrical engineering. Yet, all are experts in the boundary layer. In some sense, that diversity can make a common language difficult. When a weather forecaster hears the world "stable," they understand that means a specific atmospheric state that will affect air movement, when a civil engineer hears "stable" they may think about something that doesn't change. Both are true, but as you will see in Chapter 6, a stable boundary layer is not one without change. In this section we define the boundary layer and its structure. In later sections, we introduce the key influencers of both and other relevant meteorological terms.

Webster's Dictionary defines boundary as "something that indicates or fixes a limit or extent" and layer as "one thickness, course, or fold laid or lying over or under another." From that, we can deduce that a boundary layer is a certain thickness of the atmosphere. Intuitively, one might interpret that to imply that the entire atmosphere constitutes a boundary layer. The layer of air between earth and space. In the strictest sense one would be right. In fact, there are many boundary layers, and as you will see in Chapters 8 and 9 even boundary layers within boundary layers.

To apply the terminology to the subject at hand, we must shift our thinking about the atmosphere and move from the macro to the micro. Instead of viewing the atmosphere as a single layer between earth and space, we need to think of it as individual molecules moving around in response to external factors. We also need to remember

---

[2] For a wonderful and comprehensive history of boundary layer meteorology as a discipline and research area I highly recommend reading LeMone et al. (2019).

the atmosphere is constantly changing. Even in the time I typed this sentence, the air around me has moved. When my finger bridged the space between itself and the key the molecules in that space were either displaced or compressed. That brings us to our first fundamental concept in all meteorology:

Concept 1: Air is a compressible fluid

What that means is the density of air (mass per unit volume) can change in response to a change in pressure. Almost every fluid is compressible at high enough pressure; air is a compressible fluid at the pressures that exist in the natural world. What that means for boundary layer meteorology is that the density of air is variable.

Now that we understand the highly inconstant nature of air, let us get back to our definitions. When searched as a phrase, *Webster's* defines "boundary layer" as "a region of fluid (such as air) moving relative to a nearby surface (such as that of an airplane wing) that is slowed by the viscosity of the fluid and its adhesion to the surface." That is getting a little closer to what this book is about. Now we see that a boundary layer relates to *moving* fluids, a surface, and the properties of the fluid. One can imagine from this definition that boundary layers are everywhere: air over the airplane wing, water over a rock, or even a spill of cooking oil on your kitchen countertop.

But we aren't talking about **a** boundary layer, we are talking about **the** boundary layer. A layer that is a subset of the whole atmosphere, moving relative to a nearby surface. We need a definition that is more specific than the layer between earth and space and less specialized than the air over an airplane wing.

Simplifying *Webster's* definition down to the scale of the earth, we come up with *a region of air moving relative to the surface of the earth*. But that definition can still describe the whole atmosphere because all of it moves. (You may be starting to see why we felt this book was necessary.) Our definition is fixed by changing just a few words.

The atmospheric boundary layer *is the <u>lowest</u> layer of the atmosphere, which is <u>affected</u> by the surface of the earth*. That means we can assume that when we talk of the boundary layer we are talking of the just lowest layer of the atmosphere. In other texts, you will find variations of this definition. Some set spatial or temporal limits on the boundary layer. I choose here to stick with this more encompassing version to be as inclusive as possible, as many problems where the boundary layer is key may not fit exactly in those limits. My personal definition of the boundary layer is *the air near here*.

Our new definition of the boundary layer has two main parts, and it is important to dissect them a little. First, the vertical spatial component – the "lowest layer" of the atmosphere, which implies there is an upper limit where the earth no longer impacts the air. The second, is the causal component – the "affected by" part. It is this causal component that is the focus of the rest of this book. The goal of most science is identifying causal relationships, and boundary layer meteorology is no different. Identifying the causes of changes lets us predict the future states.

If we think about this definition just a little more, we do still need to put one more limit on it, because the "surface of the earth" is not a static entity. Go for walk and watch what is under your feet. For me, from my front door to my mailbox I will find concrete, grass, sand, dirt, and mulch. To further complicate things, if I walk that path at 10 a.m., and then again at 6 p.m., each of those surfaces will be exposed to sunlight differently. So, the "surface of the earth affecting the air above it" is highly variable in both time and space. What this means in practice is the surface conditions usually change too quickly for the atmosphere to get into balance. Which brings us to the next big concept we need to understand.

Concept 2: The atmospheric boundary layer is always changing

This constant changing is what makes the boundary layer both fascinating and challenging to study and understand. It is also what makes this book necessary. The constant movement of air in the boundary layer at different spatial scales is generally referred to as turbulence. Look at Chapter 2 for all the things you need to understand on how we formally define and quantify turbulence. The rest of the book deals with how to handle this constant changing compressible fluid.

1.2.1 The acronym game

Not to belabor the point, but the air near here is of interest in many different fields, so sometimes multiple terms are used to define *nearly* the same thing. You will see the following acronyms used by the various authors in this text, all of which are referring in some way to the definition above.

- PBL – Planetary boundary layer. This is a more encompassing term that reminds us the boundary layer covers the entire earth. It can also be used to define the boundary layers that can (and do) exist on other planets.
- ABL – Atmosphere boundary layer. This is often used to distinguish the atmospheric boundary layer from other smaller boundary layers relevant to the context, for example, the boundary layer that forms over the surface of a leaf as air flows over it.
- MABL – Marine atmospheric boundary layer. This is to distinguish that the author is talking about the boundary layer as it exists above the ocean.

You will also see several other acronyms referring to various parts or subsections of the boundary layer. These acronyms are generally used to refer to a part of the boundary layer that is affected by a specific part of the earth's surface.

- ASL – Atmospheric surface layer or sometimes just "surface layer." This is very close to the definition of the boundary layer above but tends to be used for situations where mechanical (shear) generation of turbulence exceeds buoyant generation or consumption (see Chapter 2).
- CRSL – Canopy roughness sublayer or roughness sublayer. The area of the boundary layer were the distribution and structure of foliage elements and plant spacing affects boundary layer structure (see Chapter 8).
- ML – Mixed layer or sometimes the convective mixed layer (CML). This is the state during a typical daytime scenario. The turbulent transport is convectively driven (heating from below), the atmosphere is well mixed, and the conditions are unstable, or is very susceptible to vertical uplift.

- SBL/SABL/SL – The stable boundary layer. This is typical nocturnal condition where there is no convection, buoyancy is negative, and vertical motion is suppressed (see Chapter 6).
- RL – Residual layer. This a layer that occurs at night. It is the leftover mixed layer from the day before. It is sandwiched between the growing stable layer below and the entrainment zone above.
- IBL – Internal boundary layer. This is a layer that forms within an existing boundary layer due to a change in some surface property (see Chapter 9 for examples).

1.3   The influencers

Now that you are clear on what the boundary layer is, it is time to start thinking about how exactly the earth's surface affects the atmosphere. This goes back to two fundamental science concepts: conservation of mass and energy. Mass and energy can be moved or changed, but they cannot be created or destroyed. There is a constant exchange of mass and energy between the earth and the atmosphere. Again, this is happening right now as you read this. Consider yourself as part of the earth's surface. When you breathe you take in and push out air. So does just about everything else on the surface of the earth. At a molecular level air moves in and out of soils, plants, and animals. That brings us to our third big concept.

*Concept 3: The earth is a source and sink of mass and energy
to the atmosphere*

That is to say that air has mass in the form of gases and particulates, and that mass comes from or returns to the earth. What drives those exchanges? Many things, at many scales of time and space. In this chapter, I will briefly introduce the largest of those influencers: the sun and the surface. Other chapters discuss the nuances of those influencers and impacts of others.

1.3.1   Radiation

The dominant changes in the boundary layer are due to the daily cycle of sunlight. The sun provides ALL the energy to the earth–atmosphere system and much of it serves to heat the earth, which in turn heats the atmosphere directly above it. In fact, it is this diurnal cycle of energy exchange that makes the boundary layer distinct from the rest of the atmosphere. In the absence of anything else, there is day (energy input to earth) and night (no energy input). This 24-h cycle is too short to allow the entire atmosphere to adapt, so the boundary layer exists as the result of the cycle of energy input.

When we think about the sun in the bigger picture – multi-day meteorology or multi-year climatology – we talk about the solar constant, the 11-year solar cycle, and the seasonal changes tied to the earth–sun geometry. There are well-defined relationships between latitude, longitude, time of day, and day of the year that define the daylength and provide a starting point for energy budgets.

When we think about the sun on the scale of interest for the boundary layer, there are other factors that modify the quantity, quality, and direction of incoming solar radiation. Composition is one. This is because the elements (gasses and solids) that make up the atmosphere can scatter and/or absorb radiation. Scattered radiation is redirected. Absorbed radiation is turned into heat. Both scattering and absorption depend on the wavelength of light and the size and type of the scatterer itself. There are different scattering mechanisms, and again the details are outside the scope of this book. Suffice it to say the interaction between radiation and the atmosphere is why the sky is blue, why sunsets sometimes appear red, and why we can use remote sensors like radar and lidar to measure the atmosphere. For a better, fuller discussion of scattering and absorption mechanisms I recommend Bohren (2007).

In the end, what this means for us is that the surface receives both direct and diffuse radiation, which is transferred into and out of the surface of the earth, as well as through the atmosphere itself. We use this to define net radiation – the sum of incoming and outgoing radiation at the surface.

1.3.1.1 Radiation laws

There are three laws that quantify how radiation works. They are important to define here. While you may never need to use them explicitly, they give us an understanding of how different surfaces can interact with radiation differently and ground our conceptual sense for energy balance and exchange. Additionally, understanding radiation laws helps us interpret how variations in the atmosphere itself, such as clouds, can alter the amount of energy available at the surface. Finally, in an even more practical sense, radiation laws control many of the ways we measure the atmosphere and land surfaces remotely (see Chapter 4). So, a base knowledge of them makes it easier for you to interpret data of that nature.

The laws of radiation are the mathematical ways to define the relationships between temperature, energy, and wavelength. Before digging into the laws, let us remember that everything emits and absorbs radiation. Everything from a single molecule in the atmosphere to the entire earth. This means that the radiation laws can be applied at different scales. It also means that absorption and emission of radiation is happening all the time.

To understand the radiation laws, we must define a theoretical construct known as a black-body. A black-body is an idealized object that is a perfect emitter and absorber of all wavelengths when it is at thermal equilibrium. Conceptually, a black-body is the starting point – it is a way of removing all other influences on the radiation budget. Black-bodies allow us to say that at specific temperature, this body will absorb or emit this much radiation at this wavelength. That is Planck's law.

Planck's law is an empirical law that defines the quantitative relationship between energy and wavelength for an object. Formally, the law defines energy radiated per unit volume by a black-body for a wavelength. The law itself is written in terms of Planck's constant, the speed of light, the Boltzmann constant, and the absolute temperature of the body.

Planck's law is related to two other radiation laws. Max Planck's work derived from the earlier work of Wilhelm Wien. Wien then built further on Planck's work and developed *Wien's Displacement law* to define the peak wavelength of black-body emission. Wien's Displacement law states that peak wavelength of a black-body is inversely proportional to the temperature. What this means is that warmer bodies have their peak emission at shorter wavelengths and vice versa.

The third of the radiation laws is the *Stefan–Boltzmann law*. This law defines the total energy of a black-body radiator as a function of its temperature. Mathematically it is the integration of Planck's law. The Stephan–Boltzmann law formally says that the total energy radiated per unit surface area of a black-body across all wavelengths per unit time is directly proportional to the fourth power of the black-body's thermodynamic temperature T. Functionally, this means hotter objects have much more energy than cooler objects.

When plotted on a graph as radiant energy versus wavelength, we see what are commonly referred to as black-body curves. Figure 1.1 shows the black-body curves for the sun and the earth. When looking at a curve like this we can see that Plank's law defines the curve, the Stephan–Boltzmann law defines the total energy, and Wien's displacement law defines where the peak is along the x-axis.

For completeness it is worth noting that all three of these radiation laws derive from *Kirchhoff's Law of Thermal Radiation*, which provides the theoretical background for understanding heat energy transfer. It states that for a body of any arbitrary material emitting and absorbing thermal electromagnetic radiation at every wavelength in thermodynamic equilibrium, the ratio of its emissive power to its dimensionless coefficient of absorption is equal to a universal function only of radiative wavelength and temperature. Planck's law is that universal function.

In reality, a black-body does not exist. Real objects are actually grey-bodies; they emit and absorb different wavelengths selectively. This value is characterized by an emissivity value. Emissivity is some number less than one that scales the Stefan–Boltzmann law to reality.

1.3.2 Energy transfer

The next fundamental concept related to energy is how it is transferred. As we stated previously, energy cannot be created or destroyed, but it can be altered or moved around (i.e., transferred). Energy is transferred in several ways. Radiation, which is a form of transfer as well as a form of energy. Conduction, which is transfer at the

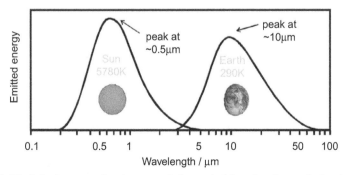

Figure 1.1 Black-body curves for the sun and the earth. Note that the vertical axis is not to scale, as the Stefan–Boltzmann law tells us that the total energy from the sun is $\sim$150,000 times that from the earth.
Image from Tuckett (2019).

molecular level. Convection, which is overturning in a fluid. And advection, which I like to think of as convection in the horizontal direction. There is also transfer of heat energy that occurs in the phase change of water. We call that latent heat. At the surface–air interface energy transfer starts with radiant energy from the sun transforming to heat energy *in* the surface. That heat is conducted back to the atmosphere where the air molecules touch the surface molecules. Within the boundary layer, convection is a major way that energy is moved upward, and advection moves it from one location to another. The scale at which convection and advection happens is explored more in Chapter 2.

A word about budgets

Air is a resource. We use it and can manage it just as we do any of the other natural resources such as water, fisheries, or forests. Air is a resource that is required for life to function effectively. Any resource can be budgeted, much like you do with your own financial resources. The atmosphere and its properties are no different. Throughout this book you will see reference to budgets. Budgets are a way to keep track of a finite resource, where it goes, and how it changes. In this chapter we introduce the radiation budget. Later chapters discuss various types of budgets. The concept of a budget is especially useful to understand interactions between the different components. Sources are the supply of that resource; sinks are a loss of the resource. Budgets have a spatial and temporal scale as well. You can create a water budget for a single watershed, or a single lake or pond within that watershed. You can compute an annual, monthly, daily, or even hourly budget. You'll see some examples of all throughout this text.

The use of a budget model for these resources stems directly from the fundamental physical laws: the conservation of mass, the conservation of energy, and the conservation of momentum. These quantities cannot be created or destroyed, but they can be altered and moved. In terms of air, water, carbon, or anything else, a budget is a convenient way to apply the conservation laws. By keeping track of what goes in, what is stored, and what goes out we are provided with a framework to understand what changes and how those changes occur.

1.3.3   The surface

You can see from this discussion that a starting point for much of the rest of this book is the exchange of energy between the land and the air. It is important here to distinguish what type of energy we are talking about. A full energy budget includes ALL the types of energy – not just radiation, but also heat energy and even exchanges of chemical energy in processes such as photosynthesis. As a side note, this is one of the places that boundary layer meteorology becomes less meteorology and more ecological or biological.

As an example of an energy budget, I will use net radiation at the surface. The terminology associated with net radiation is the same as doing an energy budget at any interface, for any type of energy. It comes down to these three statements:

- Net radiation = incoming radiation at surface minus outgoing radiation from the surface.
- If the incoming radiation is greater than the outgoing radiation, there is a positive (+) net radiation.

- If the outgoing radiation is less than the incoming radiation, there is a negative (−) net radiation.

If we think about this one step further in the context of conservation of energy, we can see that a positive net radiation means a gain of energy at the surface. That is the energy that goes to heating the surface, and basically running the rest of the world, and we can continue to budget it accordingly through different subsystems. A negative net radiation is a loss of energy from the surface to the air. This is primarily the heat that is conducted and convected into the boundary layer and the rest of the atmosphere.

You can extend this thinking to just about every concept throughout the rest of the book. By defining an interface and following the laws of conservation of energy and mass, we can better examine the time and space variations of these budgets and identify how the elements at the surface impact the air above it. In the net radiation example, we can go one step further to break down net radiation into the components of the earth's radiation (longer wavelengths) and the sun's radiation (shorter wavelengths) and make an equation that accounts for incoming and outgoing longwave and incoming and outgoing shortwave. In meteorology as a whole, we can take that further and explore how clouds impact that balance for any location. In the boundary layer specifically, we go even further and look at the differences in radiative properties between surfaces – the energy budget of a leaf is different than the energy budget of a puddle. As you progress through this book, you will see how we use this concept to understand turbulence, mass exchange, and even wind power generation. This is our next big concept.

Concept 4: To understand, quantify, and model the atmosphere,
we define boundaries and budget the changes across them

To balance a budget in the boundary layer we need to think about the other places energy can go. Radiant energy is absorbed and turned into heat energy, and it can be moved by thermal or mechanical forces. To balance it all we must consider ALL the possible inputs, outputs, and transfer mechanisms. Balancing a budget is achieved by applying the principle of energy conservation at a surface. For heat, the inputs are radiation and metabolism of organisms (i.e., you and I act as heat sources) and the outputs are radiation, convection (movement), evaporation, and conduction. The balance is achieved through adjustments of temperature.

At the surface, we can specifically examine the surface energy budget of the different components of heat flux. Latent heat flux is the energy that is absorbed by water. Sensible heat flux is the heat transfer between the surface and atmosphere by conduction and convection. Soil heat flux is the heat transfer between the surface and the underlying soil. Thus, at the surface energy balance, we can see that net radiation is equal to the sum of latent heat flux, sensible heat flux, and soil heat flux. By convention, we say that these fluxes are positive when they are upward (away from the surface, gain to the atmosphere), and negative when they are downward (towards the surface, loss from the atmosphere). Net radiation itself is defined as positive towards the surface. To be complete there is also a storage term. There is some energy stored in the parts of the earth, but it is smaller than the other components, so it is often ignored.

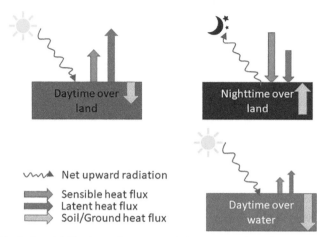

Figure 1.2 Typical variability in surface energy balance terms for different conditions. Adapted from Stull (2001).

In the daytime, the energy balance is generally dominated by solar heating, so net radiation is strong towards the surface (+), latent heat and sensible heat are strong away from the surface (+) as the surface heats the lower atmosphere, and soil heat flux is into the ground, also away from the surface (+). At night, there is no solar shortwave radiation, so there is a net loss of energy to space (−), latent and sensible heat are also lost (−) because the ground is cooling and water vapor is condensing, and soil heat flux is towards the surface (+) as heat moves upward from the deeper soil to the cooler surface. Furthermore, this balance changes over a water surface because water has a higher heat capacity than land and turbulence in the water can transport heat away faster and more efficiently. Figure 1.2 represents this variable energy balance.

Now let us add the change over time, and we can see that the surface energy budget will vary day to night and season to season. Once again, we see the boundary layer processes acting on different time scales over different surfaces. Figure 1.3 shows both the net radiation budget components and the surface energy balance components over time for warm and cool days in an urban environment. Q^* represents net radiation; you see it rise and fall over the course of a day, with negative values at night. Q_H represents sensible heat flux; you can see it is very close to zero at night and almost always positive. Q_s represents latent heat flux; it too is near zero at night, and almost always positive. ΔQ_s is the storage term; in an urban environment storage is not negligible. Also notice the dataset variability denoted by the shaded areas. While there is clearly a pattern, the data ranges within that pattern.

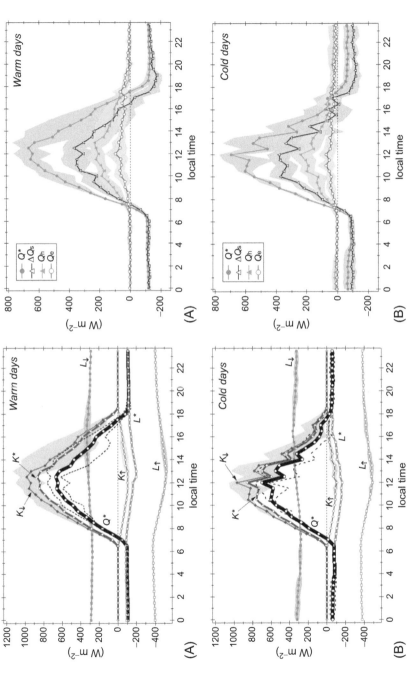

Figure 1.3 (*Left*) Average diurnal pattern of the net radiation budget for warm days (A) and cool days (B). (*Right*) Average diurnal pattern of the energy budget during warm (A) and cool (B) days. The *colored shadows* represent ±1 standard deviation of the total averages of measured fluxes and give an indication of the day-to-day variability in each phase of the daily cycle. The time scale corresponds to the local standard time. From Velasco et al. (2011).

1.3.4   The subsurface

In our conversation above we introduced soil heat flux as a component of the radiation balance. This reminds us that it is not just the surface in contact with the air, but also the subsurface that influences the exchanges of interest. The boundary layer is impacted by a shallow layer of material beneath the surface itself. We most often think of this as soil, vegetation, water, or human-made materials. The interaction is confined mostly to molecular exchange, and the speed of that exchange is due to the specific thermal properties of the subsurface material. These include thermal conductivity, heat capacity, thermal diffusivity, and thermal admittance. For the purposes of this introduction, suffice it to say that these properties control how fast and how efficiently the subsurface material can transfer heat. These properties also change with depth of the substrate, moisture content, and other physical properties such as soil porosity, which is the amount of air mixed in between the individual grains of soil.

1.3.5   Mass exchanges

Another interaction between the surface and the air is the exchange of mass. Following the principles of conservation of mass, we can budget the movement of mass in much the same way we did energy. Remembering always that these exchanges happen at multiple time and spatial scales. You may already be aware that the hydrologic cycle, carbon cycle, and nitrogen cycle are extremely important for human and ecosystem health. It turns out that the boundary layer is the critical controlling factor of exchanges of water, carbon, and nitrogen between the earth's surface and the atmosphere. Climatologists and atmospheric scientists tell us that the carbon content of the atmosphere has increased (see Chapter 12). Biologists can tell us that plants turn carbon dioxide into sugars to grow, and that excess carbon is impacting that process (Kirschbaum, 2004). Similarly, hydrologists tell us there are changes in the global water budget, decreases in water quality, and less access to clean water (UNESCO, 2015). Ecology tells us that excess nitrogen from human activities threatens the health of coastal waters (Sinha et al., 2017). Boundary layer meteorologists tell us *how* these exchanges are happening.

The big picture is clear, but as with anything else, it is the details that matter. I sometimes think of it as a Pointillism painting. When viewed from afar the scene is evident, but if you zoom in, it becomes apparent that each individual dot is required to be in just the right place, with just the right color for the whole image to work. In our case, we want to know exactly how the mass (e.g., water, carbon, or nitrogen) moves between the air and surface. Beyond that, can we predict when, where, and how that might happen?

Ask yourself, will a pine forest exchange mass in the same way as a grassy field? They are both green surfaces covered with thin, skinny leaves. So, from a geometry perspective they might. But, from a meteorology perspective they do not. They don't because turbulence (Chapter 2) moves the mass from the leaf to the atmosphere, and turbulence is different over a rough (trees) surface and smooth (grass) surface

(Chapter 8). From a biological perspective, pine trees and grass transpire differently, but that is (just a little) outside the scope of this book. So, while at first blush, it might seem obvious that a pine forest and a grassy knoll are different at an individual level, it is less obvious when scaled up (Chapter 3). It is even more complicated when we account for measurement footprints (Chapter 4) and stability concerns (Chapter 6).

1.3.5.1   Water – The necessity for life

Conceptually, water is something we all know and understand because it is impossible not to. We drink it, we cook with it, we swim in it, we even watch it fall from the sky or condense on a glass. If asked, almost everyone can come up with some basic facts about water. It supports life on Earth. It is made up of two atoms of hydrogen bonded to one of oxygen. It can evaporate, condense, and freeze. It covers most of the earth and makes up most of the human body. It moves around as part of the hydrological cycle, evaporating from the surface, condensing to form clouds, and returning to the lakes, rivers, and oceans through precipitation.

Water is part of the mass exchange, and it must be conserved as it moves through the earth–atmosphere system. More specifically, water is one of the masses exchanged between the surface and air. But it is not only that exchange that is part of the picture. Soil, storage, and throughfall are also part of the picture. Because to the atmosphere, a wet surface is different than a dry surface, thus how the land itself interacts with water is also a component of understanding how the land interacts with the atmosphere. From a surface–air interface, the properties of the surface such as soil type, storage capacity, and porosity all affect how the surface is a source or sink for water to and from the atmosphere. That means understanding the water cycle is crucial for the development of good land surface models (LSM) and climate models.

It is also important to remember that water movement does not only impact mass balances, but it also plays a role in the energy exchanges we discussed previously. This is because of the latent heat associated with water phase changes. Additionally, density and temperature are crucial factors in determining boundary layer structure and dynamics. Water is part of that as well, as water content will change air density.

We present here some definitions related to moisture in the atmosphere that you will see used in other parts of the text.

- Humidity – the amount of water vapor in the atmosphere. It can be measured in both absolute and relative terms.
- Water vapor flux – the movement of water vapor from one location to another. Most often we consider this as from a surface to the air or vice versa.
- Transpiration – the process by which plants exhale water through their stomata (tiny holes on their leaves) after the process of respiration.
- Evapotranspiration – the process by which water is transferred from the land to the air by both evaporation from surfaces and transpiration from plants.
- Latent heat – energy required for a change of water phase (gas, liquid, and solid) that occurs without changing the water's temperature. Condensation releases heat, while evaporation absorbs heat.

1.3.5.2 Carbon – The building block of life

Whereas water is the necessity of life, carbon is the stuff life is made of. Carbon is less conceptually understood by most, yet it is equally as important. Without carbon, there is no life. It is a non-metal element that readily forms compounds with many other elements. In the atmosphere, carbon (C) is a key component of two gasses of interest: carbon dioxide (CO_2) and methane (CH_4). Much like water, carbon continuously cycles through the earth–atmosphere system.

Now you may be wondering, why do we have carbon dioxide in the atmosphere? Well, the primary reason for CO_2 to be present in the atmosphere is to absorb energy emitted by the sun and the earth. CO_2 is a greenhouse gas and therefore helps provide a livable climate on Earth through the greenhouse effect. Thus, the amount of carbon in the air directly impacts the net radiation budget we discussed earlier. CO_2 is also needed in the atmosphere for plants to conduct photosynthesis. Like most life on Earth, plants are made up of carbon and thus play a crucial role in the carbon cycle. This is because they conduct photosynthesis by taking carbon from the atmosphere along with water and sunlight to create sugars or glucose. Plants are autotrophs, so they create their own food and energy. To use the sugars they have created, they perform the process of respiration. Respiration is the process that uses oxygen and sugar to consume their self-made energy and create biomass. During this process plants release water and CO_2 to the atmosphere through their stomata. CO_2 and water are consumed during the daytime through photosynthesis and CO_2 and water are released into the atmosphere as a byproduct of respiration at night (Bonan, 2016).

So why is CO_2 so important to micrometeorology? Photosynthesis is not constant over an entire ecosystem, creating a gradient across an ecosystem. Also, plant canopies and ecosystems affect wind flow, which then alters the canopy processes such as photosynthesis. Each affects the other, and therefore we cannot understand the exchange of carbon without micrometeorology and vice versa (Schurgers et al., 2015).

We have several different ways that we express carbon fluxes in the atmosphere and biosphere. These different expressions include:

- Gross primary production (GPP) – The net photosynthesis of an ecosystem expressed in $gC m^{-2} yr^{-1}$.
- Net primary productivity (NPP) – The biomass produced annually; it is quantified as GPP minus Plant Respiration.
- Net ecosystem productivity (NEP) – Similar to NPP, it is the difference between GPP and ecosystem respiration.
- Net ecosystem exchange (NEE) – The CO_2 exchange between the atmosphere and the land surface or ocean. Unlike GPP, NEP, and NPP, the values of NEE are negative for CO_2 coming out of the atmosphere and into the biosphere.
- Respiration (R) – The process that converts glucose into CO_2, water, and energy. Typically occurs at night for plants. Can be used to talk about plant respiration, animal respiration, or microbial respiration (Chapin et al., 2011).

In the big picture, carbon in the atmosphere is part of climate change (see Chapter 12). In the boundary layer, carbon is part of the mass exchange.

1.3.6 The wind

Remembering that the atmosphere is a fluid of mostly gas particles is a key component to understanding how air moves. Since air moving was a fundamental part of defining the boundary layer, we must also think about how to define that motion. To do this we must know that atmosphere also follows the ideal gas law. The ideal gas law is derived from the laws of motion, and it relates variables we can observe (pressure and temperature) to the mass and velocity of molecules in the atmosphere. The ideal gas law is a combination of several other laws that tell us the following:

- At a constant volume, pressure is proportional to temperature.
- At a constant pressure, volume is proportional to temperature.
- At a constant temperature, volume is *inversely* proportional to temperature.

Thus, at a constant energy, pressure volume, and temperature can all change. The real take-home message of the ideal gas law is that under different conditions, different things will happen. It is the meteorologist's job to predict what.

When pressure and temperature changes occur, volume changes must also occur. This is what initiates motion or drives the wind. Pressure differences drive the motion, but it is altered by other forces, such as friction near the surface or the Coriolis force arising from the rotation of the earth. In introductory meteorology courses we tell students, pressure changes initiate motion, Coriolis forces bend the wind, and friction slows it down. In general, we define the wind as the dominant horizontal movement and think of turbulence as the dominant vertical motion. The equations of motion to define how changes in pressure, temperature, and density over time result in changes in velocity in space and time, are, as you would imagine, somewhat complicated. If you really want to or need to get into the equations, I suggest grabbing a good book on fluid dynamics (e.g., Schlichting and Gersten, 2017) or atmospheric physics (e.g., Andrews, 2010).

1.3.7 Hydrostatic equilibrium

In other chapters you will see the term hydrostatic equilibrium, mostly in the context of departures from that state. While it is mostly turbulence that dominates the vertical structure of the boundary layer, there are still changes in pressure as you move away from the surface. Pressure decreases with height. This is because of hydrostatic equilibrium. If we freeze any parcel of air, we can identify three forces on it, which are acting in balance: (1) pressure from the air above it is pressing down, (2) pressure from the air below it is pushing upwards, and (3) the weight of the parcel itself, which is the product of the density of the parcel, the volume of the parcel, and gravity. When these forces are in balance the non-moving parcel is said to be in hydrostatic equilibrium. In reality, air in the boundary layer is always moving (Concept 2). That movement is because of the thermal or mechanical forces that cause turbulence. However, defining the state of balance allows us to mathematically diagnose other variables we may not otherwise be able to measure.

1.4   Some other concepts and definitions

As mentioned, meteorology has its own language, and in this section we present a few more concepts and definitions that you may find useful when reading the rest of the book.

Log wind profile: You will see the word "profile" throughout this book. This simply means how a variable changes with height away from the surface. Many graphs will have height on the *y*-axis and the variable of interest on the *x*-axis. That is considered a profile. One that is used often is the log wind profile. In a perfect world, wind speed decreases logarithmically with height, starting from a value of 0 m/s right at the surface.

Potential temperature, θ: This the temperature of an air parcel if it were moved adiabatically to sea level. This is a theoretical construct. Potential temperature is a composite variable that makes the math easier. This is because by normalizing the effects of pressure, potential temperature becomes a conserved quantity.

Virtual temperature: This is the temperature that dry air would have if its pressure and density were equal to those of a given sample of moist air. This is another composite variable that helps the math. By defining virtual temperature one can use the dry-air equation of state for moist air.

Virtual potential temperature: This is the potential temperature of the virtual temperature. It is a useful quantity for comparing potential temperatures of air at different pressures. This is most relevant in understanding buoyancy and stability, and profiles of virtual potential temperature are often used to show the diurnal variation in boundary layer structure.

Turbulence: Chapter 2 is all about this, but in case you do not want to read it: Turbulence is the three-dimensional motion that transforms mechanical energy to internal energy (heat) through a cascade of rotating eddies of diminishing size.

Fetch: This is the distance across a uniform surface that generates turbulence by shear at the surface. This is sometimes called the generating area. You can think of it as the upstream area of uninterrupted flow.

1.5   Beyond the air

As previously mentioned, boundary layer meteorologists come from a variety of disciplines. Many now work in the area of biogeosciences. This chapter briefly discussed the water and carbon cycles as an introduction to the larger group of cycles known as biogeochemical cycles or the flow of gasses and their chemical changes through the earth–atmosphere system. While water and carbon are the two most important, other biogeochemical cycles are also of interest. These include non-carbon gasses such as nitrous oxide (N_2O), ammonium (NH_3), mercury (Hg), and others. Some have coined the term eco-climatology to represent the work being done in this area. While water, carbon, and nitrogen are extremely important in the boundary layer we have chosen not to include chapters on them explicitly; for the interested reader I recommend Bonan (2016).

1.6 Moving forward

The rest of this book is written in such a way that each chapter can stand alone. The assumed background knowledge is minimal; this chapter should suffice, and if it does not, the references noted should help. The text does however assume that you have a general interest in science, so common terms, such as energy and force, will not be defined.

1.6.1 A word about the math

While this book is designed to be introductory and accessible to the non-specialist, measuring the environment is inherently a quantitative activity. Thus, the primary way we try to explain the atmosphere is with mathematical models to predict atmospheric behavior. Given that, it is impossible to avoid equations within this text. Wherever possible, equations have been kept to a minimum. When equations are necessary, we do our best to explain them textually or graphically so they can be understood in the broader context. If after reading this text you are going to continue as a meteorologist, I highly encourage you to take a calculus course, because calculus is about representing CHANGE, and we just learned that the boundary layer is all about change.

1.7 Summary and review

In this chapter we have provided a definition of the atmospheric boundary layer and introductions to the components of the earth system that can influence what happens in the boundary layer. The four big concepts presented here are fundamental to any meteorology and apply throughout the remainder of the text. The variation of the boundary layer influences our day-to-day lives because it is the air near here that we breathe, feel, and directly impact. In summary we now know:

- The structure, depth, and dynamic properties of the boundary layer change over the course of the day.
- During the daytime, when solar heating is active there are two sublayers: the convective boundary layer and the surface layer that is directly attached to the earth.
- During the night, when solar heating is gone, there are three sublayers: the surface layer, the stable boundary layer (Chapter 6), and the residual layer, which contains the atmosphere's properties from the day before.
- There are three main ways the surface influences the atmosphere: sensible heat flux, moisture flux, and radiation. Molecular conduction transfers heat, moisture, and momentum to the air from the surface and then turbulence takes over.
- We can use composite variables such as potential temperature to show the changes in the boundary layer through the day.
- Other natural cycles also impact the boundary layer; these include photosynthesis and the hydrologic cycle.

References

Andrews, D.G., 2010. An Introduction to Atmospheric Physics. Cambridge University Press.

Bohren, C.F., 2007. Atmospheric Optics. The Optics Encyclopedia. Wiley, https://doi.org/10.1002/9783527600441.oe004.

Bonan, G., 2016. Ecological Climatology. Cambridge University Press.

Chapin, F.S., Matson, P.A., Vitousek, P.M., 2011. Principles of Terrestrial Ecosystem Ecology. Springer, New York.

Foken, T., 2017. Micrometeorology, second ed. Springer. 362 pp.

Garrett, J.R., 1992. The Atmospheric Boundary Layer. Cambridge University Press. 396 pp.

Kirschbaum, M.U., 2004. Direct and indirect climate change effects on photosynthesis and transpiration. Plant Biol. 6 (3), 242–253. https://doi.org/10.1055/s-2004-820883. 15143433.

Lee, X., 2018. Fundamentals of Boundary-Layer Meteorology. Springer International Publishing.

LeMone, M.A., Coauthors, 2019. 100 years of progress in boundary layer meteorology. Meteorol. Monogr. 59, 9.1–9.85. https://doi.org/10.1175/AMSMONOGRAPHS-D-18-0013.1.

Schlichting, H., Gersten, K., 2017. Boundary-Layer Theory. Springer-Verlag. 805 pp.

Schurgers, G., Lagergren, F., Molder, M., Lindroth, A., 2015. The importance of micrometeorological variations for photosynthesis and transpiration in a boreal coniferous forest. Biogeosciences 12, 237–256. https://doi.org/10.5194/bg-12-237-2015.

Sinha, E., Michalak, A.M., Balaji, V., 2017. Eutrophication will increase during the 21st century as a result of precipitation changes. Science 357, 1–5. https://doi.org/10.1126/science.aan2409.

Stull, R., 2001. An Introduction to Boundary Layer Meteorology. Kluwer Academic Publishers. 670 pp.

Tuckett, R., 2019. In: Worsfold, P., Poole, C., Townshend, A., Miró, M. (Eds.), Greenhouse Gases, third ed. Academic Press, pp. 362–372.

UNESCO, 2015. International Initiative on Water Quality: Promoting Scientific Research, Knowledge Sharing, Effective Technology and Policy Approaches to Improve Water Quality for Sustainable Development.

Velasco, E., Pressley, S., Grivicke, R., Allwine, E., Molina, L.T., Lamb, B., 2011. Energy balance in urban Mexico City: observation and parameterization during the MILAGRO/MCMA-2006 field campaign. Theor. Appl. Climatol. 103, 501–517. https://doi.org/10.1007/s00704-010-0314-7.

Always in flux: The nature of turbulence

Alexandria G. McCombs[a] and April L. Hiscox[b]
[a]Department of Geography and Geosciences, Salisbury University, Salisbury, MD, United States, [b]Department of Geography, University of South Carolina, Columbia, SC, United States

2.1 Introduction

Chapter 1 discussed what the boundary layer is; in this chapter we will discuss the feature of the boundary layer that makes it so challenging: turbulence. The boundary layer is almost continuously turbulent, and that turbulence is responsible for a large portion of the exchanges we are interested in. Micrometeorology is the study of the atmosphere of near-surface atmospheric phenomena and processes at spatial scales less than 3 km and time scales of 1 h or less (Foken, 2008; Lee et al., 2004; Stull, 1988). In the traditional scale organization, phenomena at these scales are considered to be microscale (see Chapter 3 for more on scale). Micrometeorology studies attempt to quantify the nature of turbulence in the surface layer (Lee et al., 2004; Stull, 1988). Turbulence and micrometeorology were first discussed in the mid-1920s, when Schmidt (1925) and Geiger et al. (2009) published their books transitioning hydrodynamics to meteorology. Some of the earliest atmospheric turbulence investigations were reported in Lettau (1939) in the book *Atmospheric Turbulence*. These studies document the research that led to the flourishing of the field after the Second World War (Leclerc and Foken, 2014). In subsequent years, scientists have developed quantitative solutions for understanding the atmospheric boundary layer and atmospheric turbulence. This research continues today since there is still much we do not know about microscale phenomena. In this chapter we define turbulence and present an overview of the ways it is quantified and understood in the atmospheric boundary layer. It should be noted that in the spirit of an accessible, conceptual approach, this presentation is broad. If after reading this chapter you are still curious about the finer points of turbulence we recommend other texts such as Wyngaard (2010), which are entirely dedicated to unraveling the intricacies of overturning fluids.

2.2 What is turbulence?

Within the atmosphere there are typically two types of air flow: **turbulent** and **laminar flow**. Laminar flow occurs when all parts of the fluid are moving along the direction of mean motion. However, when we talk about turbulent flow, we are referring to non-laminar flow, where the motions of molecules are irregular (Oke, 1978; Wyngaard, 2010).

Conceptual Boundary Layer Meteorology. https://doi.org/10.1016/B978-0-12-817092-2.00005-9

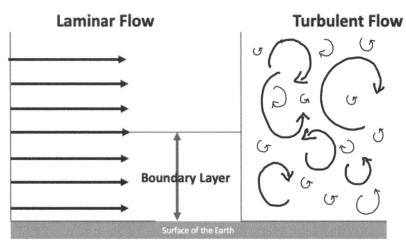

Figure 2.1 As air moves through the boundary layer, there are two types of flow. Organized consistent flow known as laminar flow (*left*) and disorganized, random flow known as turbulent flow (*right*).
Credit: A. McCombs.

Figure 2.1 shows the difference. In both laminar and turbulent flow, wind speed increases with increasing altitude, but at some altitude far from the surface, turbulence itself decreases. Reynolds number is used to determine whether the flow is turbulent or laminar. It is a non-dimensional ratio of the forces acting on a fluid. A low Reynolds number indicates laminar flow and a higher number indicates turbulence.

We can define turbulence as the variability or randomness of the wind, separate from the mean wind speed over a set time period. We describe turbulence either qualitatively or statistically because turbulent flow is chaotic and unpredictable. Turbulence is often visualized as irregular swirls that are called eddies. Eddies are generated within the boundary layer and are caused by forcings found at the surface of the earth. These eddies will often occur in different sizes. For instance, when the sun heats the surface of the earth on a sunny day, the warmer air at the surface will rise in the atmosphere because warm air is less dense; these are called thermals. Turbulence can also be caused by obstacles found at the surface of the earth, such as buildings, trees, and mountains. The frictional drag of the air flowing across the surface of the earth will cause wind shears, which often cause eddies in the atmosphere. Wind shear is the change in wind speed and direction with altitude (Stull, 1988).

When there is no frictional drag or other external force on the atmosphere, the flow of the atmosphere will remain constant due to the three conservation laws in fluid dynamics. The atmosphere is a fluid, which is why meteorologists use fluid dynamics to describe the physics of how it flows. The three conservation laws in fluid dynamics are conservation of mass, conservation of momentum, and conservation of energy. **Conservation of mass** stipulates that the fluid mass cannot be created or destroyed, so given a volume of space, the fluid flowing into the volume must be the same as the fluid flowing out of the volume. **Conservation of momentum** is Newton's second law of motion, which states that any change in the product of mass and velocity (called momentum) of a fluid will be equal to the external force acting on the fluid (Anderson, 2007). Finally, the **conservation of energy** states that the total energy of a fluid will remain constant in a volume of space given no external forces. Total energy is conserved over time, but the type of energy can be converted from kinetic energy to potential energy, and vice versa (White, 1974). We used these three laws in Chapter 1 to introduce the states of the atmosphere. For the three laws of fluid dynamics to hold true, we are assuming that the atmosphere is a closed system and there are no external forces acting on it.

Unfortunately, the atmosphere is not a closed system and there are many external forces acting on it. Therefore, turbulence is not constant. For example, a volume of air cannot go through a building, so turbulence is created as the volume of air goes around the building. You can read more about flows in the built environment in Chapter 9. Abrupt changes in the surface roughness, which is a measure of the obstacles or obstructions in the atmospheric flow, often cause an increase or decrease in turbulence (Wyngaard, 2010). Surface roughness is not the only variable that can change turbulence; other causes for change in turbulence include changes in the heating of the earth's surface, changes in air density, changes in topography, or the presence of vegetation or waterbody (Leclerc and Foken, 2014). An example of changes in turbulence can be seen in Figure 2.2 during a total solar eclipse, when solar heating is shut off for just a few minutes during the day. You can see how variability decreases as the solar radiation is removed and increases again post eclipse.

2.2.1 Defining turbulence mathematically

Turbulent flow is thought of as a two-dimensional or three-dimensional flow; both are of interest for different reasons. Two-dimensional flow is typically thought of as the horizontal wind or transport of air horizontally across the surface of the earth. Three-dimensional flow is the more common conceptualization because it is more realistic to quantify eddies in all three dimensions (x, y, and z) (Wyngaard, 2010).

Statistically we would describe the variability as the variance of the wind. Variance is a measure of how far a set of numbers is spread out from their average value. So, given a sample of wind data we can compute the mean wind speed as the average wind velocity. The variance is the average deviation from that mean. Because eddies are three-dimensional and can move in all three dimensions, we break the wind into three spatial dimensions (Figure 2.3). Mathematically, we also remember that any variable can be split into two components the mean and the perturbation or deviation from the mean. Figure 2.4 depicts this process. Therefore, the instantaneous wind speed can be partitioned as u, v, and w in the following manner:

$$u = mean(u) + u', where\ v\ is\ wind\ speed\ in\ the\ x\text{-}direction \qquad (2.1)$$

$$v = mean(v) + v', where\ v\ is\ wind\ speed\ in\ the\ y\text{-}direction \qquad (2.2)$$

$$w = mean(w) + w', where\ w\ is\ wind\ speed\ in\ the\ z\text{-}direction. \qquad (2.3)$$

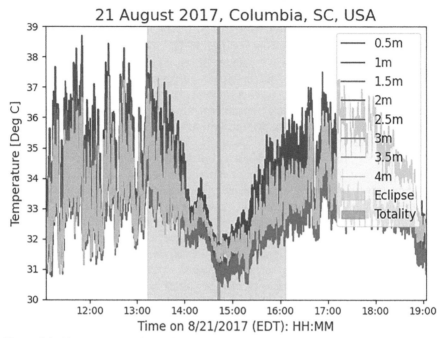

Figure 2.2 Air temperature data collected every 1 s on August 21, 2017, during the Great American Eclipse at the University of South Carolina campus weather station in Columbia, SC, USA. Data was measured at 8 heights above the ground surface (0.5–4 m).

The entirety of the eclipse event is shaded in light gray, and eclipse totality at the station location is in dark gray. The rapid changes in air temperature indicate turbulence due to heating of the surface from the sun. When the sun is covered by the moon, the changes in temperature are significantly smaller than before and after the total eclipse event.

Credit: A. McCombs, A. Hiscox, M. Stewart.

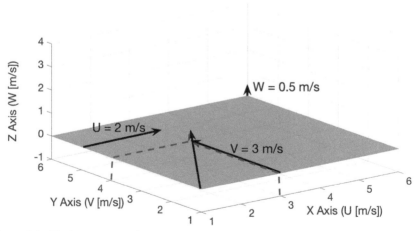

Figure 2.3 Wind speed varies in the x, y, and z directions, x is notated as U, V, and W. Wind speed can vary in each direction and the three directions are used to create a wind vector.
Credit: A. McCombs.

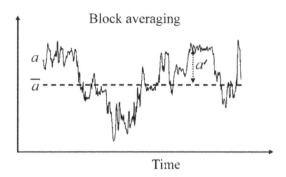

Figure 2.4 Mean and turbulent components of a time series.
From Lee (2018).

The terms u', v', and w' indicate the instantaneous wind gust or deviation from the mean. Each of these terms, the mean and the perturbation, can vary in space and time. Additionally, the three-dimensional wind speed can be negative or positive to indicate the direction in which the wind is blowing (Stull, 1988). In a practical sense, the instantaneous wind speeds are often measured every 0.1 s to every 1 s, being limited by measurement technology (Chapter 4).

The reason we notate wind speed in three dimensions is because it allows meteorologists to observe the rotational pattern that occurs in turbulent flows in the atmosphere. These rotational motions are dissipative and unpredictable. In order for turbulence to remain constant the atmosphere requires a constant source of energy, or else the energy dissipates quickly with time and space. This means that larger eddies that occur over longer time scales and spatial scales dissipate to smaller eddies that occur on smaller time and spatial scales. Most of the energy of motion, also called kinetic energy, is found in the large eddy structures. The energy found in these large structures is transferred to smaller structures over space and time. The smaller the eddy the less kinetic energy it has because the motion is not as large (Stull, 1988).

There is a set of rules for working with the mean and turbulent parts defined in Eqs. (2.4)–(2.9). This is known as Reynold's averaging, and it follows these rules.

1. The average of the turbulent value is equal to zero. This makes sense since u' is defined as the departure from the mean, so one half of the u' values would be above the mean and one half would be below, thus averaging to zero.

$$\overline{u'} = 0 \tag{2.4}$$

2. The mean of a constant times a variable is equal to the constant times the mean of the variable. This too makes sense since the mean of any constant is the same value as the constant.

$$\overline{cu} = c\overline{u} \tag{2.5}$$

3. The mean of the sum of two variables is equal to the sum of the means. This one is a little less obvious, but proving it to yourself is a straightforward exercise.

$$\overline{u + v} = \overline{u} + \overline{v} \tag{2.6}$$

4. An average value acts like a constant when averaged a second time over the same period. You need some calculus to prove this one, so just remember.

$$\overline{(\overline{u}v)} = \overline{u}\,\overline{v} \tag{2.7}$$

5. The average of local changes with time (slope) equals the slope of the averages.

$$\overline{\left(\frac{\partial u}{\partial t}\right)} = \frac{\partial \overline{u}}{\partial t} \tag{2.8}$$

Combing these Reynolds averaging rules, we can derive Eq. (2.9):

$$\overline{uv} = \overline{u}\,\overline{v} + \overline{u'v'} \tag{2.9}$$

This tells us that the average of u times v is equal to the product of the averages plus the average of the product of the perturbations or turbulent part. This last term does not necessarily equal zero because they are two different variables (Stull, 1988). The last term is equivalent to the covariance of the perturbations; in some cases this represents flux. In the case of $u'v'$ it is the horizontal momentum flux. In statistics, covariance indicates how much two variables move together. To physically measure turbulence, the eddy covariance method is utilized as a way to apply the statistical approach discussed earlier. Using Reynolds averaging we can capture the effects of turbulent motion when taking measurements using the eddy covariance method. Eddy covariance methodologies are discussed in more detail in Chapter 4.

2.2.2 Turbulence scales

Ideally, meteorologists would be able to measure the size and shape of every eddy in the atmosphere. Unfortunately, that is both impractical and impossible. When we measure turbulence in the atmosphere, we can only take a snapshot of the eddies. What we really measure is an atmospheric property or variable, such as wind speed, humidity, or temperature, as it changes over time in one location. To solve this conundrum, we assume, using **Taylor's hypothesis**, that turbulent eddies remain "frozen" as they pass the measurement sensor. This means the eddy moves past the sensor at the mean windspeed without changing. Taylor's hypothesis also says that changes in temperature as a result of the eddy movement are negligible. Without these assumptions we are unable to solve for the governing equations that tell how air temperatures change as each eddy goes past the sensor (Aubinet et al., 2012; Leclerc and Foken, 2014; Oke, 1978; Stull, 1988; Wyngaard, 2010).

According to Taylor's hypothesis, we are seeing a spectrum of turbulence. That is to say eddies exist at multiple sizes and any given measurement will capture multiple eddies. In a practical sense, this means we can measure and interpret turbulence in both the time domain *and* space domain. Much like light, which can be referred to by its wavelength *or* its frequency, this is done through time series analysis techniques, which recognize the patterns of periodic peaks. The spectrum of scale and

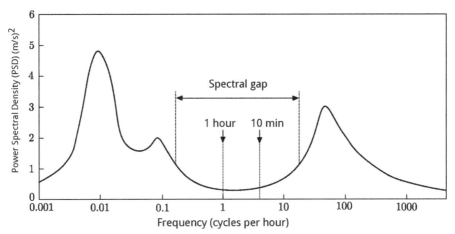

Figure 2.5 Relative spectral energy (i.e., energy per unit frequency) as a function of eddy cycling frequency or time period. As eddies generated at the PBL scale by convection and shear break down into smaller eddies, the energy is dissipated.
From Martin-Martinez et al. (2012) based on a study of I. Van der Hoven (1957).

transformation between the time and space domains is discussed in Chapter 3, so we will not dig into it deeper here. For an even more complete discussion of time series see Wilks (2019).

What we know from all this is there is a spectrum of turbulence (Fig. 2.5). Most energy is contained in the large (synoptic) scale motions *or* the smaller (turbulent) scale motions. The shape of this spectrum will show what size eddies dominate the flow. We also see a gap in the spectrum. This spectral gap is a scale where there is little to no turbulence (or deviations from the mean), and it is used to separate the larger scale motions from the turbulence scale motions (Figure 2.5).

In broad terms we can say turbulence occurs at multiple scales. Large-scale turbulence tends to be of the same scale as the dimensions of the system – in the boundary layer that is the mixing depth. Large-scale turbulence tends to be driven by buoyancy. The inertial subrange of turbulence is where energy falls off as kinetic energy and is transferred from the larger to smaller scales. And the small-scale turbulence is influenced by viscosity; the small stuff is where energy is dissipated (Figure 2.5).

2.2.3 Types of turbulence

Since turbulence is defined as a deviation from the mean, it is natural that turbulence is further classified by its statistical properties. There are three characteristics of the turbulence that underpin most of the current theories of boundary layer meteorology: stationarity, homogeneity, and isotropy. As a side note, it is the fact that these properties do not hold true all of the time that means these theories are constantly evolved and refined by boundary layer meteorologists.

Stationarity: Turbulence is assumed to be stationary. This does not mean eddies are not moving, it means that the statistical properties (i.e., mean, standard deviation, variance, and autocorrelation) stay the same over time. The variable itself changes over time, but how it changes stays the same. Stationarity is an underlying assumption of most time series techniques. A truly stationary time series has no periodic fluctuations. That statement is in direct contradiction to the discussion of the turbulence spectrum above. The spectrum is many different periodic fluctuations happening in the same space. These two concepts work together in turbulence analysis: we choose a subset of the turbulence that is stationary to perform analysis. That subset is the length of time used to compute the mean component.

Homogeneity: We also often consider turbulence to be homogeneous, meaning that while it is a random process, the average turbulent properties are independent of the position in the fluid. Homogeneity is stationarity in the spatial sense. It means that the turbulence is invariant in space. In homogeneous turbulence the fluctuations are random, but the average value is the same over the entire flow field. Quantitatively the structure is the same everywhere.

Isotropy: The third assumption is that turbulence is isotropic. Isotropic turbulence is defined by the American Meteorological Society *Glossary of Meteorology (2022)* as "Turbulence in which the products and squares of the velocity and their derivatives are independent of direction, or, more precisely, invariant with respect to rotation and reflection of the coordinate axes in a coordinate system moving with the mean motion of the fluid." Isotropic turbulence is homogenous by definition; isotropic turbulence has the added restriction of no directional preference in the statistics. Three-dimensional isotropic turbulence is never encountered in real atmospheric flows because rotation and buoyancy suppress vertical motions, so an eddy is never perfectly round.

2.3   What is a flux?

When we say something is "in flux," we understand that to mean that things are changing or impermanent. In the environment, we define a flux as a change over an area. More formally a flux is a transfer of some quantity per unit area per unit time. In other words, how quickly and how much of something crosses some flat plane. To understand fluxes, we must remember that movement in the air can transport a property of the air with it. That is, molecules, particles, or eddies can transport heat, water vapor, or gases. We can define a flux for any one of these quantities or some combination of them. For example, mass flux can include *all* the material that moves through the plane, or it could be the mass of *just* the water vapor. Therefore, mass flux refers to the transfer from one location to another of anything that has mass; it could be air itself or any other quantity of something in the air, such as a pollutant or other gas. The radiation budget discussed in Chapter 1 is actually a budget of radiant fluxes, even though we don't always refer to it that way. The units we used were watts per square meter. That is energy (the property) moving per unit area (the flat plane). Boundary layer meteorology is mostly concerned with fluxes of mass, heat, moisture, and/or momentum.

Starting with the earth–air interface as our flat plane means it is vertical fluxes with which we are most often concerned. It is, after all, the exchange of these quantities from surface to air or air to surface that are at the heart of any practical question and make the boundary layer what it is. However, a flux can happen in any direction if there is a gradient of the variable of interest. So horizontal fluxes are also important, especially if we are tracking something that is already in the air (see Chapter 10 for more information on the specific case of plume tracking). For convenience's sake, and the remainder of this discussion, we define a cartesian coordinate system, with the x–y plane parallel to the surface as the reference frame to compute fluxes (Figure 2.3). This gets tricky over hills or other topographic features, which is discussed in Chapter 7.

Thinking about fluxes in terms of the units helps us use them in a practical sense. Since it is difficult to measure heat or momentum directly, we instead measure temperature or wind speed, so this yields a different form of a flux used in practice. These are known as kinematic fluxes and they are found by dividing a standard flux by the density of moist air. This is valid because air density is nearly constant in the boundary layer. Table 2.1 presents fluxes, their SI units, and the more commonly used kinematic form.

The next step to understanding a flux would be to define it mathematically; this is done through the derivation of a transfer equation taking the general form of flux = change in property/change in time. We recommend the derivation and discussion of Monteith and Unsworth (2013) to see how this concept turns into practice. They start with the reminder that "carriers" can transport a property. Here we end with the main points from that concept:

- In the boundary layer, the carriers are molecules, particles, or eddies and the properties are heat, mass, and momentum.
- Heat, mass, and momentum move from one place to another so the net transport can occur in any direction that the property decreases.
- The carriers can unload any excess property where the property's value is less than the starting point.

Table 2.1 Fluxes.

	SI units	**Kinematic units**	**Example application**
Mass	$kg\,m^{-2}\,s^{-1}$	$m s^{-1}$	Carbon flux
Heat	$J\,m^{-2}\,s^{-1}$	$K\,ms^{-1}$	Energy balances
Moisture	$kg_{(H2O)}\,m^{-2}\,s^{-1}$	$kg_{(H2O)}\,kg_{(air)}^{-1}\,m^{-2}\,s^{-1}$	Evapotranspiration
Momentum	$(kg\,m\,s^{-1})$ $m^{-2}\,s^{-1} = N\,m^{-2}$	$m^2\,s^{-2}$	Turbulence partitioning
Scalar quantity	$kg_{(q)}\,m^{-2}\,s^{-1}$	$kg_{(q)}\,kg_{(air)}^{-1}\,m^{-2}\,s^{-1}$	Pollutant movement

Note: Moisture and scalar fluxes are both fluxes of mass, but the mass of the air is removed from the formal definition but retained in the kinematic units because of the inclusion of moisture in the density.

In relationship to turbulence, eddies transport heat, mass, or momentum in the horizontal and vertical directions, so the quantities we are actually concerned with are turbulent fluxes.

2.3.1 Momentum flux

One of the trickiest fluxes to understand conceptually is momentum flux. This is because unlike heat that we can feel, or mass that we can weigh, momentum is less tangible. We cannot feel it in the same way as we do heat or mass, so we just have to understand it. Momentum is the tendency of something to continue moving once it is already moving. Momentum is first defined by Newton's laws of motion, which says something will keep moving in the direction it is going unless some other force acts on it. So, if we give a molecule of air a push it will keep moving until something stops it. Momentum transfers on a molecular level via viscosity. Viscosity is an expression of the internal friction of the fluid (air).

Momentum transfer results from a gradient between two surfaces. Consider Figure 2.6, which depicts a stream of air (gray area) *flowing* over a solid surface. In this depiction, horizontal wind speed increases with height, let's assume linearly, even though we know that is not true in the actual atmosphere. Let us also assume that the temperature does not change with height, so there is no additional kinetic energy associated with the molecules (i.e., temperature) at higher heights than lower. There is molecular movement in the flowing air and therefore there is a constant interchange of momentum in all directions. Some of that momentum transfers downward layer by layer in the air, and eventually from the air to the non-moving surface below it, where it will experience friction and the drag, so it will slow down. The other two things to notice in Figure 2.6 are μ and τ. These are calculus-derived versions of dynamic viscosity and shearing stress, respectively. See Monteith and Unsworth (2013) for those explanations.

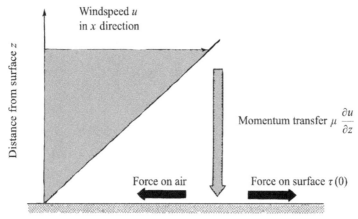

Figure 2.6 Transfer of momentum from moving air to a still surface.
Credit: Monteith and Unsworth (2013).

2.3.2   Heat and mass flux

Because heat and mass are transferred by the same molecular agitation process as momentum, we could draw the same picture for both. In the case of heat, we would conceptualize it as a layer of warm air over cold air, and in the case of mass, higher concentration over lower (think humid air to dry surface). The difference in these two cases is that it is molecular conduction and diffusion transferring the property rather than stress, so the horizontal forces in Figure 2.6 don't exist. Following a similar thought (and mathematical) process we can derive transfer equations for heat and mass. In all three cases the resulting transfer equations take the form:

$$transfer = coefficient \ x \ change \ in \ property/change \ in \ height. \tag{2.10}$$

The coefficient in these equations is dependent on the temperature and pressure of the air.

2.3.3   How does the surface influence the flux?

Now let's bring this back to the boundary layer itself. Flux is an exchange or transfer of a property. So far in this discussion we have been thinking about it as a transfer from the air to the surface (downward arrow in Figure 2.6). But it works the other way too. Heat flows from areas of more (warm) energy to areas of lower (cool) energy. Similarly, momentum and mass move from high to low. As we saw in Chapter 1, the properties of the surface itself influence the air. In general, during the day the ground warms faster than the atmosphere causing a temperature gradient. Heat transfers from the surface to the air. Similarly, when dry air is over a wet surface and the temperature allows evaporation, water transfers from the surface to the air. Carbon also transfers to air from the surface. The processes act in both directions, making the surface a source *or* a sink for all these properties.

2.4   Energy production

When it comes to energy production in the boundary layer there are two main types of energy generation. These two types of energy production include mechanical energy also known as shear, and thermal energy production or buoyancy.

   Mechanical energy production is caused by an increase in frictional drag at the surface often due to an increase in roughness at the surface. For instance, if air is moving over a bare agricultural field to a forest, there would be mechanical energy produced in that transition. This type of energy production typically generates high-frequency and low-magnitude turbulence, which means that the eddies are quite small (Stull, 1988; Oke, 1978; Wyngaard, 2010).

   Thermal energy production is dominated by a temperature gradient from one surface to the next or at different altitudes within the boundary layer. Another name for thermal energy production is buoyancy or convection. Convection is defined as vertical motions in the atmosphere that are dominated by buoyancy forces. Buoyancy force is the vertical force exerted on a parcel of air due to the density differences between the parcel of air and its surrounding atmosphere. The thermal indifferences cause instability in the atmosphere, which then can create large eddies. So, large eddies in turbulence are caused by thermal energy production. Typically, you will see the buoyancy force dominate during the daytime and decrease at night.

2.4.1 Richardson number

Richardson number is used to quantify the ratio of mechanical turbulence to thermal production or destruction of turbulent energy. Basically, it is another way to tell us something about the stability of the atmosphere at any given moment, and characterizes the change of turbulent energy to laminar flow (Foken, 2008; Leclerc and Foken, 2014). The Richardson number is a dimensionless value and can take different forms. The flux Richardson number can be computed as the ratio of the buoyant production term to the mechanical production term in the fully expressed TKE equation. It is sometimes thought to have a critical value of 1 indication if flow is turbulent or becoming laminar (Stull, 1988). Unfortunately, since we need turbulent components to define the flux Richardson number, it is not always possible to compute it. The gradient Richardson number defines the ratio in terms of vertical gradients of potential temperature and wind speed. It too has a critical value, although the exact value is under debate (e.g., Grachev et al., 2013). Finally, a bulk Richardson number can be defined by approximating the gradients with differences at discrete points in space. This version is the most practical in use since truly continuous gradients are not measurable in the real world.

2.4.2 Stability

As we saw in Chapter 1, the boundary layer is ever changing; during the day the convective boundary layer dominates, during the night, the stable boundary layer dominates. These two conditions are defined by the size, shape, and frequency of turbulence. The boundary layer is considered neutral when the surface buoyancy flux is exactly zero, in other words, shear production dominates. This condition rarely occurs because when the surface is heating, buoyancy is positive, and when the surface is cooling, buoyancy is negative. An unstable boundary layer is one where there is an upward heat flux from the surface and turbulence production is enhanced. A stable boundary layer is one where vertical motion is suppressed and there is no radiation/heat input from the surface. We can define stability in terms of the change in potential temperature with height.

$$\frac{\partial \theta}{\partial z} \begin{cases} > 0 & stable \\ < 0 & unstable \\ = 0 & neutral \end{cases} \qquad (2.11)$$

Chapter 6 provides an in-depth view of the stable boundary layer and its implications.

2.5 Turbulent kinetic energy

To quantify the intensity of turbulence we use a derived value of turbulent kinetic energy or TKE. TKE is the full expression of *all* the transport of momentum, moisture, and heat. The equation itself is the sum of the velocity variances for each turbulent component divided by 2. It can be derived from decomposing kinetic energy into the mean and turbulent parts.

$$\bar{e} = 0.5\left(\overline{u'^2} + \overline{v'^2} + \overline{w'^2}\right) \tag{2.12}$$

When fully defined through the derivative and prognostic equations, each component of the TKE equation is quite long and pretty math heavy. Stull (1988) provides a comprehensive breakdown in that way. Here we conceptually describe all the components of TKE.

The definition of TKE comes from the same concepts of budgeting discussed in Chapter 1. For a particular time and space and parcel of air, we account for all the things that contribute or subtract from the turbulence kinetic energy. It looks like Figure 2.7.

In Figure 2.7 the left two terms make up what is happening to the parcel – it is storing energy and that energy is being moved by the mean wind (advection). On the right there are all the other physical processes that can generate or dissipate turbulence. Buoyancy can be a production or loss term (positive or negative depending on the heat flux). It is generally positive during the day and negative at night. Mechanical forces are momentum flux, which occurs in the opposite direction of the mean wind, so it is a production term. Turbulence transport is how TKE is moved by eddies, so it is also generally a production term. Pressure perturbation is the term that represents how pressure redistributes TKE; this suppresses turbulence, so it is a loss of energy term. The conversion of TKE to heat energy is a loss of kinetic energy that is due to viscous dissipation. It is the relative balance of these terms that indicates how well the flow can either maintain turbulence or become more turbulent.

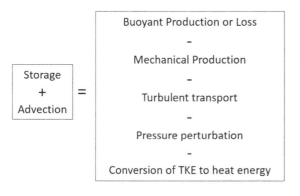

Storage + Advection =

Buoyant Production or Loss

-

Mechanical Production

-

Turbulent transport

-

Pressure perturbation

-

Conversion of TKE to heat energy

Figure 2.7 Breakdown of the components of the turbulent kinetic energy equation. Credit: A. Hiscox.

A key component to understanding turbulence and TKE is that turbulence is not a conserved quantity. Turbulence is a dissipative process that moves kinetic energy to heat energy. So, if no additional generation produces turbulence, then the kinetic energy associated with it will also decrease. In most real cases turbulence never goes to zero because a little is generated locally or transported into a location by the mean wind.

When we look at each of the terms of the TKE individually, we see some general trends, but it is important to understand that we are talking about things that are **always** changing, so general trends are exactly that – a general starting point.

- **Storage** is a rate of change term. Over land there is a strong diurnal variation in how TKE is changing; its value is small in the morning and grows to a peak later in the day. Typical values can range over two orders of magnitude. There is also a height dependency to the storage term. TKE changes faster nearer to the surface, but not touching it. Storage over the ocean is near zero, and there is no diurnal change there.
- **Advection** is a term that can be ignored over large areas but plays a large role in small areas.
- **Buoyancy** is the term that is directly related to heat flux (discussed above). Near the surface of the convective mixed layer the buoyancy term is large and it generally decreases linearly with height. Buoyancy is positive when it is contributing to TKE production – convection, unstable conditions. It is negative when it is consuming TKE; static stability suppresses TKE. Buoyancy is stronger in the vertical than in the horizontal, therefore it is not isotropic.
- **Mechanical** production is also known as the shear term. Like storage, it is also a rate of change term, but it is a rate of change with respect to height. It is generally positive and generally the maximum is near the surface. This is because at the surface we see the maximum shear. Shear is a change of wind speed over some distance. If you think back to the log-wind profile introduced in Chapter 1, wind speed is zero at the surface and increases logarithmically above. The biggest change (i.e., shear) is therefore near the surface. As you get further away, the change in wind speed with height is small, so mechanical production goes nears zero. Shear production is greatest on windy days, and stronger in the horizontal than the vertical, so it is also not isotropic.
- **Turbulent** transport is a flux divergence term. Mathematically it is also a derivative (like storage and shear). It represents the quantity of flux moving into or out of a layer. It is probably the most difficult of the terms to understand conceptually. Overall, its impact is to move turbulence around. It does not create or destroy it like the other terms, it just shuffles it from one place to another. So, it acts as a production or loss at the local scale, but it is zero over the whole mixed layer. Analysis and observation of turbulent transport tells us that some TKE forms at the ground is moved upward before it is dissipated.
- **Pressure perturbation** causes a redistribution of TKE. This goes back to the physics of fluids; we know it happens, so it must be included in the budget. However, in reality, the pressure perturbation term is extremely hard to measure because p' is very small, so this term is normally calculated as a residual. What is left over after determining the values from all the other terms is attributed to pressure. In recent years more focus has been put on this term (He et al., 2020; Finnigan et al., 2020).
- **Dissipation** is always a loss of TKE. When the kinetic energy is changed to heat energy it is no longer TKE. Dissipation is largest near the ground. The smallest eddies dissipate turbulence, so if there are more smaller motions, the dissipation term is larger. Dissipation is the end of an energy cascade where large eddies generate smaller eddies and smaller eddies support viscous dissipation.

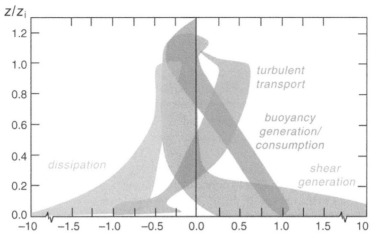

Figure 2.8 Typical ranges of terms in the TKE budget equation during the daytime, composited from observations and numerical simulations.
Figure from Markowski and Richardson (2010), adapted from Stull (1988).

Figure 2.8 shows an example of all the terms and how much they vary. It is a composite of many studies and drives home the point of just how complex turbulence is. Unlike the ideal gas law, which allows us to easily see the connection between temperature and pressure, the variability depicted in Figure 2.8 does not lead itself to a simple nor even a higher order mathematical relationship.

We can combine and compare the relative magnitudes of terms to classify states and give a qualitative understanding of turbulence. If buoyancy is much larger than shear, we can assume "free convection." If shear is greater than buoyancy, we are in "forced convection." Classifying states can determine which parameterizations or approximations are most appropriate for models or other uses. We can also use the TKE equation to tell us about what is happening to the boundary layer. If TKE is increasing, the boundary layer is likely to get more turbulent over time. If TKE is decreasing, the boundary layer is likely to get less turbulent. All the individual contributions depend on the time of day and the properties of the boundary layer and underlying surface.

2.6 The turbulence closure problem

Now that we have defined turbulence and the initial ways we quantify it, we need to take a step back and remind ourselves of the why and the what. Let us remember that we can talk about turbulence in two ways: qualitatively and quantitatively. Qualitatively is not necessarily without numbers, but more about comparing those numbers to get the relative importance of one term vs another. Quantitatively is about really getting down to the math to predict exactly how the atmosphere will change on small time and spatial scales. What we have done earlier in this chapter is identify the things we want to predict (turbulence) and put them in a form and set of units we could possibly measure (i.e., wind speed, temperature, and moisture content). We then defined how these variables work in the mean and turbulent parts. The next steps would be to define equations from the underlying physical principles (conservation of mass, momentum, energy). When we do this, that is, put equations to the concepts discussed above, we always have a leftover term that relates to how those mean variables fluctuated. So, we would then define them in terms of variance and flux – adding the rate of change and adding more terms to the budget. When all is said and done, after you've mentally apologized to your math teacher because you realize that people really do use calculus, we have a big, long equation with a lot of terms that can predict TKE. Unfortunately, if we look a little closer at that equation, we have a problem. We have more variables than we do equations to define them. Having more unknowns than defining equations is a mathematical problem. It means our description of turbulence is not closed – there is always a variable that we cannot predict. This is called the turbulence closure problem. It is a consequence of the non-linear nature of turbulence. The only way to make it work is to parameterize the highest order terms. To parameterize a term means to define it in terms of things we do know and make some approximation for those we do not know. It's a way to handle things where we don't yet know the true physics, or the physics are too complicated to use because of computer/time limitations (Stull, 1988; Lee et al., 2004; Wyngaard, 2010).

The turbulence closure problem is important because it tells us there is always something we have left to figure out. It is also important because it defines the usefulness of any of the models we may develop or use. The turbulence closure problem is also mathematically advanced, so a more detailed explanation is far outside the scope of this book. The turbulence closure problem is also the reason you will see similarity theory applied a lot throughout the rest of the book. It is also the reason the field of boundary layer meteorology continues to exist. We need to make assumptions to create closure – and testing and refining those assumptions is what boundary layer meteorologists are always striving to do.

2.7   Key concepts – The takeaway

This chapter is intended to give you a general understanding of turbulence, why it matters to the air near here, and how we quantity turbulence. The list below is a summary of the topics we believe to be most important in this chapter and hope that you use them to further study micrometeorology.

- The boundary layer is almost continuously turbulent, and the main sources of turbulences are thermal forcing from the surface or vertical wind shear due to friction at the surface.
- We use turbulent kinetic energy (TKE) to measure turbulence intensity.
- Reynolds number tells us something about the level of turbulence. Higher Reynolds numbers mean more turbulence.
- Richardson numbers tells us something about the type of turbulence: mechanical or buoyant, and the shift between the two.

- To maintain turbulence, a supply of energy must be maintained. In the boundary layer that is thermal differences between the air and surface that are always present.
- Turbulence occurs over several orders of magnitude for time and space scales.
- There are two ways to analyze turbulence: (1) statistics and (2) simplified descriptions to find empirical relationships (experiments).

References

Anderson, J.D., 2007. Fundamentals of Aerodynamics, fourth ed. McGraw Hill.

Aubinet, M., Vesala, T., Papale, D., 2012. Eddy Covariance: A Practical Guide to Measurement and Data Analysis. Springer, p. 451.

Finnigan, J., et al., 2020. Boundary-layer flow over complex topography. Bound. Lay. Meteorol. 177, 247–313. https://doi.org/10.1007/S10546-020-00564-3.

Foken, T., 2008. Micrometeorology. pp. 1–306, https://doi.org/10.1007/978-3-540-74666-9.

Geiger, R., Aron, R.H., Todhunter, P., 2009. The Climate Near the Ground: Geiger, Rudolf, Aron, Robert H., Todhunter, Paul: 9780742518575: Amazon.com: Books, seventh ed. Rowman & Littlefield, p. 623.

Glossary of Meteorology, 2022. https://glossary.ametsoc.org/wiki/Welcome. (Accessed March 12, 2022).

Grachev, A.A., Andreas, E.L., Fairall, C.W., et al., 2013. The critical Richardson number and limits of applicability of local similarity theory in the stable boundary layer. Bound. Lay. Meteorol. 147, 51–82. https://doi.org/10.1007/s10546-012-9771-0.

He, J., Cohen, Y., Lopez-Gomez, I., Jaruga, A., Schneider, T., 2020. An Improved Perturbation Pressure Closure for Eddy-Diffusivity Mass-Flux Schemes., https://doi.org/10.1002/ESSOAR.10505084.1.

Leclerc, M.Y., Foken, T., 2014. Footprints in Micrometeorology and Ecology. Springer, pp. 213–214.

Lee, X., 2018. Fundamentals of Boundary-Layer Meteorology. Springer International Publishing.

Lee, X., Massman, W., Law, B. (Eds.), 2004. Handbook of Micrometeorology: A Guide for a Surface Flux Measurement and Analysis. Kluwer Academic Publishers.

Lettau, H., 1939. Atmosphärische Turbulenz. Verlagsges.

Markowski, P., Richardson, Y., 2010. Mesoscale Meteorology in Midlatitudes. John Wiley & Sons, Ltd.

Monteith, J., Unsworth, M., 2013. Principles of Environmental Physics: Plants, Animals, and the Atmosphere, fourth ed., pp. 1–401, https://doi.org/10.1016/C2010-0-66393-0.

Oke, T.R., 1978. Boundary Layer Climates, second ed. Routledge.

Schmidt, W., 1925. Der massenaustausch in freier luft und verwandte erscheinungen. Henri Grand Verlag, p. 118.

Stull, R.B., 1988. An Introduction to Boundary Layer Meteorology. Kluwer Academic Publishers, p. 11.

White, F.M., 1974. Viscous Fluid Flow. McGraw Hill.

Wilks, D.S., 2019. Statistical Methods in Atmoshperic Sciences, 4th. Academic Press, pp. 617–668.

Wyngaard, J.C., 2010. Turbulence in the Atmosphere. Cambridge University Press.

Here, there, and everywhere: Spatial patterns and scales

Sreenath Paleri[a], Brian Butterworth[b,c], and Ankur R. Desai[a]
[a]University of Wisconsin-Madison, Madison, WI, United States, [b]Cooperative Institute for Research in Environmental Sciences, University of Colorado-Boulder, Boulder, CO, United States, [c]NOAA Physical Sciences Laboratory, Boulder, CO, United States

3.1 What is scale

Take a look at Figure 3.1. Do you see any patterns? One or more than one? A quick glance shows that the atmospheric boundary layer (ABL) is characterized by motions and features that occur at multiple spatial and temporal intervals. The concept of "scaling," as it is conventionally referred to, is a fundamental technique and aspect of atmospheric sciences and the fluid dynamics that it rests on. This chapter is meant to provide a general introduction to the broad ideas involved therein with respect to the ABL. While reading this chapter it can be helpful to remember that the atmosphere is a fluid. Therefore, many of the concepts presented here draw from fluid dynamics. It is also helpful to remember the basic linkage between time and space as mediated through a notion of speed: Wind speed is defined as a length per unit time, how much distance can a particular disturbance in the air cover in unit time. We begin with a discussion of scale in general and then present the two key areas where scale is important to the ABL: "scale-invariance," which allows the identification of key relationships as a function of scale, and "scale-dependency," which allows those relationships to vary with conditions external to the ABL.

3.1.1 Range of scales in the atmospheric boundary layer

Atmospheric motions and fluctuations of the state variables span a wide range of spatial and temporal scale from millimeter-sized whorls to the entire planet.

We can define or categorize these scales of motion by their spatial extent or wavelengths involved. Consider winds moving past a rough surface, like a tree or a patchy piece of land or a small hill. It can produce small eddies as it flows past the object. The wavelength of motion involved here would be the distance between an eddy's upward moving arm and downward moving arm.

We can come up with a similar definition for the scales of motion based on their time period or frequency of occurrence. This can help us differentiate between processes that last only a few minutes from the ones that may take up entire days. Also note that we can bring in some sense of periodicity to the atmospheric motions by talking about their frequency of occurrence. There could be hourly, daily, or monthly

Figure 3.1 Open-cell cumulus cloud convection seen by geostationary satellite imagery. Cold air moving over relatively warmer ocean water generates cells of convection with clouds generating in areas of rising air, and clear areas in descending air, following a regular pattern that scales with scales of turbulence in the marine boundary layer.
Credit: From https://cimss.ssec.wisc.edu/satellite-blog/archives/4402.

cycles for the motions concerned. Finally, because wind moves air across space, we can sometimes represent a scale in both time and space scales, with the difference between the two a function of the wind speed (Figure 3.2): spatial scale (meters) = time scale (seconds) × wind speed (meters per second).

In the boundary layer, we generally talk about the eddy-resolving scale or energy-containing scale, from tens of meters to kilometers; the inertial subrange, from centimeters to tens of meters; and the microscale or dissipating scale, from millimeters to centimeters. This hierarchy of motion is connected by a cascade of eddies that transfers energy and momentum across scales from the largest to the smallest.

The largest, energy-containing or eddy-resolving, scale motions are generated by variations in the friction of Earth's surface acting on larger-scale planetary motions and by the differences in heating of the air by surfaces. These "input" energy into the boundary layer. The energy contained in the large scales are transferred to smaller and smaller "scales" of motion that can span from centimeters to the depth of a typical boundary layer, up to 1–2 km. This range of scales is referred to as the inertial subrange where energy is just transferred from large-scale motions to the smallest, microscale motions. Finally, the microscale or dissipating scale motions are the smallest and span from length scales of the order of a few millimeters to a few centimeters. These

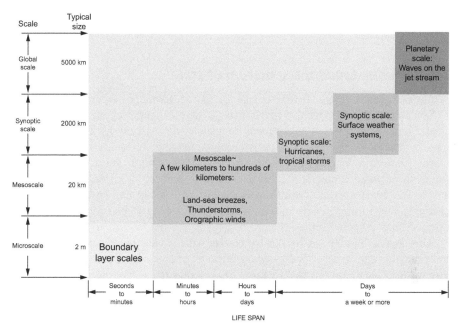

Figure 3.2 Scales of motion in the atmosphere as a function of time scale and space scale. Credit: Meteorology Today, C. Ahrens.

tiny eddies come about from the breaking apart of the larger inertial subrange eddies. Eddies spun up by a gust of wind around a street corner or over a tree branch would fall into this category. They are formed by thermal convection or winds blowing past rough surfaces and can last a couple of minutes. Because so much of boundary layer meteorology is connected to the study of microscale motions, it is also sometimes called micrometeorology.

Each of these three scales have characteristic time scales, which define how fast an eddy of that size will turnover, or how quickly variations occur in wind speed, temperature, and so on. *Generally, the smaller the spatial scale, the shorter the time scale.* Microscale motions generally take tens of milliseconds, inertial subrange motions take seconds to minutes, and eddy-resolving motions occur on scales of minutes to an hour.

Together, these three scales make up the boundary layer. There are larger scales happening too, such as the mesoscale, ranging from a few kilometers to about hundreds of kilometers. This regime is very diverse, including features such as land-sea breezes, thunderstorms, and orographic winds. The next scale, the synoptic scale, is the scale of surface weather systems. Large scale circulations span from hundreds to thousands of kilometers and can live for weeks. The largest scale of motion is planetary scale, with global features such as the waves on the jet stream.

The space and time ranges within these definitions for each class of motion are not very precise, as can be seen in a space time figure (Figure 3.2). Mesoscale motions can last from a couple of minutes to several days and span a couple of hundred meters to kilometers. Despite the lack of strict cutoffs, this classification helps us pick up patterns of behavior and study how they interact with each other to result in the observed features.

3.1.2 Understanding a spectrum of scale

While we can look at scale in time or space, a more informative approach to scaling is to look at the spectral space. By spectral space, we mean representing a given time or spatial series of data into its constituent frequencies or wavelengths. This is generally done by applying a mathematical transformation to the given measurements that represents the same data as values per frequency (Hertz or seconds^{-1}), wavelength (meters), or wave number ($2 \times$ pi/wavelength, in meters^{-1}) instead of by time or space. The transformation from time or spatial series to spectral space can be interchanged, so a spectral representation of data or a signal can be inverted back to its original time or spatial series.

This approach builds on Fourier's theorem, which notes that any periodic, continuous function can be described by an infinite sum of sine and cosine waves. The definition does have certain qualifications, like requiring a periodic and continuous signal. However, we can choose measurement windows such that the data under investigation can be considered as periodic and continuous for the appropriate physics being investigated. There are several mathematical ways to apply Fourier's theorem to observed data. One of the most common is the Discrete Fourier Transform (DFT) and its inverse. DFT works on discrete (non-continuous) data and outputs both the power at any wavelength and its phase. As an example, consider the following time series and related DFT power spectrum (Figure 3.3):

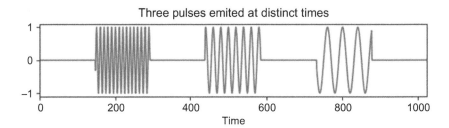

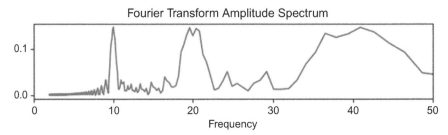

Figure 3.3 Example of a synthetic time series of some signal (*top*) and its representation in frequency-space based on a discrete Fourier transform.

The original time series has three harmonics of different wavelengths (short, medium, long). When the DFT is applied, the wavelengths of each stand out (10, 20, 40 s^{-1}). The peaks vary in sharpness because representing them with continuous sine and cosine functions require that multiple waves are superimposed. More advanced spectral methods, such as wavelet decomposition, allow for finer-grained or more data-specific spectral analysis. Fortunately, many computer programs can perform these transforms easily, but they should still be used with caution, as automated filtering may eliminate time scales of interest.

Generally, in the boundary layer, we are interested in plots of power vs frequency on time series such as wind speed, temperature, or concentration of a gas, usually plotted in log–log or log–linear space, called a power spectrum. The shape of the spectrum (straight line, flat line, non-linear) tells us about how different scales contribute to the signal. The peaks in the power spectrum allow one to identify which frequencies have higher contribution to signal power and are thereby the dominant mode of variability of the signal (an important piece of information to understand and predict geophysical signals). Sometimes it is also useful to plot the covariance or correlation between two variables such as between vertical wind speed and temperature, called a co-spectra. The co-spectra can be used to diagnose which scales contribute to joint variations in two variables. In the case of wind speed and temperature, this would provide the scales that most contribute to sensible heat flux (which is driven by the covariance of vertical wind and temperature in a turbulent boundary layer).

3.1.3   Why scale matters

- Are there universal laws that govern what scales of motion turbulence can generate and how those relate to large-scale parameters of the flow?
- How do scales of variability in surface or atmospheric properties such as surface roughness, land surface temperature, or patterns of turbulent motion influence patterns in boundary layer growth and variation?
- Do these variations influence the rate at which clouds or convection develop over space?

These are all fundamental questions we might want to ask to better understand how weather and climate interact with the surface and the air above it. These types of questions all depend on how scale is defined and identified. The remainder of this chapter discusses how scale is used to understand the boundary layer.

As previously mentioned, scale invariance refers to situations where a variable of interest varies in a consistent way across scales, such that it can be described by a simple relationship. We are getting a little ahead of ourselves here, but one example of this can be seen in Figure 3.5, which shows the power spectrum of a turbulent signal with the frequency content in the signal. The axes have been transformed to their logarithmic values. That means any simple power law relationship between energy and frequency, where one of them varies exponentially as another, would show up as a linear relationship in this log–log plot. Once you've had this idea firm in your minds, have a look at section B of this plot. You can see that this section is seemingly linear. That means, for these ranges of frequency, if we know the values at some scales, we can interpolate them to larger or smaller scales because we know the power law relationship across these scales.

This approach is essential to numerical modeling of the atmosphere, where we discretize the equations of momentum, mass, and energy onto a fixed spatial grid to solve the differential equations through numerical methods. Although the governing equations of motion are the same for a gust of wind, a tornado, and a cyclone, the scales of their physics differ vastly from each other. If the variables of interest are scale invariant, then we don't have to resolve them at all scales to have a realistic numerical representation. We can in fact simplify the governing equations of motion by looking at what are the dominant force balances and including just them in our analysis and adjusting the grid size accordingly and parameterizing all smaller scales.

For synoptic scale motions, we don't need to consider the effects of viscosity, and by scale analysis, it can be dropped from our governing equations. However, anything smaller than the grid resolution needs to be represented in some other way to account for turbulent dissipation by the boundary layer. For example, a mesoscale weather model might be run with 3×3 km grids, so all motion that varies at scales <3 km needs to be represented in the conservation equations. Otherwise, the model may miss key features of how mass and energy are dissipated or generated. Theories that allow us to relate the power spectrum to power–law relationships form the basis of parameterizing the unresolved scale, also known as turbulence closure (Chapter 2) or Kolmogorov scaling, or similarity theory.

There are other times when similarity theory or power–law relationships do not provide an adequate explanation. That is, there is no universal way to describe how scale influences a flow at all scales and situations. In these cases, scale-dependent theories arise. Often the effects of scale-dependence can be subtle, non-linear, or overlapping with other scales. For example, the scale at which patch of ground varies in roughness (from forest to ocean, for example) might generate variation in the depth of the boundary layer, but those variations might disappear at a certain wind speed, which might in itself influence temperature gradients or the roughness of the ocean surface. The patterning of surface properties can generate certain types of features in the atmosphere, where the homogeneity or heterogeneity of surface fluxes might lead to generation of horizontal rolls, open or closed cell convection, areas of mesoscale convergence or divergence, all of which influence cloud development, gravity waves, convective initiation, and other key weather generation phenomena.

3.2 Scale invariance

The movement of winds and meteorological variables in the ABL are governed by the equations of motion. These are a complex set of equations built upon the fundamental conservation laws that account for all the ways in which motions can be affected at a given point, including advection, production/loss by buoyancy or shear forces, turbulent transport, storage, and dissipation. Unfortunately, these partial differential equations cannot be solved analytically. Attempts to solve the unknowns in the equations produces additional unknowns to infinity. This is known as the closure problem in turbulence (Stull, 1988).

Because motions can't be determined analytically there has been much focus in micrometeorology on finding ways to characterize the unknowns. By simplifying the equations with parameterizations for the unknowns we can then solve the equations, as in the case in large eddy simulations (LES). Chapter 5 discusses LES in more detail. Using parameterizations means the solutions are estimates, which may not hold in all situations. However, they are estimates based on empirical evidence from measurement campaigns.

In many cases these parameterizations rely on the fact that the behavior of turbulence and other meteorological variables in the boundary layer repeat across multiple scales – are scale invariant as discussed earlier. If a meteorological feature is scale invariant it means that the pattern of that feature does not change as the scale (e.g., length, time, energy) changes or, more commonly, changes by a common factor across scales. By finding the common factor through empirical studies, those processes can then be calculated by simple equations or parameterizations for use in models (such as numerical weather forecasting or global climate models). Scale invariance has uses for distinguishing atmospheric stability, turbulent vs laminar flow, and predictions of wind speed at specific at heights above the ground.

To begin a description of scales in the ABL it helps to characterize the common processes. One key characteristic of the ABL is that most flows are turbulent (Tennekes and Lumley, 1972). Turbulence refers to the chaotic motions of air, in which eddies of multiple scales swirl simultaneously in three dimensions (as discussed in Chapter 2). While the disorder of turbulence makes it impossible to describe precisely how motions will behave at a given time/place, its behavior in the aggregate is predictable. To illustrate this, we can consider the example of how fluid flows and turbulence develops over a plate or out of a pipe (Figure 3.4). If you followed the molecules of liquid that passed across the plane or out the pipe, you would find that at a given distance down the entrance turbulence has caused them to mix rapidly in an apparently random fashion. However, if you run the experiment over many iterations you would find that the behavior tends toward a mean flow shape, in this case of the flat plate, a logarithmic shape with slower flow near the surface, while out of the pipe, a parabolic shape, with faster flow in the center and slower flow at the edges.

This same repeatability occurs in the ABL, where ensemble averages of turbulent motions show distinct and predictable behavior. For example, if we take measurements of three-dimensional winds from a tripod over a field of grass we would find that at any given moment we might have an updraft or a downdraft. We cannot predict what will happen instantaneously. But if our measurements are taken at a sufficiently fast rate (10–20 Hz) and we average over a period of time (e.g., 30 min) we find that certain regular behavior exists in the winds.

To "see" how this collection of data points behaves in the aggregate, micrometeorologists often observe spectra of the turbulent atmosphere. As described earlier, spectra refer to the use of Fast Fourier Transforms to determine which frequencies are contributing the most to the fluctuations about the mean for an individual variable (power spectra) or the flux of some vector or scalar quantity (co-spectra). Over many field experiments, it has been found that these spectra have a similar shape (Figure 3.5). The spectra show three ranges of motion.

(a)

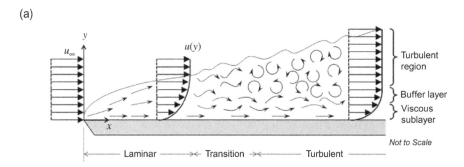

(b)

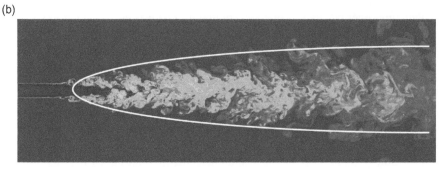

Figure 3.4 (A) Turbulence develops as a fluid passes over a flat plate, showing both the motion of individual molecules (*curved lines*) and the mean flow (*solid arrows*). Similarly, in (B), a numerical simulation of a turbulent jet flowing out of a pipe shows the highly dispersive effects of turbulence on the concentration of the fluid, depicted by the color, as the mean flow (*white lines*) fans out in a parabolic arc.
Panel A: COMSOL, www.comsol.com/blogs/which-turbulence-model-should-choose-cfd-application/; panel B: adapted from Dept. of Energy, Lawrence Livermore National Laboratory, Turbulence Analysis and Simulation Center, https://tasc.llnl.gov/galleries/image-gallery.

1. The majority of energy is found in the larger eddies at the lower frequencies. This region of the spectrum is the energy-containing range. It provides the system with turbulent energy produced from buoyancy and shear forces. The size of the eddies at the peak of this region is known as the integral scale of turbulence (Λ) and is typically around 10–500 m. It is calculated as the product of the mean wind speed and integral time scale of turbulence, or the time scale over which turbulence is correlated (Kaimal and Finnigan, 1994).
2. The middle of the spectrum is the inertial subrange. Here energy is neither produced nor dissipated but transferred from larger to smaller eddies (Kaimal and Finnigan, 1994). Interestingly, this transfer occurs at a consistent rate, whereby energy decreases by 5 decades for corresponding increases in frequency of 3 decades (Kolmogorov, 1941b; Foken, 2017). This is a great example of scale invariance in the ABL.
3. The region of the spectrum at the highest frequencies is the dissipation range. Here the eddies are of a sufficiently small size that they can no longer be maintained due to the viscosity of the air and are converted into internal (or heat) energy (Kaimal and Finnigan, 1994). The length scale of these eddies is typically around 1 mm.

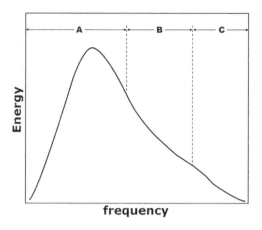

Figure 3.5 Schematic diagram of energy co-spectrum in the atmospheric boundary layer. The spectrum is divided into separate regions where (A) represents the energy production region of larger eddies at lower frequencies, (B) the inertial subrange where energy is passed from larger to smaller eddies, and (C) the frequencies at which dissipation occurs. Adapted from figure 2.1 in Kaimal and Finnigan (1994).

3.2.1 Similarity theory

From spectral analysis we know that there are characteristic length scales of the different processes related to turbulence in the form of its generation, energy cascade, and dissipation. But how do we know at what length scale these processes transition? This is where scaling and similarity theory comes in.

Similarity theory in atmospheric sciences refers to the assembling of sets of variables into dimensionless groups that can be used to relate the behavior of the atmosphere across various scales (Stull, 1988). It is based on a 100-year-old theorem called Buckingham's Π-theorem, in which variables that are important to the process in question are grouped so that their fundamental units (e.g., length, mass, time) cancel out (Buckingham, 1914). The difficulty in the analysis is with hypothesizing which variables are relevant to the process (Foken, 2017). Similarity theories help us establish empirical relationships among groups of these variables, which could be power laws or straight-line fits, and so on. The exact relationship between the groups needs experimental data, physical insight, or even theoretical modeling. However, the final result would be a family of lines or curves that show the same behavior or are similar to each other, hence the name.

By applying similarity theory, we develop dimensionless parameters that do not have any physical dimensions assigned to them, but just real numerical values. Most often, in atmospheric sciences and fluid dynamics, non-dimensional parameters are used to understand force balances or scaling regimes. The values of these numbers can be interpreted to derive conclusions about what forces are driving the physical process being observed.

Two such parameters used widely in boundary layer applications are the Kolmogorov microscales and the Reynolds number. Here we present the definitions and use of these two important parameters.

The Kolmogorov microscale, also known as the dissipation length scale, describe the length scales at which viscous forces resist the continued energy cascade from larger to smaller eddies, and instead turn those small eddies into heat. In a low viscosity fluid like air the Kolmogorov scale (η) is typically around 1 mm. This means you should be able to observe eddies that have lengths about 1 mm or greater. Smaller than this the viscosity of the air impedes their motion and the kinetic energy of those small eddies contained gets converted to heat energy. However, the exact length of this scale does vary. The dissipation rate is governed by the speed of the air (u) and the length scale (l) of the surface over which it is moving (Foken, 2017). So, while the viscosity of air remains relatively constant, η will change based on the flow characteristics. By being a dimensionless relationship between the relevant variables, the value of η accounts for that change. For example, when wind speeds are weaker and average length scale of the eddies is longer, η will increase.

This relationship relates directly to another example of similarity: the Reynolds number. The Reynolds number is defined as the ratio of inertial forces to viscous forces for a fluid in motion. The Reynolds number addresses the question of when flows will become turbulent. This is an important question in the ABL because turbulence affects atmospheric stability, ABL development, and dispersion. It turns out to be a balancing act between two different types of forces acting on the air: inertial forces and viscous forces. Inertial forces can be thought of as the forces arising from the air being in motion, while viscous forces represent the resistance of the liquid to deformation. A fluid with low viscosity (e.g., air) does not resist deformation strongly, while a thicker liquid (e.g., honey) has a high viscosity and more strongly resists deformation.

Reynolds number is derived from the Navier–Stokes equations for the conservation of momentum (three equations for the three Cartesian directions) where each term of the equation represents one of the different forces at play. After taking the ratio of inertial to viscous forces and canceling the units the Reynolds number is:

$$Re = \frac{UL}{\nu} \tag{3.1}$$

where U is a characteristic velocity scale of the wind, L is a characteristic length scale of the wind, and ν is the kinematic viscosity of the air (Garratt, 1992). Studies have found that at Reynolds numbers above a threshold value (critical Reynolds number, Re_c) the fluid will become turbulent. If it drops too far below Re_c, it will tend to transition from turbulent flow back to laminar flow. An interesting feature of turbulence is that once the flows become turbulent, they become Reynolds number independent. This means that after turbulence initiates, it behaves by the normal characteristics of turbulence whether the Reynolds number is just above Re_c or several orders of magnitude above it. This is specifically true at larger scales. However, at smaller scales the Reynolds number will continue to impact the characteristic length scale of motion (i.e., the Kolmogorov microscales). Higher Reynolds numbers, which represent turbulence with more energy at the top of the energy cascade, will support smaller scales of motion before energy is dissipated by viscosity (Stewart, 1968).

The more exciting conclusion one can draw is that flows with the same Re behave similarly because we have condensed all the dimensional information into this one scaled parameter. So, flow in a scaled model like a laboratory experiment can be used to represent a real, physical flow. Such a relationship is referred to as dynamical similarity.

A key point to keep in mind is that similarity relationships have an implicit assumption of steady state or equilibrium since they have no time tendency terms in relationships. In the ABL, they can be used to describe the equilibrium profiles of mean and turbulent variables as a function of height or position. Time-dependent variables, such as the ABL height, have to be measured independently and then included as part of a dimensionless group.

3.2.1.1 Similarity in the neutral surface layer

The atmospheric surface layer is considered to be the lowest 5% ($\sim$100 m) of the boundary layer, where shear due to the surface is the dominant producer of turbulence. In a neutral surface layer, there is no buoyant turbulence production or consumption, and shear production (or consumption) dominates. The surface layer is also called a constant flux layer, where surface–atmosphere fluxes vary by less than 10% of their magnitude with height. In more mathematical terms, the vertical flux divergence is small and negligible. This assumption simplifies the description of the layer by using the flux at just one height. This "height" or level can be chosen as the surface, thus making life and the science simpler. Hence, height above the surface can be included as a likely scaling parameter for any non-dimensional groups being prepared.

A widely used scale in surface layer studies is friction velocity (u_*), a velocity scale that represents the wind stress on the surface as:

$$u_* = (|\tau/\rho|)^{1/2} \tag{3.2}$$

where τ is wind stress and ρ is air density (Sutton, 1953). Since, in the surface layer, fluxes are typically constant with height, friction velocity is related to changes in the mean wind with height by:

$$\frac{\partial U}{\partial z} = \frac{u_*}{kz} \tag{3.3}$$

where U is mean wind, z is height, and k is the Von Karman constant, experimentally found to be $\sim$0.4. Integrating this equation over z gives

$$U(z) = \frac{u_*}{k}\ \ln\left(\frac{z}{z_0}\right) \tag{3.4}$$

where z_0 is the roughness height or height at which wind speed goes to zero near the surface. Such an equation is extremely useful as it allows one to predict what the wind speed is at a chosen height above the ground by measuring turbulence at the surface.

The above relationship gives a functional form for the mean wind profile. Scaled by u_* the mean wind varies as the natural log of z. All profiles, measured on different days, over different surfaces (with different z_0) will behave "similarly." This equation holds for neutrally stable boundary layers, where the turbulent kinetic energy generation is mechanical, by wind shear and surface stresses. Under such stability conditions, the wind speed increases linearly with height on a log scale.

Our constant of integration under these assumptions, z_0, becomes the height above the surface where the wind speed reaches zero. It is referred to as the aerodynamic roughness length. It is a property of the underlying surface and does not depend on atmospheric properties such as wind speed, shear stress, and so on, and in turn is determined by the individual components that make up a surface, which are referred to as "roughness elements." For example, a tree is a roughness element and so is a building, while a forest or a city becomes a surface. Each surface types have their own values of z_0, which have to be inferred from targeted field measurements, by extrapolating the log wind data. A common empirical relationship between z_0 and the height of the roughness elements is $z_0 = 0.15\,h$, with h being the height of the individual element, like the height of the tallest tree. More specific cases of z_0 and its applications can be found in Section 3.3 of this text.

A dense forest canopy or a cityscape can act like a displaced surface, higher than the actual underlying surface. To correct for such effects, the idea of displacement length (d) is used. We then introduce d into the log wind profile, such that the mean U goes to zero at $z = d + z_0$.

$$U(z) = \frac{u_*}{k} \ln\left(\frac{z-d}{z_0}\right) \tag{3.5}$$

3.2.1.2 Non-neutral surface layers

Unfortunately, neutrality is not the norm in surface layers. Buoyant eddies can play a role in the turbulent kinetic energy production on a clear sky day with a lot of solar radiational heating incident at the surface or when you have a land surface interspersed with forests and waterbodies, introducing patchiness, and so on. In such cases, buoyant effects can become important quite easily. Shear instabilities could remain as the dominant turbulence production terms in non-neutral surface layers too since we are in the "surface" layer. In either case, whether shear driven or buoyancy driven, the ABL can become stable or unstable. In such cases we would need to modify our empirical similarity relationships with the additional physics that comes into play. As a result, wind and other profiles behave differently in both stable and unstable environments (Figure 3.6).

Fortunately, Eq. (3.5) can be extended for stable and unstable cases by adding an additional, empirically found correction, that is broadly called Monin–Obukhov similarity theory (MOST), so named for the two scientists who came up with it.

Under MOST, the quasi steady-state turbulence immediately above a flat, horizontally homogeneous land surface is governed by the following parameters:

1. l a length scale of the turbulence, taken as z, distance from the surface
2. velocity scale of the turbulence, taken as u_*, the friction velocity

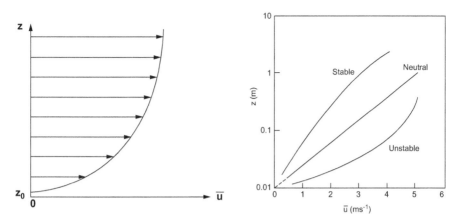

Figure 3.6 Theoretical wind profile showing winds going to zero at roughness height (z_0) and log wind profile under stable, neutral, and unstable conditions.
Left: Adapted from figure 9.4 in Stull (1988); *Right*: figure 1.6 from Kaimal and Finnigan (1994).

3. mean surface fluxes of
   a. temperature, Q_0
   b. conserved scalar, C_0
4. a buoyancy parameter g/θ_0, θ_0 being the near-surface potential temperature.

These parameters govern dependent variables such as the mean vertical gradients of wind, potential temperature, and the mixing ratios of conserved scalars as well as the turbulent statistics. Just as u_* was used as a scaling variable for neutral surface layers, under MOST, the relevant scaling parameters for temperature and scalar transport turn out to be:

1. $\theta_* = \dfrac{-Q_0}{u_*}$

2. $c_* = \dfrac{-C_0}{u_*}$ (3.6)

3. $L = \dfrac{-u_*^3 \theta_0}{kgQ_o}$

L is referred to as the Obukhov length, the characteristic length scale of interest here.

With these scaling parameters, non-dimensionalized mean vertical gradients of wind, potential temperature, and so on are diagnosed as functions of the non-dimensional Obukhov parameter, $z/L = \zeta$. Unlike a neutral surface layer, where one could construct empirical relationships using normalized ratios of governing variables, in the more complex non-neutral surface layer, these relationships become functional relationships and the observational data to do so are collected from targeted field experiments. The Obukhov length is often interpreted to be related to stability in the atmosphere:

From Eq. (3.6),

for neutral conditions, the surface heat flux, $Q_0 = 0$ and $L \to \infty$,
for unstable conditions with a positive heat flux at the surface, $Q_0 > 0$ and $L < 0$.
and for stable conditions with a negative heat flux at the surface, $Q_0 < 0$ and $L > 0$.

Under MOST, if one knows velocity, temperature, or scalar gradients at multiple heights or grid levels, and the surface fluxes, one can estimate near surface meteorology such as 2 m temperature. This is the conventional way of calculating surface meteorology in most operational weather and climate models. However, now you can see why this is not an ideal method since MOST remains a largely idealized treatment over flat, homogenous surfaces and expects strong wind speeds and shear, by including u_* in the parameterizations.

For the wind profile, this equation can be integrated to find wind speed with height as:

$$U(z) = \frac{u_*}{k} \left[\ln\left(\frac{z}{z_0}\right) + \psi\left(\frac{z}{L}\right) \right] \tag{3.7}$$

where $\psi\left(\frac{z}{L}\right)$ represents the height integrated correction factor (called the Businger–Dyer factor) as a function of z/L, which is a measure of the atmospheric stability (Businger et al., 1971).

The Richardson number is an alternative stability parameter. It is derived from force balances of the vertical component of Navier–Stokes equations of momentum. We can do a very similar analysis to the Reynolds number, which gives a turbulent Richardson number, Ri_t:

$$Ri_t = \frac{\text{buoyancy force/mass}}{\text{inertia force/mass}} \sim \frac{g\theta l}{\theta_0 u^2} \tag{3.8}$$

With higher values of Ri_t, turbulent buoyant forces become more significant. Ri_t can also be expressed in terms of L.

Above the surface layer, in the mixed layer or for a convective boundary layer, eddies tend to scale with the whole ABL height and not just height above the surface. That would mean instead of z, height above the surface we will have to consider z_i, height of the ABL.

In that case, another common scale used to describe the ABL is the convective velocity scale (w_*). This is similar to friction velocity except it is used to characterize vertical motions in the entire boundary layer. It is calculated by combining gravitational acceleration, temperature, boundary layer height, and surface heat flux into a dimensionless group (Deardorff, 1970). It can be used to generate a time scale calculated as

$$t_* = \frac{z_i}{w_*} \tag{3.9}$$

where z_i is ABL height. It describes roughly the time it takes for free convection to force one complete overturning of air from the top to the bottom of the ABL (Stull, 1988). Diurnal velocity and time scales are shown in Figure 3.7. You can see that in the morning around sunrise the time scale is short because the ABL height is low. In the midday it lengthens as the ABL rises. And then in the evening it continues to increase, even as the w_* decreases, because the boundary layer height is still at its maximum, but the strength of upwards convection (represented by w_*) decreases, meaning there is less energy for these overturning eddies (Figure 3.7).

When viewing these types of dimensionless parameters we can see that often the behaviors collapse along specific lines, slopes, or curves. For example, if we view the mean variance in vertical winds $(\overline{w'^2})$ with height at different times we would expect to have curves of varying magnitude, rising to differing boundary layer heights (Figure 3.8). However, if we use our convective velocity scale (w_*) squared as a scaling value and we plot it against normalized height (z/z_i), we find that the relationship

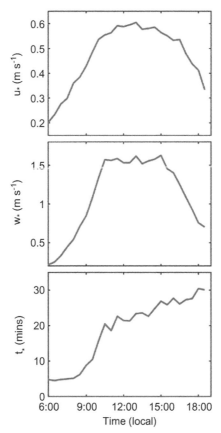

Figure 3.7 Mean daytime friction velocity ($u\_^*$), convective velocity scale ($w\_^*$), and free convection time scale ($t\_^*$) measured during the CHEESEHEAD19 field campaign in northern Wisconsin in 2019.

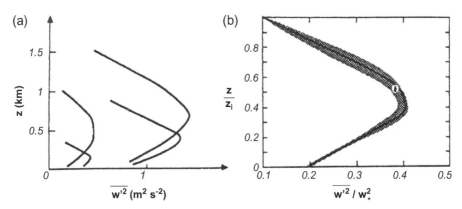

Figure 3.8 (A) hypothetical vertical profiles of $(w^\wedge('^\wedge2))^-$ and (B) the same profiles from subplot (A) normalized (or scaled) with w_*. *Shading* refers to the range of the curves. This plot is figure 9.3 from Stull (1988).

collapses along a common curve. The implications for prediction are great because we can then estimate with reasonable certainty $\overline{w'^2}$ whenever we have measures of ABL height and w_*.

3.3   Scale-dependence

By now you can see that there is a lot of idealization involved in the similarity theory approach. This is justifiable because of the complexity of real-world flows and situations. The important question to keep in mind is when are the idealizations not enough or where is scale invariance not valid. The idealized theories start to break down when the assumptions of neglecting the impacts of surface heterogeneities start breaking down over real surfaces. When mesoscale variations of land surface properties start influencing atmospheric properties at heights such that the fluxes at these heights are no longer equal to the surface fluxes (i.e., there is no constant flux layer because there is a divergence of surface fluxes in the vertical) then the previously formulated similarity theories such as MOST fail at estimating the fluxes (Mahrt, 2000). In such cases no universal scaling law would be applicable to describe the ABL response to surface forcings. We will have to come up with scale-dependent theories.

3.3.1 Blending length formulations

One of the earliest attempts to diagnose the impacts of surface heterogeneity on the vertical structure of the atmosphere uses the concept of a blending height (Mahrt, 2000; Raupach and Finnigan, 1995; Wood and Mason, 1991). Blending height theories ask at what height above the surface does the influence of surface heterogeneities stop impacting changes in the large scale and turbulent nature of the atmospheric variables. For heights larger than such "blending heights," the surface can then be considered as homogeneous and uniform, as is assumed in the earlier similarity theory approaches. The general form of such a blending height formulation is as follows:

$$z_{\text{blend}} = C\left(\frac{u_*}{U}\right)^2 L_{\text{hetero}} \qquad (3.10)$$

These heights would depend on the length scales of the underlying heterogeneities, L_{hetero}. So, building from such considerations, we can infer a minimum length scale of these heterogeneities that can influence atmospheric properties at the heights. In the ABL literature, researchers have come up with broadly three length scales for surface heterogeneities, depending on the stability conditions:

(1) In case of near-neutral or stable conditions, when the shear production of turbulence by local diffusive mixing is dominant

$$L_{\text{blend}} = C_{\text{blend}} z \left(\frac{U}{u_*}\right)^2 \qquad (3.11)$$

This length depends on the mean wind speed, U at height z, and the surface shear, u_*. C_{blend} is a drag coefficient to describe the surface.

(2) For unstable conditions, when buoyancy effects are important (Wood and Mason, 1991)

$$L_{wm} = C_{wm} z \frac{U\Theta_0}{\overline{w'\theta'}_{sfc}} \qquad (3.12)$$

Referred to as a thermal blending height formulation, for L_{wm} the absolute value of the spatially averaged surface heat flux, $\overline{w'\theta'}_{sfc}$ and the spatially averaged potential temperature Θ_0 also enter into the formulation to account for their influences.

(3) For the convective boundary layer, when there are eddies that scale the entire depth of the boundary layer and cause bulk mixing (Raupach and Finnigan, 1995)

$$L_{Rau} = C_{Rau} \frac{U z_i}{w_*} \qquad (3.13)$$

where w_* is the previously introduced velocity scale for the convective boundary layer.

The takeaway from these theories is that, at any height z, there is a minimum length scale, L_{hetero}, for horizontal heterogeneities to induce spatial variations in atmospheric properties. So, to infer the vertical impact of differing scales we can use a non-dimensional parameter like L_{hetero}/L_x, where L_x is one of the abovementioned scales as appropriate.

   If we calculate L_x using z_i, the height of the boundary layer, and find that the length scales of surface features are much smaller than the calculated L_x values, we can infer that $z_{\text{blend}} \ll z_i$, that is, the blending heights are much smaller than ABL heights. Then we can use MOST to predict atmospheric properties including surface fluxes. Such heterogeneities are called microscale heterogeneities.

But, if the L_{hetero} length scales are larger such that their vertical influence affects a significant fraction of the boundary layer, then the flow above the blending heights would be too high to estimate surface fluxes.

$$0.05z_i < L_{blend} < z_i \tag{3.14}$$

where $0.05z_i$ is an approximate surface layer height.

These are mesoscale heterogeneities. At these heights, there is significant changes in surface–atmosphere fluxes from surface values and MOST cannot be used to estimate spatially averaged values.

If the observed length scales are much larger than the predicted ones, $L_{hetero} \gg L_{blend}$, then we have macroscale heterogeneity. In such scenarios the boundary layer establishes an equilibrium with the local surface type and similarity theory can be applied without considering the changes in surface conditions far ahead. This concept is conceptualized in Figure 3.9. However, macroscale heterogeneity can lead to large variations in boundary layer properties across space, which may then interact and mix in complex ways as the large-scale wind mixes them, generating what are called "internal boundary layers" and/or mesoscale circulations. This concept forms the basis for generation of phenomena like sea breezes or urban heat islands.

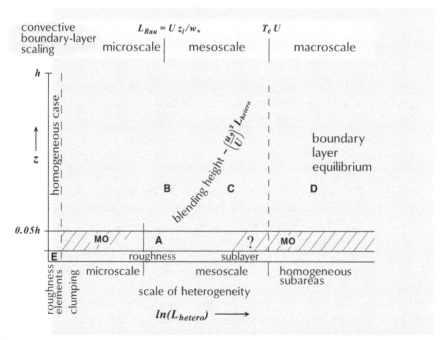

Figure 3.9 Scale of heterogeneity and relationship with blending height, from Mahrt (2000).

3.3.2   Representing scale dependence in the atmosphere

We can see that the blending height approaches to quantifying the effects of surface heterogeneities cannot help us with heterogeneities of the order of mesoscale. Also, these approaches do not give an estimate of the impact of the amplitude of variations. There have been efforts to understand the quantitative impacts of heterogeneity amplitude alongside scales on the ABL properties starting with the early modeling works of Avissar (1995). These works showed how variations in surface properties such as topography, surface roughness, surface soil moisture, stomatal conductances, and so on can affect surface fluxes and the turbulent characteristics. These studies also demonstrated that such variations can induce mesoscale circulations under appropriate environmental conditions. Consequentially, these studies pointed out that scaling of exchange processes and lower atmospheric properties appear to be strongly non-linear and cautioned against scale invariant theories. An approach or a model developed for one particular scale and surface parameters cannot be applied for processes at another scale. They also pointed out the previously overlooked factor of the spatial distribution of variations. The modeling studies specifically sought to investigate the ABL response to spatial variations of surface properties and found that they are strongly linked as well.

Through similarity theories we've seen attempts at trying to relate the mean or larger scale motions with the turbulent characteristics (such as u_*, θ_*, etc.). These approaches help us formulate relationships between surface turbulent fluxes and measurable or observed mean gradients. Models use discretized equations and have to be integrated. Having terms representing the effect of the smaller scales on the "resolved" larger scales of motion is all well and good, but how do we account for them in the more realistic scenarios of operational weather and climate models? This is one of the most active and long-standing research questions of boundary layer meteorology, an area of active research and field experiments!

As discussed at the beginning of this chapter, numerical models discretize the governing equations and solve them numerically on a discrete grid. That means only some scales are resolved and some become too small to be "seen" on this numerical grid. Subgrid parameterizations and closure schemes for these turbulent variance and covariance terms in the budget equations require linking them to the known, resolved values or properties of the mean flow or other physically significant parameters. This leads to the development of simple models where the turbulent fluxes or properties are related to the mean flow, just like similarity approaches. Eddy viscosity approaches assume that the growth and decay of turbulent eddies is a "flow" property and can be represented by an "eddy diffusivity" for momentum or temperature. The assumption is that, just as the dynamic viscosity parameterizes the resistance to an external flow through a fluid, the eddy diffusivities parameterize how eddies decay for the given flow and conservation equations. These can then be used to quantify the unresolved scale variables and "close" the time-integrated equations.

In contrast, large eddy simulations resolve turbulence, but not entirely. In an LES, the turbulence in the larger, energy-containing scale is explicitly simulated and the smaller scales of turbulence are parameterized with some flow properties like turbulent kinetic energy (kinetic energy of turbulent eddies).

These modeling forays and experiments showed conclusively that the smaller scale turbulent processes have a significant influence on the mean properties and nature of the ABL and by extension the free atmosphere. Pioneering modeling studies called for more extensive observational data to benchmark and inform future numerical experiments. We will go over a couple of experiments, designed to understand the nature of surface atmosphere exchange over heterogeneous surfaces and how scale plays a role in such exchange processes.

The Boreal Ecosystem-Atmosphere Study (BOREAS) (Sellers et al., 1995) was initiated as a large-scale international investigation focused on improving the understanding of the exchanges processes between the boreal forest and the lower atmosphere. The experiment was aimed to improve process models and develop methods to implement those process models over large scales in global climate models. An intensive measurement campaign was conducted that involved the first generation of towers and aircrafts that measured the turbulent surface fluxes. The researchers came up with a nested experiment design, taking measurements at a wide range of scales, since some of the important governing processes can only be studied at very small spatial scales (e.g., the links among leaf biochemistry, spectral properties, and photosynthesis) but had to be linked to global climate processes.

The International H_2O Project 2002 (IHOP) (Weckwerth et al., 2004) took place in the Southern Great Plains (SGP) of the United States from May 13 to June 25, 2002. The key research focus of the measurement campaign was to better understand and predict ABL processes, convection initiation, and precipitation, and their relationship to atmospheric water vapor. IHOP 2002 data provided an unprecedented look at the interplay of larger-scale moisture evolution with the land surface, vegetation, and terrain across a region with a strong east–west rainfall gradient that is reflected in changes in land use.

LITFASS-2003 (Beyrich and Mengelkamp, 2006) was a mesoscale field experiment carried out in the heterogeneous landscape around the Meteorological Observatory Lindenberg (MOL) of the German Meteorological Service in May and June 2003. It was carried out as a field component of the Evaporation at Grid/Pixel Scale (EVA_-GRIPS) project. The primary research goal of EVA_GRIPS was to determine the area-averaged evaporation over a heterogeneous land surface at the scale of a grid box of a regional numerical weather prediction or climate model, and at the scale of a pixel of a satellite image. Basically, the goal was to scale evaporative fluxes from a sub-grid level to the grid level. LITFASS-2003 pushed this agenda a bit further, trying to include fluxes of heat and momentum as well (all three turbulent fluxes of relevance in the ABL).

Experimental determination of surface fluxes on a variety of spatial scales was achieved by employing micrometeorological flux stations and an array of relevant instrumentation. Surface energy fluxes were also derived from satellite data. Modeling works using different Soil–Vegetation–Atmosphere Transfer schemes, an LES model, and mesoscale atmospheric models clearly demonstrated how boundary layer circulations can develop over heterogeneous land surfaces.

The Land Atmosphere Feedback Experiment (Wulfmeyer et al., 2018) was a field experiment designed to investigate the complex feedbacks at play in surface–atmosphere processes. The key objective was to develop and test scalable combinations of the parameterization of surface fluxes and boundary layer turbulence in heterogeneous terrain since current parameterizations have suboptimal representations of L–A feedback. This can lead to erroneous representations of the diurnal cycle of ABL dynamics and thermodynamics including the simulation of convection initiation. The experiment involved development and deployment of new instrumentation and a novel synergy of these instruments. LAFE demonstrated that insights about L–A interaction and feedback would require the application of new observations, as the whole L–A system needs to be investigated from the land surface throughout the PBL higher up into the free troposphere (including the effects of entrainment!) One key takeaway is that the combined measurements of dynamical and thermodynamical fields are important.

The Idealized Planar-Array Study for Quantifying Surface heterogeneity (IPAQS) focused on the atmospheric surface layer. A high-resolution field experiment was conducted over the playa in Utah's West Desert, where the terrain is characterized by a relatively uniform rough surface with differences in both space and time of surface temperature and soil moisture. Such measurements can help isolate the effects of spatial gradients in surface temperature and moisture and come up with relationships between these and surface–atmosphere exchange processes. The measurement and complementary modeling efforts seek to answer questions about representing finer scale heterogeneities in the Atmospheric Surface Layer (ASL) where spatial averaging of these features does not result in realistic representations of the underlying processes.

The Chequamegon Heterogeneous Ecosystem Energy-balance Study Enabled by a High-density Extensive Array of Detectors (CHEESEHEAD) (Butterworth et al., 2021) is an intensive field campaign designed specifically to address long-standing puzzles regarding the role of ABL responses to scales of spatial heterogeneity in surface–atmosphere heat and water exchanges. An extensive network of flux towers, along with an array of instruments were deployed within a 10×10 km domain in the Chequamegon – Nicolet National Forest in Northern Wisconsin. The high-density observing network is coupled to LES and machine learning scaling experiments to better understand sub-mesoscale responses and improve numerical weather and climate prediction formulations of sub-grid processes. The key research objective is to advance spatiotemporal scaling methods for heterogeneous land surface properties and fluxes and theories on the scales at which the lower atmosphere responds to surface heterogeneity.

As you can see, there is a lot we know about scales in the boundary layer, but several key frontiers in the role of representing complex real-world surfaces and relationships are still being explored.

References

Avissar, R., 1995. Scaling of land-atmosphere interactions: an atmospheric modelling perspective. Hydrol. Process. 9, 679–695. https://doi.org/10.1002/hyp.3360090514.
Beyrich, F., Mengelkamp, H., 2006. Evaporation over a heterogeneous land surface: EVA_GRIPS and the LITFASS-2003 experiment—an overview. Bound. Lay. Meteorol. 121, 5–32. https://doi.org/10.1007/s10546-006-9079-z.

Buckingham, E., 1914. On physically similar systems: illustrations of the use of dimensional analysis. Phys. Rev. 4, 345.

Businger, J.A., Wyngaard, J.C., Izumi, Y., Bradley, E.F., 1971. Flux-profile relationships in the atmospheric surface layer. J. Atmos. Sci. 28, 181–189. https://doi.org/10.1175/1520-0469(1971)028<0181:FPRITA>2.0.CO;2.

Butterworth, B., Desai, A.R., et al., 2021. The Chequamegon ecosystem energy-balance study enabled by a high-density extensive array of detectors (CHEESEHEAD19). Bull. Am. Meteorol. Soc. 102, E421–E445. https://doi.org/10.1175/BAMS-D-19-0346.1.

Deardorff, J.W., 1970. Convective velocity and temperature scales for the unstable planetary boundary layer and for Rayleigh convection. J. Atmos. Sci. 27, 1211–1213. https://doi.org/10.1175/1520-0469(1970)027,1211:CVATSF.2.0.CO;2.

Foken, T., 2017. Micrometeorology, second ed. Springer-Verlag, Berlin, Germany, https://doi.org/10.1007/978-3-642-25439-0.

Garratt, J.R., 1992. The Atmospheric Boundary Layer. Cambridge University Press, Cambridge, UK. ISBN 0-521-38052-9.

Kaimal, J.C., Finnigan, J.J., 1994. Atmospheric Boundary Layer Flows. Oxford University Press, Oxford. ISBN 9780195062397.

Kolmogorov, A.N., 1941b. Rassejanie energii pri lokolno-isotropoi turbulentnosti (Dissipation of energy in locally isotropic turbulence). Dokl AN SSSR. 32, 22–24.

Mahrt, L., 2000. Surface heterogeneity and vertical structure of the boundary layer. Bound. Lay. Meteorol. 96, 33–62. https://doi.org/10.1023/A:1002482332477.

Raupach, M.R., Finnigan, J.J., 1995. Scale issues in boundary-layer meteorology: surface energy balances in heterogeneous terrain. Hydrol. Process. 9, 589–612. https://doi.org/10.1002/hyp.3360090509.

Sellers, P., Hall, F., Margolis, H., Kelly, B., Baldocchi, D., den Hartog, G., Cihlar, J., Ryan, M.G., Goodison, B., Crill, P., Ranson, K.J., Lettenmaier, D., Wickland, D.E., 1995. The boreal ecosystem–atmosphere study (BOREAS): an overview and early results from the 1994 field year. Bull. Am. Meteorol. Soc. 76, 1549–1577. https://doi.org/10.1175/1520-0477(1995)076<1549:TBESAO>2.0.CO;2.

Stewart, R.W., 1968. National Committee for Fluid Mechanics Films Movie: Turbulence. Education Development Center, Waltham, MA. http://web.mit.edu/hml/ncfmf.html.

Stull, R., 1988. An Introduction to Boundary Layer Meteorology. Kluwer Academic Publishers, Dordrecht, Netherlands, p. 9027727686.

Sutton, O.G., 1953. Micrometeorology. McGraw-Hill, New York, NY, p. 333.

Tennekes, H., Lumley, J.L., 1972. A First Course in Turbulence. MIT Press, Cambridge, MA, p. 284. ISBN 0-262-20019-8.

Weckwerth, T.M., Parsons, D.B., Koch, S.E., Moore, J.A., LeMone, M.A., Demoz, B.B., Flamant, C., Geerts, B., Wang, J., Feltz, W.F., 2004. An overview of the international H2O project (IHOP_2002) and some preliminary highlights. Bull. Am. Meteorol. Soc. 85, 253–278. https://doi.org/10.1175/BAMS-85-2-253.

Wood, N., Mason, P.J., 1991. The influence of static stability on the effective roughness lengths for momentum and heat transfer. Quart. J. Roy. Meteorol. Soc. 117, 1025–1056.

Wulfmeyer, V., Turner, D.D., Baker, B., Banta, R., Behrendt, A., Bonin, T., Brewer, W.A., Buban, M., Choukulkar, A., Dumas, E., Hardesty, R.M., Heus, T., Ingwersen, J., Lange, D., Lee, T.R., Metzendorf, S., Muppa, S.K., Meyers, T., Newsom, R., Osman, M., Raasch, S., Santanello, J., Senff, C., Späth, F., Wagner, T., Weckwerth, T., 2018. A new research approach for observing and characterizing land–atmosphere feedback. Bull. Am. Meteorol. Soc. 99, 1639–1667. https://doi.org/10.1175/BAMS-D-17-0009.1.

The known unknowns: Measurement techniques

Marc Aubinet

TERRA Teaching and Research Centre, University of Liege, Gembloux, Belgium

4.1 Introduction

With the development of new measurement techniques, notably in anemometry and gas concentration measurement, surface flux measurements have widely extended in the last 20 years and are now at the heart of ecosystem and atmospheric science. As a confirmation, thousands of papers based on flux measurements or referring to them have been published recently (see a.o. the review by Baldocchi, 2020).

The term "surface fluxes" describes fluxes of *tracers* exchanged by the *terrestrial surface* and transported to or from the atmosphere by *atmospheric processes*. By tracer, we mean any scalar that can be transported in the air. This includes, in addition to sensible heat, all gases (notably water vapor, carbon dioxide, methane, nitrous oxide, volatile organic compounds, ammonium, and others discussed throughout this book) or particles. Potentially all terrestrial surfaces may be investigated, including forests, savannas, peatlands, grasslands, crops, pastures, bare soils or even ice packs, lakes, seas, or urban areas. The common point is that all these tracers, wherever they come from or wherever they go, are transported by atmospheric processes, which are dominated by *turbulent transport*. All the measurement techniques described herein (but one) measure directly or indirectly this turbulent transport.

Section 4.2 reviews the mechanism by which turbulent transport operates and Section 4.3 provides some suggestions relative to the design of measurement setups. Sections 4.4 and 4.5 present different flux measurement techniques, including *eddy covariance (EC)*, which is the most widely used. The chapter ends with an overview of the main micrometeorological measurements that are generally needed to complement flux measurements. The descriptions of the techniques in this chapter were deliberately kept simple avoiding complicated mathematics, but references are given for those interested in more details. The reader may refer to the general textbooks or papers on this topic: Aubinet et al. (2012a), Burba (2013, 2021), Rebmann et al. (2018), and Mauder et al. (2021).

4.2 Transport in the surface layer

The main transport process in the surface layer is *turbulent transport*, which is driven by eddies. When the surface is homogeneous and horizontal and when turbulence is well developed, this is practically the sole transport process at work in the surface

Conceptual Boundary Layer Meteorology. https://doi.org/10.1016/B978-0-12-817092-2.00011-4

layer. All but one of the flux measurement techniques presented here are focused on turbulent transport.

Figure 4.1 presents an example of turbulent transport. Let's consider the vertical turbulent transport of water vapor from a wet surface to the atmosphere. Due to their rotary movement, eddies move the air upward and downward between the surface and the air above. If a tracer, water vapor in this case, is present at the surface, eddies may capture some molecules there and transport them to the air aloft where they release them. As a result, the air transported by eddies will be more humid when going upward than when going downward. This suggests that, in presence of turbulent transport, the covariance between the vertical component of velocity (w) and the tracer concentration is not zero. More rigorously, it can be demonstrated that the turbulent flux may be deduced from this covariance. It is equal to the flux exchanged by the surface provided that other transport (advection) or storage processes are negligible, which is the case when homogeneity and stationarity hypotheses are fulfilled. For more details see Chapter 2 and for a complete demonstration see Foken et al. (2012a).

Ideally, a system measuring turbulent fluxes of a tracer should thus be able to capture the covariance between the vertical wind, w, and the tracer concentration (in the case of sensible heat, the tracer concentration would be replaced by the air temperature). In addition, this system should be able to capture high-frequency fluctuations as eddy sizes cover a large range, going down to centimeters and the succession of downward and upward movements operating at frequencies up to a dozen Hertz. The method based on this direct measurement of the turbulent transport is referred to as the *eddy covariance* method (Section 4.4). Progress in anemometry, notably due to the development of sonic anemometers and the development of high-frequency gas analyzers have made this approach suitable for more and more tracers. However,

Figure 4.1 Air flow in the surface layer above the canopy.
From Burba (2013).

as analyzers are not available (or too expensive) for all gases, other approaches have been carried out to estimate indirectly the turbulent transport. In particular, we will consider the *flux gradient* and the *relaxed eddy accumulation* (REA) methods (Section 4.5).

Mass balance methods constitute a notable exception: these methods were devised to capture the flux emitted by isolated buildings (barns) or by herds confined in a paddock. The situation is thus completely different, as the flux is not issued from a (supposedly) homogeneous surface but rather from a source constrained on a small area. Atmospheric transport is then no more dominated by turbulent transport but rather by advection and the measurement method estimates this transport. This method is described in Section 4.5.3.

4.3   Experiment design

4.3.1   Why? Defining the objectives

First, let us recall that measuring tracer fluxes is not a scientific objective by itself but only a means to reach a scientific objective. In view of the variety of tracers as well as ecosystems that could be investigated, these objectives may vary widely.

As an example, here is an incomplete list of the most frequently pursued goals in published studies.

– Methodological objectives: validation of the measurement method, better assessment of measurement representativeness, improvement of its accuracy, reduction of the uncertainties, and adaptation of the method to specific situations (small footprint, moving sources, very small gas concentrations, reactive gases, etc.)
– Validation of vegetation models or of meteorological models
– Identification, understanding, or quantification of the mechanisms driving the fluxes at hourly, seasonal, or interannual scales
– Budgeting of CO_2, carbon, or greenhouse gas at local or regional scales
– Detection of the impact of climate change on gas exchanges and ecosystem functioning
– Detection and evaluation of the impact of extreme events or disturbances (drought, heat waves, insect attacks, fires) on the ecosystem functioning
– Analysis of the impact of ecosystem management (forest clearings, surface tillage or cover cropping in crops, use of mulch, wetland drainage) on gas emissions
– Quantification of tracer emission by cattle
– Evaluation of the potential of carbon storage and global change mitigation in cultivated ecosystems
– Evaluation of pollution levels in urban environments

No matter the why, it is important that the how follows sound principles of scientific measurement and experimental design as presented later. The why should always drive the how, but there is also a practical balance between the how and why when measuring the physical world because the environment does limit where you can place instruments.

4.3.2 How? Designing the setup

The design depends notably on the choice of the *ecosystem(s)* and of the *tracer(s)* that will be investigated. According to the research objectives, the *duration of the experiment* (from some weeks to decades), the choice of the *measurement technique* and the number of *complementary measurements* should be defined.

4.3.2.1 Setup design in relation to ecosystem specificity

One of the first questions when designing the setup is to determine the *vegetation height*. Indeed, as the measurement system needs to be placed well above the surface, the investment will be very different in forests, where erecting a tower would be probably necessary, than over short canopies like crops or grasslands where a simple mast could be sufficient. In fast-growing canopies, the possibility of changing the measurement height during the experiment should also be considered. See Chapter 8 for more details on vegetation and turbulence. In addition, some methods (e.g., aerodynamic method) are impracticable over tall vegetation because it would require too-high measurement heights. It is also worth noting that the higher the measurement height, the wider the investigated zone. A compromise is thus often necessary to focus measurement on the target ecosystem (see Section 4.4.1.3).

In sites that are frequently managed such as crops and pastures, it is necessary to provide a measurement system some protection against livestock or accidents with agricultural machines. One possibility to avoid this could be to dismantle the setup during agricultural activities. In addition to the fact that this is labor intensive, this would be at the price of a loss of continuity in measurements and possibly the loss of critical events (e.g., loss of strong emissions just following a cropping activity). Alternatively, the system could be surrounded by an enclosure, which is difficult to avoid in long-term studies. Problems in this case are that the presence of the enclosure may perturb the site management and constitutes an island where vegetation cover differs from that of the target area. Unmanaged areas can have similar concerns with interference from wildlife or extreme weather conditions and events such as periodic floods or hurricanes.

All flux measurement methods except the mass balance approaches (Section 4.5.3) rely on the hypothesis of *surface homogeneity*. The system positioning should thus be chosen to fulfill these conditions. However, this is impossible in heterogeneous environments like cities or ecosystems with irregular vegetation cover or with moving punctual sources (e.g., cattle). In these conditions, some data processing using footprint models (Section 4.4.2.2) will be necessary to treat heterogeneity problems. When the target source is constrained and located at a fixed position, mass balance approaches could be an alternative.

4.3.2.2 Setup design in relation to the investigated tracer

The choice of tracers to investigate determines the set-up design. Indeed, if many analyzers are now available and allow investigation of a large range of tracers, their price, technical requirements, and ease of use in the field vary widely.

Commercial analyzers for *water vapor, carbon dioxide* (Leuning and Moncrieff, 1990), and *methane* (Detto et al., 2011; Nemitz et al., 2018) are now readily available and are robust enough to allow long-term continuous measurements (on the order of several years). Both *closed-* and *open-path systems* exist (Section 4.4.1.2) and the latter have low energy consumption, allowing measurements to be made at remote sites with light infrastructure and without a power line. Closed-path systems, however, are more energy consuming and require access to a power grid. Several multi-year campaigns (up to 25 years) using these analyzers have now been carried out (see notably for CO_2 the review by Baldocchi et al., 2018), and there are many standard practices and guidelines available for both making measurements and processing the resultant data.

Analyzers for volatile organic compounds (Ammann et al., 2004; Rinne et al., 2016), nitrous oxide (Nemitz et al., 2018), ammonium (Erisman et al., 2002), and many others (the list grows every year) are also commercially available, but they are more expensive (Section 4.4.1.2). These devices are of the closed-path type (sometimes with a low-pressure measurement chamber) and require generally stronger power lines. Their handling and calibration (i.e., frequency of human intervention) is more delicate, which makes long-term measurements more difficult to achieve. To our knowledge, measurement campaigns focusing on these gases mostly cover one or two vegetation seasons and rarely extend continuously for several years.

Studies concerning other tracers may be found in the literature (mineral dust, particles, black carbon, elemental mercury, peroxy acetyl nitrates, sea spray aerosols, dimethyl sulfite, carbonyl sulfides, and others), but they generally require very specific analyzers that had been set up in the frame of specific research and are not commercially available (at least not at present).

4.3.2.3 Determination of the temporal scale

The length of the measurement campaign and the need for *continuous measurements* also depends on the research objectives. Establishing an annual gas budget (carbon dioxide, methane, nitrous oxide) at a site and studying its inter-annual variability or its response to extreme events requires campaigns that extend for several years of continuous measurements. However, for other objectives, shorter campaigns may be considered. For example, studying the impact of one methical option or seeking for exchange mechanisms by analyzing functional relationships, determining the emission of methane or ammonium by cattle may be operated on seasonal campaigns. In addition, comparing the impact of some management operation on fluxes could also be carried out over shorter periods but could require the use of multiple measurement systems. As stated previously, it is important for the question to drive the choice of measurements.

4.3.2.4 Choice of the spatial scale

Typically, the methods presented here provide measurements at the *"ecosystem" scale*. The setups are indeed placed on a "flux tower" of several to several tens of meters high over the target ecosystem, and these systems allow investigating the fluxes on a surface of about 1 ha. This will be the focus of the remainder of this chapter, as it can be considered the staple scale for boundary layer studies. However, as discussed in Chapter 3, both smaller and larger spatial scales may be considered. At *lower scales* (some squared centimeters), chambers methods may be devised to study fluxes at the leaf (Long and Bernacchi, 2003), branch, or soil sub-plot (Pumpanen et al., 2004; Pavelka et al., 2018) scale. At *larger scales* (regional or continental) fluxes may be investigated by developing tower flux networks (Baldocchi et al., 2001; Yamamoto et al., 2005; Yu et al., 2006; Dolman et al., 2006; Franz et al., 2018), very tall towers (Berger et al., 2001), aircraft flux measurements (Crawford et al., 1996), or remote sensing (Lees et al., 2018). These methods are out of the scope of this chapter, and new methods will appear as technology advances.

4.3.2.5   Choice of the flux measurement technique

The ***eddy covariance technique*** (Aubinet et al., 2000, 2012a; Burba, 2013) is the most relevant method to measure gas fluxes at the surface because it is the sole method measuring the real turbulent transport process at work between the surface and the atmosphere. Due to the development of fast response analyzers, it is applicable to more and more gases and is therefore recommended when practicable. It is described in detail in Section 4.4.

The ***disjunct eddy covariance method*** (Rinne and Ammann, 2012) is a variant of the eddy covariance method but allows the use of slower analyzers or the study of multiple gases in parallel. It is notably used to study emissions of volatile organic compounds.

However, these methods may be difficult to apply to some gases because of the difficulty to capture high-frequency concentration fluctuations. This could be because fast response analyzers are not available or too expensive but also because the gases themselves tend to stick to the tube surfaces through which they are flowed (e.g., ammonium). In these conditions, alternatives using only slow response gas analyzers (***flux gradient method***, Section 4.5.1; ***relaxed eddy accumulation***, Section 4.5.2) should be contemplated. These alternatives only constitute approximations of the eddy covariance method and the limitations that apply to the eddy covariance method generally apply to these methods as well.

Finally, when the target source is spatially confined (a cattle herd constrained in a paddock, a single building, a barn), the homogeneity conditions are not required and thus ***mass balance methods*** could be an alternative (Section 4.5.3).

4.3.2.6   Supporting measurements

The interpretation of the fluxes often requires the measurement of complementary variables. The amount and the frequency of these supporting measurements again depends on the research objectives, but there is a general agreement on a minimum set of micrometeorological measurements taken at the same frequency as the fluxes. This includes: ***air temperature and humidity, short wave, infrared and photosynthetically active radiation, soil temperature and humidity, precipitation, and air*** *pressure*. In sites with tall vegetation, a ***vertical profile of CO_2 concentration*** is also necessary to take account of CO_2 storage in the air below the canopy at night (Section 4.4.2.2; Nicolini et al., 2018).

Other complementary measurements may be desired. For example, the European Integrated Carbon Observation System (ICOS) requires additional ***biomass measurements*** (leaf area index, biomass, chlorophyllian fluorescence; Gielen et al., 2018; Loustau et al., 2018), ***soil characteristics*** (carbon content, pH, C/N ratio … Arrouays et al., 2018), or ***phenological survey*** (Hufkens et al., 2018). The rationale for these measurements, implementation details, and required frequencies are described in the references above.

4.4 Eddy covariance

Eddy covariance in a nutshell

The eddy covariance technique consists of directly measuring the covariance between *w* and tracer concentration (Montgomery, 1948; Swinbank, 1951; Obukhov, 1951).

It measures the turbulent transport through the surface layer and is thus representative of the flux exchanged by a surface as far as other transport processes are negligible, which is the case when some hypotheses are met, the main ones being stationarity and a horizontal and homogeneous surface (Foken et al., 2012a).

Eddy covariance systems are made of a high-frequency three-dimensional anemometer (Section 4.4.1.1) and a gas analyzer (Section 4.4.1.2). These systems must be placed above the surface in the surface layer.

They may provide flux measurements on the scale of every hour (usually, most measurement periods last for one half-hour), allowing the flux daily dynamics be captured. They may provide continuous measurements for long-term periods (25-year datasets are available at some sites) allowing flux inter-annual variability and long-term trends to be obtained (Baldocchi, 2014).

The technique may be applied over any surface, including forests, grassland, crops, tundra, peatlands, ice core, urban areas, or lakes.

It may potentially be applied to any tracer, the limit being the availability (and the price) of the corresponding high-frequency gas analyzers. However, this limit continuously extends thanks to advances in gas analyzer technology. At present, routine measurements are made for velocity (momentum flux, turbulence kinetic energy (TKE)), temperature (sensible heat), water vapor (and latent heat), carbon dioxide, and methane.

The measurements integrate all the processes generating a flux at a given site. When the flux results from counteracting contributions (e.g., photosynthesis and respiration, both contributing to the net flux at a vegetated site), it directly provides the net flux. This is an advantage as far as gas budgets are required but makes the flux analysis complicated, especially if emission mechanisms are sought.

The measurements are integrated over a surface of several thousands of m^2, depending on the measurement height. This provides a larger spatial representation of the measurements than chamber or sampling methods but limits its significance over heterogeneous ecosystems or over spatially limited sinks and sources.

Despite the availability of data treatment software packages, the data treatment for flux calculation is not straightforward and always requires a specific expertise (Aubinet et al., 2012a).

4.4.1   Measurement system

A typical eddy covariance setup includes a *3D anemometer* and a *gas analyzer* that are placed on a tower or a mast. In addition, a shelter may be needed to place the analyzers, the power supply system, the data loggers, and the communication systems.

4.4.1.1   3D anemometer

Historically, hot wire 3D anemometers were used in the first eddy covariance systems (Swinbank, 1951). However, their difficulty of use in the field made them impracticable for long-term measurements. It is the development of *sonic anemometers-thermometers* (SAT) in the 1990s (Zhang et al., 1986; Foken and Oncley, 1995) that stimulated the development of the eddy covariance technique.

The basic principle of SAT (Kaimal and Businger, 1963; Munger et al., 2012) is to measure the *forward and return transit time* of an ultrasonic pulse between two transducers placed at a fixed distance apart (Figure 4.2). These times depend on the speed of sound and of the wind speed parallel to the line separating the two transducers. If the distance between the transducers is known, wind speed may be deduced from the difference of the inverse of the forward and return transit time. In addition, the average of these inverses provides the speed of sound, from which temperature is deduced.

Typically, 3D SATs are made by three pairs of transducers separated by tens of centimeters and placed according to three axes (Figure 4.3). Generally, these axes do not follow an orthogonal configuration but are positioned to minimize wind perturbations in the anemometer volume and to reduce the impact of rain because the introduction of moisture in the path will alter the transmission speed for sound.

Figure 4.2 SAT working principle: composition of velocities in the SAT volume: wind speed *(blue)*, speed of sound in still air *(purple)*, speed of the forward and return signals *(red)*. The transit times are equal to the ratio of the distance between the transducers and the transit speeds.

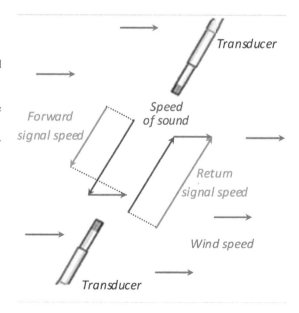

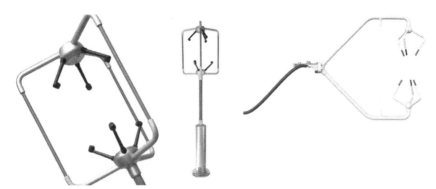

Figure 4.3 Two examples of 3D sonic anemometers. Left: Gill WindMaster HS (http://gillinstruments.com/new-images/applications/sonic-anemometer/sonic-anemometer-1.jpg); right: Campbell CSAT3A (https://www.campbellsci.fr/csat3a).

Transformation of transit times in three orthogonal components of velocity and in sonic temperature is made by the software built into the anemometer.

There are many different SATs with different qualities and prices. For eddy covariance measurements, research grade SATs are recommended with velocity resolution greater than $0.01 \, \mathrm{m \, s^{-1}}$, temperature resolution above $0.01 \, \mathrm{K}$, and acquisition frequency of the order of $20 \, \mathrm{Hz}$. Although only the vertical component of velocity is used in the computation of the eddy covariance term (Section 4.4.2.1), the use of a 3D SAT is recommended (Rebmann et al., 2018).

The absence of moving pieces in SATs makes them very robust and allows continuous long-term measurements without failure (Rebmann et al., 2018). However, several possible problems can affect sonic measurements.

- *Lateral wind flows* distort the sonic path of pulses, which may affect mainly sonic temperature estimates. Corrections have been proposed for this effect by Schotanus et al. (1983) and Liu et al. (2001).
- *Anemometer structure* may *distort* and shadow the velocity field inside the anemometer volume. Manufacturers try to optimize anemometer geometry to limit shadowing and distortion and introduce corrections to the anemometer's built-in software. However, there is still discussion about the correction accuracy and the resulting uncertainty (Munger et al., 2012; Foken et al., 2012b; Gash and Dolman, 2003).
- *Pulse transfer time* is perturbed by the presence of water in the sonic path. Errors may thus appear in presence of *rain or dew* on the transducers. This error is more important on temperature than on velocity (an important part of this error is suppressed by the transit time subtraction). Manufacturers optimize transducer shape to favor rain drain off. However, a quality control is always needed during and shortly after rain events (Munger et al., 2012; Rebmann et al., 2018).
- For the same reason, anemometers stop working in presence of *frost or ice* on the transducers. Some manufacturers propose heating the transducers to remove frost and ice. However, during the heating, natural convection from the heating element may perturb the wind field inside the anemometer space and introduce additional errors. Measuring velocity during these periods is therefore not recommended (Munger et al., 2012; Rebmann et al., 2018).

Finally, it is important to note that the sonic temperature measured by an SAT is not exactly equal to air temperature (but approaches **virtual temperature**; see Chapter 1), so these devices are not recommended for absolute accurate temperature measurements. However, the difference between sonic and air temperature is mostly constant so that the SAT provides unbiased measurements of air temperature fluctuations (Munger et al., 2012; Aubinet et al., 2000).

4.4.1.2 Tracer measurement

When the tracer is the temperature or another component of velocity, it can be measured by the SAT. As a result, a single SAT is enough to capture momentum and sensible heat flux as well as all TKE components and higher-order turbulence components (velocity and temperature skewness and kurtosis).

For gas concentrations, specific analyzers are required. The most current analyzers are based on *far-infrared spectrometry* techniques. In some cases, *mass spectrometry* or *time-of-flight spectrometry* is also used.

In addition to the *accuracy* and *robustness* that is expected from all gas analyzers, the analyzers used for eddy covariance must capture gas concentration fluctuations at high frequency ($\sim 10\,Hz$), they must be *synchronized with velocity measurements*, and they must measure the concentration at a position *as close to the SAT volume as possible*, without perturbing the wind field inside it. These constraints are sometimes contradictory, and a trade-off is then necessary.

Gas analyzer types

The current technique to measure concentration of greenhouse gases is far-infrared absorption spectrometry. An analyzer consists of an infrared light source covering a wavelength range that spans absorption lines for the gas of interest, a filter, and a detector. Light is absorbed by the gas in the light path, and the reduced intensity observed by the detector is related to its molar concentration. Intensity reduction is evaluated by comparing the detector signal with a reference signal that could be obtained by measuring the intensity of light in a second cell with zero gas concentration or at an adjacent non-absorbing wavelength.

There are two types of analyzers. In the first type (*open path*), the light path is directly in the open air, and in the second type (*closed path*), the light path is in an enclosure through which the air to be analyzed is continuously flowed (Figure 4.4). The flux data treatment differs according to the analyzer type and each type has specific advantages and drawbacks. They are discussed below.

Nondispersive infrared gas analyzers (*IRGA*) are the most currently used for the measurement of water vapor and carbon dioxide. In IRGA, broadband light sources and band-pass filters are used to select the wavelength range of interest. Nondispersive analyzers are well fitted for measurements of water vapor and carbon dioxide but are inapplicable when the gas concentration is very small or when absorption lines are confounded with those of other gases. For these situations, analyzers rely on *laser spectroscopy* or *mass spectrometry*.

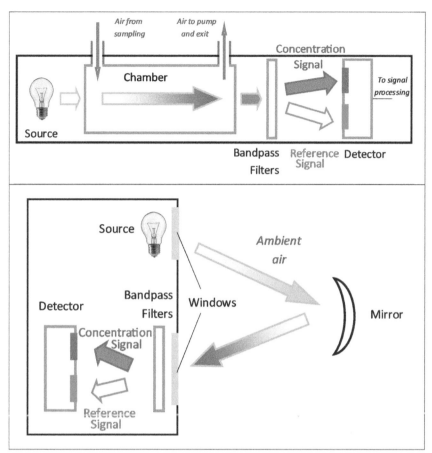

Figure 4.4 Working principle of non-dispersive infrared gas analyzers. Above: closed path system; below: open path system.

Concentrations vs fractions

All analyzers measure *gas molar (or mass) concentrations*. However, fluxes depend on *dry molar (or mass) fractions*. Both variables are related by the ideal gas law and Dalton's law. Relation between molar concentration and fraction of a non-H_2O gas thus depends on temperature and water vapor concentration. As a result, spurious tracer concentration fluctuations could be observed in the absence of any flux, due to fluctuations of air density or water vapor concentration. This needs a data treatment, generally called *density* (for the effect of temperature) *and dilution* (for the effect of water vapor) *corrections* (Webb et al., 1980). This treatment is necessary to avoid any bias in fluxes and applies differently for open- and closed-path analyzers. Both need a continuous measurement of temperature and water vapor fluctuations inside the analyzer volume. This is notably why an analyzer must always measure H_2O concentration fluctuations in addition to the investigated gas.

Specificities of open-path IRGAs

In *open-path analyzers*, the light path is directly in open air: infrared light is transmitted from the source to air and from air to the detector through windows.

The main advantage of the system is that the concentration fluctuations are directly measured *in-situ* and are not perturbed by any gas transport. In addition, the absence of a pump reduces the power needed to run the system, which makes the system more adapted to remote sites not equipped with main power. The main drawbacks are that the windows may be dirty because of dust and rain, leading to calibration drifts, which are difficult to correct in the absence of continuous maintenance. In addition, when the windows are exposed to the sun, they heat, which generates natural convection in the measurement volume and induces measurement biases. Corrections for these biases have been proposed but are not exact as they are generally based on empirical relations (Burba et al., 2008). To minimize these impacts, the IRGA may be tilted to promote rapid runoff of rain and opposite to the zenith to minimize window solar heating.

Another problem is the physical size of the device that may perturb the wind field into the SAT volume and bias velocity measurements. Avoiding this requires putting the IRGA and SAT at a minimum distance apart. However, by doing this, the two measurements are no longer taken at the same place, which translates into a loss of signal in the high frequencies. Sensor separation is the main source of high-frequency losses for open-path systems; some newer systems are developed to avoid this problem (Helbig et al., 2016).

In open-path systems, both density and dilution corrections must be made to the fluxes. They can be applied either on the fluxes by using the so-called WPL correction (Webb et al., 1980; Leuning, 2007) or by transforming instantaneous molar concentrations in dry molar fractions before flux calculation. In both cases, these corrections rely on the water vapor fluctuations measured by the IRGA and on the temperature fluctuations measured by the SAT.

Closed-path IRGAs

In *closed-path systems*, the light path is in a chamber through which air is continuously pumped. Analyzers of this type are more flexible and may be used for other purposes than eddy covariance.

Advantages of closed-path systems are that the chamber is less prone to dirtying, which greatly reduces the uncertainty due to calibration drift; the possibility of putting the sampling tube close to the SAT without perturbing the wind field, which reduces the error due to sensor separation; and the fact that it is not subject to natural convection bias. In addition, the analyzer does not need to be placed close to the measuring point, which could facilitate its maintenance (in case of measurements on high towers, e.g., the analyzer may be placed at the tower bottom).

The main drawback of the closed-path analyzer is that it requires a pumping system to transport air from the sampling point to the measurement chamber. This requires at least tubing, filters, and a pump, which is power consuming (Rebmann et al., 2018). All these devices introduce a lag between the air capture and its passage in the measurement chamber, which acts as low-pass filter by damping concentration fluctuations in the tube (Leuning and King, 1992; Leuning and Judd, 1996). This could reduce the system half power frequency to much lower values than the acquisition frequency and constitutes generally the most important limiting factor of high frequency fluctuations detection. In addition, pumping creates a depression in the chamber, which could be problematic for some analyzers. A careful dimensioning of the pumping system (tube length and diameter, flow rate) is thus needed to comply with these different constraints. There is no standard system (but there are standard constraints) and, generally, the pumping system is fitted to each analyzer. Finally, condensation may occur in the tube, which biases water vapor measurements and affects other tracers. This may be avoided by heating the tubes (Metzger et al., 2016; Rebmann et al., 2018).

Despite these cautions, errors remain, and the flux post processing must include two corrections: one that synchronizes measurements of gas concentration and wind velocity (Lee and Black, 1994; Moncrieff et al., 1997; Aubinet et al., 2000; Peltola et al., 2021) and one that corrects the loss of high-frequency fluctuations due to the tube (Leuning and Moncrieff, 1990; Massman and Ibrom, 2008).

In closed-path systems, if the dilution correction must be brought to the fluxes, this is not necessary for density correction. Indeed, as soon as tube length is sufficient, the temperature fluctuations are damped (Rannik, 1997) so that they are no longer observed in the measurement chamber. The correction may be applied on the fluxes by using the water vapor term of the WPL correction (Webb et al., 1980; Leuning, 2007) or by transforming instantaneous molar concentrations into dry molar fractions. In both cases the correction relies on the water vapor fluctuations measured by the IRGA.

In some intermediate situations (namely, in so-called *enclosed systems*), the tube length is not sufficient to totally suppress temperature fluctuations. In these cases, both temperature and water vapor fluctuations must be taken into account in the correction, but if the water vapor correction still relies on the fluctuations measured by the IRGA, temperature correction cannot be based on sonic measurements, as temperature fluctuations are larger in the open air than in the measurement chamber. Thus, this system needs an additional high-frequency thermometer placed in the chamber to perform this correction.

Details on analyzer calibration and system maintenance can be found in Munger et al. (2012), Rebmann et al. (2018), and references therein.

Other gas analyzers

When gas concentration is too small or when gas absorption lines are confounded with those of other gases, non-dispersive IRGAs cannot be used, and other techniques are needed.

In *tunable diode laser absorption spectroscopy (TDLAS)*, a laser is used as a source to emit radiation in a narrow band. In addition, this band is tunable to match an individual absorption line of a particular molecule of interest (Zahniser et al., 1995). This allows discrimination between absorption lines of different components in presence of gas mixture and avoids interference with other gases. This technique is notably suitable for measurements of ammonium (Whitehead et al., 2008), N_2O (Neftel et al., 2010) or stable isotopes (Saleska et al., 2006). A better separation between different gas species requires low pressure in the chamber and thus high-power pumps. Detectability of TDL spectrometers may be increased by extending the light path length: light is trapped in an optical cavity (cavity ring-down spectroscopy or integrated cavity output spectroscopy) and its intensity decay rate is related to the gas concentration. These methods also rely upon vacuum pumps to draw down pressure in the optical cell and require high power.

Proton transfer reaction (PTR) systems have been developed by Lindinger et al. (1998) (de Gouw and Warneke, 2007) and are notably used to measure volatile organic compounds or ammonium fluxes (Sintermann et al., 2011). These analyzers are of the closed-path type. The air is pumped through a drift tube reactor and a fraction of the tracers is ionized in proton-transfer reactions with hydronium ions. After this, the tracer masses are measured by a quadrupole mass spectrometer (PTR-MS). A major disadvantage of the PTR-MS is that it only determines the mass of product ions, which is not a unique indicator of the tracer identity. This problem may be solved by using a time-of-flight mass spectrometer (PTR-TOF-MS; Jordan et al., 2009) whose higher resolution in mass allows discrimination of isomers.

4.4.1.3   Setup

The setup should be carefully designed to capture the target ecosystem, avoid perturbing the measurements, allow maintenance, and ensure system resistance to environmental extremes like strong winds, lightning, thunder, or damages by animals. Some general indications are given below, and more practical details may be found in Munger et al. (2012) and Rebmann et al. (2018).

Site selection

The choice of the EC system location often results from a compromise between scientific objectives and practical considerations.

First, the location should be chosen to provide measurements that are *representative of the target ecosystem*. This supposes to place the system downwind to or at the middle of the target ecosystem and to allow an adequate fetch in the desired directions. Knowledge of the dominant wind directions and the use of footprint models may help to evaluate the location representativeness prior to the location determination (Rannik et al., 2012).

The surface around the measurement point should ideally be as *flat* and *homogeneous* as possible to comply with hypotheses underlying the eddy covariance technique. In practice, however, many sites would not be investigated if these conditions had to be strictly satisfied. Data treatment should therefore often accommodate non-ideal conditions.

Presence of *natural obstacles* (isolated trees, rocks, forest edges) that can cause flow distortion must be avoided.

Caution must be taken that the measurement system does not affect the *ecological integrity* of the site.

In case of managed ecosystems, regular contacts with the managers are recommended to ensure that *management activities* will not *perturb measurements and vice versa*. At some sites, it could be necessary to protect the site against theft and vandalism.

Finally, in many cases (long-term campaigns or experiments using closed-path analyzers) the presence of a *power line* is unescapable.

Tower requirements

Tower choice results from different compromises; it must be high enough to capture the fluxes in the well-mixed surface layer above the plant canopy and low enough to ensure that the footprint does not extend beyond the fetch of interest. Over short vegetation, a small mast of a few meters could be sufficient, while over tall vegetation (forests), a bigger tower is needed, which requires a stronger infrastructure. In addition, the structure should be lightweight enough to *avoid perturbing the wind field around the sensors* but also robust enough to ensure safe access for maintenance and to resist environmental extremes over the lifetime (Munger et al., 2012).

Relative positioning of SAT and analyzer

Flow distortion around the SAT should be as small as possible. This can be done by placing it at the end of a boom, also supporting the IRGA or its sampling point, facing away from the tower structure. Except for very strong slopes, the sonic anemometer must be *oriented horizontally*. The relative position of the IRGA (for open paths) or of its tube inlet (for closed path) and the sonic path sets a lower limit on the system time response. To maintain this high limit, the *distance should be as small as possible* by meeting the constraint to minimize flow distortion however possible (Figure 4.5).

Figure 4.5 Two examples of eddy covariance systems configurations: left: open-path IRGA (LI-COR 7500) with an omnidirectional sonic anemometer (Gill R3) at the Takayama site (Japan) (photo: M. Aubinet); right: enclosed path IRGA (LI-COR 7200) with a directional sonic anemometer (Gill Wind Master HS) at the Vielsalm site (Belgium) (photo: Jean Louis Wertz).

4.4.2   Data treatment

4.4.2.1   Flux computation

Flux inference from instantaneous velocity and concentration measurements is a complicated procedure that treats hundreds of thousands of instantaneous data. There are software packages that perform this computation, but they need to be correctly parameterized. As such, expertise is required. We will summarize here the main treatment steps. More detail may be found in Aubinet et al. (2000), Rebmann et al. (2012), and Sabbatini et al. (2018).

Conversion of concentrations into dry molar fractions

As previously stated Concentrations vs fractions, gas analyzers measure concentrations, while dry molar fractions are needed to compute fluxes. The conversion may be performed directly on instantaneous (high-frequency) measurements, which avoids the need to apply density and dilution corrections later. However, high-frequency temperature and H_2O concentration inside the measurement volume must be available, which is not the case (at least for the former) in all systems. If the conversion is not made at this stage, density and dilution corrections musty be applied later (see below).

Fluctuation computation by removing the mean

Computing the covariance implies removing the mean from wind velocity and concentration data series. This may be done by different methods such as block average, running mean, or linear detrending (Figure 4.6).

Historically, the running mean was first proposed mainly for technical reasons (Moncrieff et al., 1997; Aubinet et al., 2000), but the block average, which consists in simply computing the time series mean and removing it from each individual value, is more exact (complies with Reynolds rules) and is thus generally recommended (Rebmann et al., 2012; Sabbatini et al., 2018). It may be impractical when the analyzer

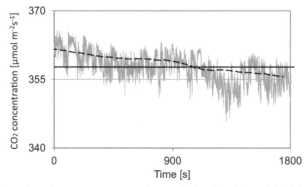

Figure 4.6 Illustration of two mean computation methods. *Gray line*: original time series; *black dotted line*: running mean with 500 s time constant; *black full line*: block average. N.B.: To stress the difference between the methods, an extreme example (during sunrise) was chosen here. This half hour would probably be rejected by the stationarity test.

is submitted to an important calibration drift. In these cases, a running mean may be applied. Linear detrending is given here for the record but is generally not recommended. Differences between block averaging and running mean are generally small provided that the running mean time constant is chosen high enough (about 500 s) and that stationarity conditions are fulfilled.

Coordinate rotation

The eddy covariance technique generally refers to the "vertical" component of velocity. More exactly, the considered component should be perpendicular to the surface or, better, perpendicular to the average wind streamlines above the surface. Despite all cautions taken during the SAT installation, this condition cannot be exactly matched (generally the average streamlines cannot be predicted before the measurements, especially on complex terrains; see Chapter 7). Because it is critical to obtain correct measurements (because horizontal velocity components are often several orders of magnitude larger than the vertical component), a coordinate rotation is necessary. Two approaches are possible: *2D rotation* or *(sector wise) planar fit*.

In *2D rotation*, rotations are applied to nullify the lateral (yaw angle) and the vertical (pitch angle) components of average velocity. Rotation angles are computed on each flux measurement period. This approach is easy to apply, does not require any long-time data series, and can give a fair estimate of the flux if the SAT alignment is not far away from the perpendicular to main streamlines (which are not difficult to predict if the site topography is smooth). However, the rotation is incomplete as it ignores the roll angle rotation, which is considered negligible. In addition, it supposes that average vertical velocity is always zero, which is not the case if the site is subject to vertical advection.

Some propositions have been made in the past (Kaimal and Finnigan, 1994; Aubinet et al., 2000) to estimate a third rotation by nullifying the lateral component of momentum flux. However, this rotation is not recommended anymore (Finnigan, 2004).

The *planar fit* approach has been proposed by Wilczak et al. (2001) who define a mean streamline plane based on measurements made on long periods (1 month minimum) that encompass all wind directions. The z-axis is then fixed perpendicular to this plane, the x-axis is fixed as the normal projection of the mean wind velocity on this plane, and the y-axis is fixed as the normal to the two other axes. This method was later refined by replacing the plane with a more complex surface (sector wise planar fit; Mammarella et al., 2007). The advantage of this method is that it may consider non-zero average vertical velocities. More detail concerning these different approaches and their implementation may be found in Rebmann et al. (2012) and Sabbatini et al. (2018).

Synchronization of velocity and time series

Velocity and concentration time series are generally shifted to each other due to differences in electronic treatment, sensor separation or, more importantly in closed-path systems, to gas transfer through the sampling tubes. These shifts induce covariance

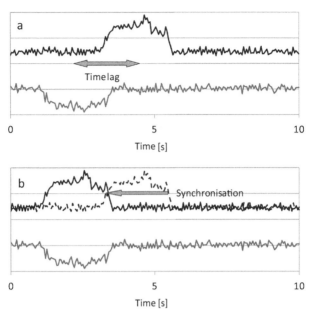

Figure 4.7 Illustration of the time lag problem and the signal synchronization. Above: The two artificial signals represented here represent two physical variables in arbitrary scales (e.g., wind velocity and CO_2 concentration) that are simultaneous but lagged because of the sampling. Below: The shift of the first signal allows restoring the simultaneity.

and thus flux losses. Time series should thus be synchronized before covariance computation (Figure 4.7).

The classical methodology (Aubinet et al., 2000; Rebmann et al., 2012) proposed to determine the time lag is *covariance maximization*. This consists in computing the covariance several times using different time lags and choosing the lag that maximizes the covariance, considering that any shift in time series would always induce a flux loss. However, this procedure can give unrealistic values when the signal is weak and the covariance maximum is not well defined. For this reason, time lags are always chosen in a limited range that complies with technical characteristics of the system, and when no maximum clearly appears, a default value corresponding to the most current time lag is used.

This approach has been recently criticized by Peltola et al. (2021) who pointed out that high-frequency concentration fluctuation damping could also generate some time lag. As a result, the simultaneous use of time lag determination by covariance maximization and of high-frequency correction Frequency corrections would overlap and lead to an "overcorrection." Thus, they recommend determining the time lag on a physical basis, on the basis of the knowledge, on one hand of tube dimensions and pumping rate and on the other hand of sensor separation distance and wind velocity and direction. This approach could become problematic, however, if all this information (notably the real pumping rate, which could fluctuate) is not available.

Density and dilution corrections

If conversion from concentrations to dry molar fraction has not been made on high-frequency measurements, flux would be biased due to temperature or water vapor concentration changes. To correct this, one uses the density and the dilution correction, respectively. Both constitute the WPL corrections (Webb et al., 1980; Leuning, 2007).

The *density correction* needs a record of temperature fluctuations in the measurement volume. The way these fluctuations are obtained and the difficulties inherent to some analyzer types have already been discussed.

The *dilution correction* needs a record of water vapor concentration fluctuations in the measurement volume, which is provided by the H_2O analyzer that uses the same chamber as the non-H_2O gas for any analyzer type. More practical details are given by Foken et al. (2012b).

Frequency corrections

High-frequency losses due to sensor separation, fluctuation attenuation in tubes, and some other, less critical features (Moore, 1986) induce an underestimation of fluxes.

The correction that is used to rectify this problem is based on spectral analysis. Poor high-frequency response of the measurement system causes observed gas cospectra to decay more quickly than expected with frequency. This effect may be quantified by computing a correction factor as the ratio of the integrals over the whole frequency ranges of a degraded cospectrum to an ideal one (Figure 4.8). This factor is only dependent on the setup (sensor separation distance, gas sampling system characteristics), wind velocity (which affects the sensor separation error), stability and, for water vapor, air humidity.

The correction procedure consists of (1) computing the *system transfer function* by computing the ratio of a real to an ideal spectra or cospectra, (2) computing a *correction factor* as the integral ratios of the ideal and the degraded cospectra (computed as the product of the ideal cospectrum and the transfer function), and (3) applying this correction on each measurement period.

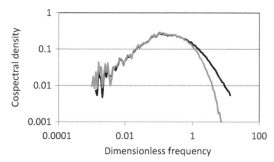

Figure 4.8 Illustration of the problem of high-frequency attenuation. The *black curve* corresponds to the sensible heat cospectrum, supposed non attenuated; the *gray curve* represents the CO_2 flux cospectrum whose high frequencies are attenuated. Data from the Lonzée site (Belgium) taken in June 2019. In the present case, the high-frequency attenuation result from sensor separation.

Different modalities exist to compute the ideal cospectrum (theoretical relations or use of the sensible heat cospectrum) to determine the transfer function (a priori approach based on setup properties or a posteriori approaches based on comparison of real and ideal spectra or cospectra) or to apply the correction factors on measurement periods (application of the correction on every measurement period or establishment of regression between the correction factors and meteorological variables). These are discussed notably in Massman and Ibrom (2008), Foken et al. (2012b), Fratini et al. (2012), Sabbatini et al. (2018), and Peltola et al. (2021).

4.4.2.2   Quality control

There is no method, other than eddy covariance, that measures fluxes at the same spatial and temporal scale. It is therefore essential to evaluate the quality of the measurements and to determine their uncertainty. Quality tests may be performed by several ways at different steps. First, the quality of instantaneous raw measurements may be assessed to detect instrumental or data logging problems. Next, different verifications may then be carried out on fluxes to check the compliance of measurement conditions to general laws and hypotheses underlying the eddy covariance approach such as *stationarity, Monin-Obukhov similarity theory (MOST)*, and *energy conservation*. Then, another test evaluates the relevance of the equality between the surface flux and the turbulent transport (negligibility of non-turbulent fluxes (advection) compared to turbulent flux), which is especially challenged in night conditions. Finally, a quantification of the uncertainties affecting eddy covariance measurements is also necessary. It will be discussed later.

Raw data analysis

Statistical tests on raw data series to detect instrument and data logging errors have been developed by Vickers and Mahrt (1997) and are now integrated in most current eddy covariance computation software packages (e.g., Sabbatini et al., 2018). The different tests help detecting spikes and data whose mean, variance, skewness, or kurtosis falls outside given limits, or data series with discontinuities in mean or variance. Each test is characterized by a quantified threshold or limit. Default values are proposed in the literature or in software package manuals, but they are not always adapted to the considered site or tracer. In each case, the user is free to decide to apply or not the test or to adapt its parameters to the site or tracer specificity.

Quality control on fluxes

Three tests checking the compliance of measurement conditions to the general hypotheses supporting the eddy covariance method tests have been proposed by Foken and Wichura (1996): the stationarity test, the integral turbulence test, and the energy balance closure. All these tests also provide quality flags. The tests may be classified and organized, and a flagging system combining all flags could finally provide a general flag for easy use (Foken et al., 2004, 2012b).

The eddy covariance method relies on stationarity of average velocity, temperature, and concentrations. The stationarity test evaluates the compliance to this hypothesis by comparing covariances computed on the whole measurement period and on fractions of this period. The degree of conformity is quantified by a threshold and results in a flag characterizing the quality of the flux data.

Compliance to surface boundary layer similarity relations is performed by comparing the relation between measured non-dimensional variance and stability parameters derived from the Obukhov length with relations predicted by MOST (Foken and Wichura, 1996). Here again, the degree of conformity is quantified by a threshold and results in a flag characterizing the quality of the flux data.

Conservation of energy is tested by considering all the energy fluxes taking place at the surface (sensible and latent heat, net radiation, energy storage in soil) and comparing the two that are measured by eddy covariance with the others. In a complete energy balance, these terms should be equal. However, the comparison is not so straightforward, as radiation and energy storage in the soil are not measured at the same spatial scale as eddy covariance fluxes. In addition, especially in forests, some terms of the energy balance are often ignored or not well measured (e.g., storage in vegetation, energy consumption by photosynthesis). As a result, full energy balance closure is rarely reached. The correction of the residual of the energy balance is not considered an issue of the correction of the eddy covariance method because the missing energy is not a missing flux at the measuring point and can be, at most, measured as advection (Foken et al., 2012b).

Night flux problems

As recalled previously (Section 4.2), the eddy covariance technique measures the turbulent flux and can be equated to the surface flux only if other transport and storage processes are negligible. This assumption is often challenged at night, especially under stable conditions (see Chapter 6) when storage of gas in the air below the measurement point and exchanges by advection may become dominant. This error has been well documented for CO_2 flux measurements. It was first addressed by Goulden et al. (1996) and identified by Moncrieff et al. (1996) as selective systematic (i.e., the flux was underestimated only at night when the ecosystem was a source and not during day conditions when it behaves as a sink), which considerably biases the sink intensity in long-term CO_2 budgets.

Ideally, the best way to correct the flux should be to directly measure storage and advection. This is possible for storage by measuring changes in tracer concentration below the measurement point (Nicolini et al., 2018), however, this has been shown to not be realistic for advection (Aubinet et al., 2010).

Consequently, the sole practicable correction at present consists in detecting and eliminating data affected by this problem. Goulden et al. (1996) proposed to use a criterion based on the friction velocity. As the source/sink term is expected to not depend on friction velocity, a flux decrease with decreasing friction velocity indicates a divergence between turbulent flux and surface flux. This decrease appears at about all sites below a given friction velocity threshold. The correction consists thus in discarding the data obtained when the friction velocity is below this threshold. As this threshold is specific to the site and setup, it must be regularly evaluated at each site. Procedures implementing this approach have been proposed by Gu et al. (2005) and Barr et al. (2013). Alternative approaches exist, all based on filtering but using different criteria: many researchers suggest using the standard deviation of w rather than the friction velocity as an indicator (Thomas et al., 2013; Hunt et al., 2016), following Acevedo et al. (2009). Besides this, by considering that advection appeared only during the night, van Gorsel et al. (2009) also suggested correcting night fluxes on the basis of measurements of turbulent fluxes and storage in the beginning of the night. More details are given by Aubinet et al. (2000, 2012b).

The impact of this error on other tracers has been less investigated. It is clear that it only affects tracers that could be exchanged by the surfaces at night, and it is also probable that it affects flux measurements of tracers whose emission is controlled by production/absorption mechanisms that carry out independently of the presence or absence of turbulent transport. This is the case, for example, for CO_2 and sensible heat but not for water vapor or isoprene whose production is negligible at night. Application of a filtering to these tracers would therefore strongly bias their budget. For other tracers, like latent heat, methane, monoterpenes, methanol, nitrous oxide, ozone, or NOx, the situation should be clarified. In these cases, a careful and specific analysis is needed for each tracer to determine if the flux decrease under low turbulence (if any) is the result of a measurement artifact or of a real surface emission slowing down.

Measurement representativeness

One central question that remains when measurements are collected and their quality is guaranteed is to know the extent to which the measured fluxes are really *representative of the target ecosystem*. Answers to this may be provided by footprint models (see reviews by Schmid, 2002 and Rannik et al., 2012) and footprint climatology (Göckede et al., 2006, 2008).

Footprint models are based on an analysis of the turbulent field around the measurement point. They allow weighing at a half-hourly scale the contribution to flux of all sources and sinks distributed around the measurement point (Figure 4.9). This weight depends on wind direction (the sources/sinks are always located upwind to the measurement system), on the relative height of the measurement point and the sources/sinks

Figure 4.9 Schematic of the footprint function.
From Schmid (2002).

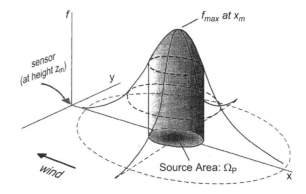

(the higher the height difference, the more distant the significant sources/sinks) and on stability conditions (larger contribution by distant (closer) sources/sinks under (un)stable conditions).

Footprint climatology is obtained by combining footprint model results with land use maps under the range of observed meteorological conditions at a site. This allows a quantification of the relative contribution of each source/sink to the flux over longer periods.

When the target ecosystem differs from the whole surface viewed by the EC system, the footprint model may be used to select the flux data to better focus it.

Models are based on either analytical or numerical techniques. Since the establishment of the footprint concept (Schuepp et al., 1990; Leclerc and Thurtell, 1990; Schmid, 1994), many models have been developed. Among the most currently used footprint models are those devised by Kormann and Meixner (2001) and Kljun et al. (2004).

Causes of uncertainty

Uncertainties on individual flux measurements result from a sensor failure to capture the right measure because of inherent imperfections, incorrect maintenance, or inadequate meteorological conditions. Uncertainties also result from imperfect setup positioning or non-fulfillment of the hypotheses underlying eddy covariance theory.

All these causes generally result in missing data (if the measurement system failed or if quality control led to data removing) or in errors affecting the measurement. Errors are generally classified as *random* and *systematic* (or selective systematic) errors. In the present case, we will consider three types of uncertainties depending on whether they result from random errors, systematic errors, or missing or screened measurements because of poor quality.

Random errors cannot be corrected and generate noise on the individual measurements (i.e., increase of their variance without affecting their mean) and translate directly in an uncertainty. Different quantifications of random uncertainties affecting half-hourly eddy covariance measurements may be based on one point sampling error (e.g., Langford et al., 2015; Kroon et al., 2010), covariance estimate at high time lag (e.g., Langford et al., 2015; Spirig et al. 2005), auto-covariance at zero-time lag (Langford et al., 2015), or day difference (Richardson et al., 2006). Uncertainties due to random errors propagate quadratically (i.e., the random uncertainty affecting a sum or a difference of two terms is computed as the square root of the sum of the squares of the uncertainties of these two terms; Taylor, 1997). As a result, the relative impact of uncertainties due to random errors on flux sums decreases according to inverse of the square root of the number of measurements.

Systematic errors don't generate noise and, for this reason, are more difficult to detect than random errors. However, they however a bias on the measurements. We may consider three possibilities.

In the ideal case (known knowns), errors are well identified; their causes are fully documented and quantified. A full correction is then possible, and it does not generate any uncertainty. This situation is unfortunately rare. It could be the case of the density and dilution corrections when there is no uncertainty on instantaneous fluctuations of temperature and water vapor measurements.

More generally (known unknowns), although the error is well identified, a rigorous correction is not possible because the processes at the base of the error are not fully documented, or adequate data are missing to proceed to the correction. In these cases, a semi-empirical correction may be applied and, if not possible, the data must be discarded. Uncertainties may then appear because of the incomplete correction or of the creation of new data gaps. This is notably the case of the high-frequency corrections.

In the worst case (unknown unknowns), unidentified systematic errors may remain. In these cases, the uncertainty is of course impossible to quantify. The sole ways to identify these errors are to screen the measurement and data treatment procedures to detect possible problems and to validate as fully as possible eddy covariance products by comparison with independent measurements. Example could be given by flux errors resulting from screening effects or flux distortion due to obstacles or homogeneities for which there is no easy evaluation approach.

The way uncertainties resulting from incomplete correction of systematic errors propagate is more complex and depends on how the correction is applied. It could propagate directly (i.e., the uncertainty on a sum – difference is the sum – difference of the uncertainties) on given time or variable ranges and quadratically from intervals to intervals or ranges to ranges.

Missing data do not constitute an error per se but may generate uncertainties in some products (budgets, notably), as far as data gap filling is needed. Uncertainties due to data gap filling depend on the length of the data gap, on the moment where it appears, and on the type of missing variables (Richardson et al., 2012). Data gaps created by removing bad quality data or incorrigible systematic errors (night flux error) may generate biases, as the removed data could be associated to specific conditions (rain, freeze, stable night conditions, specific wind direction) that are not represented in the dataset.

According to the considered "product" (net flux or identified component of the flux, flux response to a driving variable, annual or inter-annual budget, anomaly, impact of management or of extreme event on the fluxes, etc.), uncertainties will propagate differently, and uncertainty sources may impact differently the products. This question has still not been explored completely and a general scheme to evaluate uncertainty propagation through products calculation remains necessary. However, there are some examples. Random uncertainties will be the most important source of uncertainties of products based on a small number of these measurements, but their impact on long-term (annual or multi-annual) budgets will be much lower. On the contrary, uncertainties resulting from night flux corrections may be unimportant on short-term data series but more critical if budgets are considered. Finally, systematic errors may affect the budgets but are of lesser importance as far as inter-annual comparisons are considered because they disappear in the subtraction (Aubinet et al., 2018).

4.4.3   Disjunct eddy covariance

Similar to eddy covariance, disjunct eddy covariance (DEC) directly measures the covariance between w and the tracer concentration, but it differs in the way the sampling is operated. In eddy covariance, the air is continuously flowed into the chamber.

In DEC, the covariance is calculated on a subset (typically from 70 to 1800 subsamples within one-half hour) of the full continuous data series. The advantage of the method is that it allows concentration measurements by slower gas analyzers. However, it is essential that the time series are constituted of instantaneous wind velocity and concentration measurements. This is not a problem for wind velocity, for which a simple subsample of the sonic anemometer recording can be made. For concentrations, a specific sampling must be operated to allow this instantaneity (in practice, the sampling duration should last for 0.1 s or less).

Two sampling techniques are used to reach this goal. In *grab sampling*, air is sampled near instantaneously in a reservoir at intervals of 1–30 s thanks to high-frequency response valves (Figure 4.10). The air in the reservoir is then flowed into the analyzer during the interval between two samples. The system response time is the time it takes to grab the sample and can be as low as 0.1 s when using high-frequency valves. In practice, two different reservoirs may be used alternatively.

The *mass scanning method* is especially suitable for PTR-MS analyzers in which a continuous flow is generated and which provides fast response concentration measurements, but by sequencing multiple masses. Each analyzed compound (identified by a mass) is therefore recorded as a disjunct time series.

The method has been validated against eddy covariance measurements and the results show that the sample reduction does not cause systematic error on the flux but increases the random uncertainty on the fluxes. Spectral analysis also shows that if the DEC cospectra are affected by aliasing, compared to EC cospectra, this has no impact on the fluxes. More detail on DEC history, technique, validation, and pitfalls may be found in Rinne and Ammann (2012).

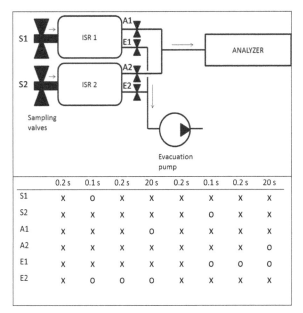

Figure 4.10 Schematic of DEC system with two grab samplers. ISR are intermittent storage reservoir. Operating sequence of a DEC system with grab samplers. o signifies open valve, x closed valve. S1, S2, A1, A2, E1, and E2 represent the different valves represented in the figure.
Redrawn from Rinne and Ammann (2012).

	0.2 s	0.1 s	0.2 s	20 s	0.2 s	0.1 s	0.2 s	20 s
S1	x	o	x	x	x	x	x	x
S2	x	x	x	x	x	o	x	x
A1	x	x	x	o	x	x	x	x
A2	x	x	x	x	x	x	x	o
E1	x	x	x	x	x	o	o	o
E2	x	o	o	o	x	x	x	x

4.5   Indirect measurements of turbulent fluxes

4.5.1   Flux-gradient method

The flux-gradient method relies on the approximation that the flux of one tracer is proportional to the gradient of its concentration. This is not always true, however, as the mechanism of turbulent transport described previously does not necessarily ensure this proportionality (Denmead and Bradley, 1987). The flux gradient method could constitute a good substitute to the eddy covariance method when high-frequency measurements are not possible.

The method is based on the measurement of vertical profiles of velocities, temperature, and tracer concentration. Its advantage is that it does not require high-frequency measurements. It thus doesn't need fast response analyzers. This was the most used method to measure fluxes before the development of eddy covariance systems and focused mainly on CO_2 and H_2O measurements. It is still used when fast response analyzers are not available or too expensive. It is also used to study sticky components whose high-frequency concentration fluctuations are difficult to capture (Kamp et al., 2020). This is the case notably for NH_3 (Nelson et al., 2019; Griffis et al., 2019), N_2O (Bai et al., 2019) or dimethyl sulfide (Tanimoto et al., 2014) as well as methane (Laubach et al., 2008; Judd et al., 1999).

The method has been extensively described in literature, notably by Prueger and Kustas (2005). It supposes that a tracer flux may be estimated as the product of the vertical concentration gradient and a turbulent diffusivity. The latter depends on friction velocity (or on the wind speed vertical gradient) and on stability conditions (Obukhov length). It can thus be deduced from measurements of wind speed and temperature profiles or from friction velocity and stability parameters obtained with a sonic anemometer. Detailed relations have been proposed by Prueger and Kustas (2005).

The method relies on similar hypotheses as eddy covariance does: stationarity of weather conditions, no flux divergence, and minimum area size with a constant, homogeneous surface. The latter requirement is more stringent than for eddy covariance, as the footprint is larger for concentrations than for fluxes. Its advantage is that it does not require the use of fast response analyzers. However, it must be operated on large heights (several times canopy heights), which increases the required homogeneous area size and becomes impracticable over tall vegetation. In addition, profile measurements require several measurements taken at different heights, implying either the use of several analyzers (which requires a careful analyzer inter-calibration) or sampling operated at different heights (which generates an error due to non-simultaneity of the concentration measurements; Kamp et al., 2020). Another limitation of the method is linked with the uncertainties affecting the similarity functions used to determine the turbulent diffusivity in stable and unstable conditions (Prueger and Kustas, 2005).

4.5.2   Relaxed eddy accumulation

Relaxed eddy accumulation method is a technique of conditional sampling; air from upward and downward eddies being sampled separately using a three-way valve commanded by a sonic anemometer (Pattey et al., 1992, 1993). The system is thus

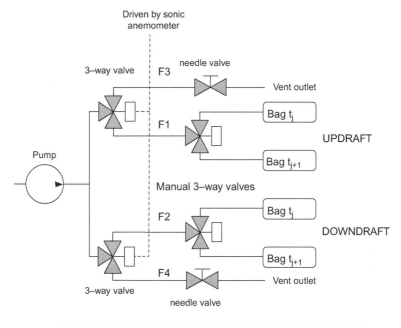

Figure 4.11 Schematic of a relaxed eddy accumulation system to measure trace gas fluxes. The status of flows F1 to F4 indicate if there is a flow (1: yes, 0: no) through the tubing depending on vertical wind velocity (W) compared to the negative and positive value of the deadband ($\pm D$). Air is collected in bags that are changed from one period to another.
Redrawn from Pattey et al. (1993).

constituted by a sonic anemometer, a pumping system with a three-way valve, and two reservoirs (Figure 4.11). According to the sign of w, the valve fills one or the other reservoir. Concentration difference between the two reservoirs is measured at the end of the measurement period and the flux is equal to the product of this difference with the standard deviation of w and an empirical constant.

The main advantage of the method is that, like the flux gradient method, it does not require fast gas analyzers. Moreover, it does not need the presence of the analyzer in the field, as air samples can be moved to the laboratory and treated for analysis. It is thus especially useful for measuring fluxes of tracers that cannot be measured by eddy covariance. The method is used notably for measurements of fluxes of herbicides (Pattey et al., 1993), ammonia (Zhu et al., 2000; Nelson et al., 2019), volatile organic compounds (Bowling et al., 1998; Rhew et al., 2017), nitrous acid (HONO) (Ren et al., 2011), and gaseous mercury (Skov et al., 2006). The method is generally implemented in campaigns that last for one vegetation season, rarely longer.

The method relies on the rapidity of the valves and especially on the accuracy of the direct measurement of w. A careful alignment of the sonic anemometer and a filtering of the velocity signal to remove any bias are necessary as no post processing is possible (Pattey et al., 2006).

The value of the empirical constant is relatively constant but may be smaller if a dead band (i.e., a w range around zero where no sampling is proceeded). It could be determined for a given setup by comparing REA and EC estimates of the flux of a scalar used as a marker (e.g., CO_2). However, this requires scalar similarity between the studied tracer and the marker, a condition that is not always fulfilled (Ruppert et al., 2006).

4.5.3   Mass balance methods

Contrary to the preceding techniques, mass balance methods don't rely on surface homogeneity. On the contrary, they are devised to capture fluxes emitted by spatially constrained sources. They are generally used to measure emissions from cattle herds (Griffith et al., 2008; Laubach and Kelliher, 2004), barns, or soils (Harper et al., 2011; McInnes and Heilman, 2005) and concern most often methane or ammonium.

In the *modified mass difference* approach, a rectangular control volume is delimited around the source and the tracer flux is measured across each its vertical faces (Harper et al., 2011). The flux is estimated as the product of horizontal wind speed normal to the face and the tracer concentration at the same level. The method requires extensive tracer measurements at many points, which requires measuring the line average concentration across the face at each height. This can be done by surrounding the control volume with porous tubes (Harper et al., 1999, 2002; Denmead et al., 1998; Leuning et al., 1999) or to use an open-path laser (Desjardins et al., 2004). One important limitation of this method is that it neglects turbulent exchanges at the control volume faces, which leads to a flux overestimation.

The *integrated horizontal flux* method is easier to implement in the field, as is only requires placing a single vertical profile of tracer concentration and wind speed downwind to the source and one tracer concentration measurement upwind to the source. The flux is then estimated as the vertically integrated product of wind speed and the difference between downwind and upwind tracer concentration. The method suffers from several shortcomings, however. First, the setup configuration is valid only for one given wind direction. One solution is to change the relative position of the setup and the source according to wind direction (Laubach and Kelliher, 2005). A second limitation, if only one single vertical profile is used, is that not all the sources may be captured by the sensors (especially in the case of moving cattle). The problem may be solved by using line averaging sensors such as open-path lasers or open-path Fourier transform infrared (FTIR) spectroscopy, with path lengths comparable to the source dimension (Desjardins et al., 2004). The problem of neglecting turbulent fluxes is the same as in the modified mass balance method and leads to a similar flux bias.

4.6   Meteorological measurements

4.6.1   Radiation

The different types of radiation that are considered at a flux site include generally shortwave radiation (range of emission of the sun – blackbody at 5778 K), longwave radiation (range of terrestrial emission – blackbody at 278 K), net radiation, which is a budget of the former two and photosynthetically active radiation (PAR, range from 400 to 700 nm) (Figure 4.12). The first two are important components of the energy budget of the ecosystem and are logically quantified in energetic units ($W\,m^{-2}$), while the third one is the driver of plant photosynthesis, which is a kind of photoelectric

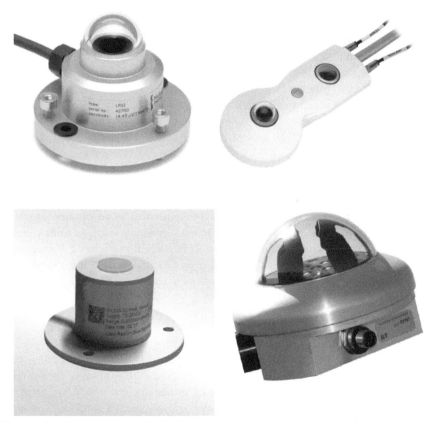

Figure 4.12 Some examples of radiation measurement devices. Top left: First-class pyranometer (SR11 from Campbell Scienctific) (https://www.campbellsci.es/lp02). Top right: Four-component net radiometer (CNR4 from Campbell Scientific) (https://www.campbellsci. es/lp02, https://www.campbellsci.fr/cnr4). Bottom left: PAR sensor (SP-RK200-02 from SmartyPlanet) (https://www.smartyplanet.com/fr/produits-smartyplanet/par-sensor/). Bottom right: Diffuse/direct radiation sensor (SPN1 from Delta T) https://delta-t.co.uk/product/bf5/.

reaction. Its rate depends on the number of absorbed photons, thus it is quantified in photonic units ($mol\,m^{-2}\,s^{-1}$).

For the shortwave emission, the radiations of interest are the incident and the reflected radiation, and for infrared, the radiations are incident and emitted radiation (the combination of these four terms providing the net radiation). For PAR, the radiation is the fraction absorbed by the vegetation. The incoming shortwave radiation may also be separated into its direct and diffuse components.

Shortwave radiation is measured by pyranometers, which can be made of thermopiles or of photo diodes, longwave radiation is measured by pyrgeometers that are made of thermopiles, and PAR is measured by photodiodes. The diffuse shortwave radiation can be measured by a specific pyranometer equipped with a disk to mask the sun. Recent systems use multiple sensors and a complex shading system ensuring that at least one sensor is always exposed to the sun and one is maintained in the shadow. More details on the measurement principles can be found in Carrara et al. (2018).

Sensors looking upward (incident pyranometer (including diffuse/direct sensor), pyrgeometer, and PAR sensors) should be placed horizontally and free from obstructions (including those due to tower or mast structure and other instruments). Sensors looking downward (reflected shortwave, infrared emitted by the surface) should be placed horizontally and, ideally, their field of view should be entirely and exclusively the targeted surface. To avoid perturbations due to the mast or to the tower structure, the sensors could be placed at the end of a horizontal boom at the top of the tower.

Finally, PAR absorbed by the vegetation, which is the relevant variable to characterize photosynthesis, is measured by difference between the incident PAR measured above the vegetation and the transmitted PAR measured below the vegetation. In view of the canopy heterogeneity, measurements of PAR below the vegetation require many sensors or, better, mobile automated systems (such as tram systems) to provide a representative spatial sampling. More details concerning measurement principles, sensor installation, accuracy and precision requirements, maintenance and calibration, and data treatment can be found in Carrara et al. (2018) and the references therein.

4.6.2 Temperature

Temperature is one of the most common and basic meteorological measurements. Temperatures of interest at tower sites are mainly the air (possibly at several heights) and the soil temperatures. There are a plenty of measurement methods. The most current methods in automatic recording systems use electrical resistances, generally made of platinum or semi-conductor thermistors. More detail on measurement techniques may be found in Brock and Richardson (2001). However, it is not so easy to obtain accurate measurements of temperature. The problem does not come from the measurement sensor, but rather from the sensor conditioning.

In the case of *air temperature*, conditioning is critical, as an air temperature sensor placed in the open air would be influenced not only by the air temperature but also by shortwave radiation (all components including direct and diffuse as well as incident and reflected radiation) and infrared radiation emitted by the sky and the surrounding surfaces. To avoid this, the sensor should be placed in a ventilated shield and ventilated (Figure 4.13). Ventilation may be natural, using appropriate devices or generated by an aspiration system. WMO (2014) recommends that the measurement uncertainty should be in the order of magnitude of 0.1 K, in the range $-40°C$ to $+40°C$ and lower than 0.3 K outside this range, and a maximum response time in the range 20–40 s. Details concerning air temperature measurements in flux tower can be found in Rebmann et al. (2018).

Soil temperature is useful as it acts as a driving variable for many processes at work in the soil (respiration, N_2O or methane emission, etc.). Soil temperature is very sensitive to depth, especially in the first few centimeters. Diurnal and annual temperature variations are indeed much larger close to the surface. For this reason, soil temperature must be measured at different depths with a decreasing spatial resolution with depth. For example, in the ICOS network, it is recommended to establish a profile with five to six sensors down to 1 m depth (when the bedrock depth allows it) with two sensors in the first 10 cm. Sensor conditioning is also important for soil probes, but the aim here is mainly to protect the probe against humidity and allow for long-term burial.

As horizontal heterogeneity is expected to be larger in the soil than in the air, it could be useful to install several profile replicates distributed in the target area. Representativeness of soil temperature profiles could be challenged in managed crops.

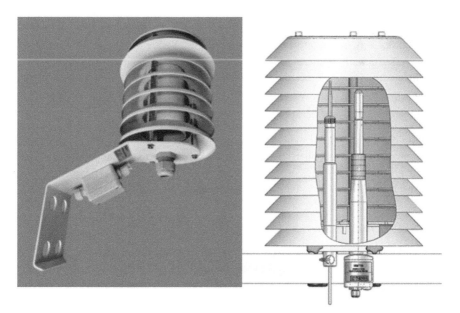

Figure 4.13 Examples of radiation shields: left: meteoclima with forced ventilation according to ISO17714:2007, image kindly provided by meteoclima; right: Vaisala DTR13 but with only natural ventilation, image kindly provided by Vaisala.
Images and legend copied from Rebmann et al. (2018).

On one hand, placing the sensors in the protected island (Section 4.3.2.2) could lead to biased estimates of temperature profiles because of the differences in vegetation and soil structure between the island and the target area. On the other hand, placing it in the target area would generate risks of destructions during cropping activities. More information on soil profile design and requirements may be found in Op de Beeck et al. (2018).

4.6.3   Humidity

There are many different variables that can be used to characterize air humidity, including water vapor concentration, relative humidity, water vapor pressure, dew point temperature, and wet bulb temperature. All these variables are interconnected and the knowledge of one of them along with those of temperature allows for computing all of them. Consequently, in sites equipped with an eddy covariance system, these variables are already accessible as they are measured by default. The meteorological station may anyway be completed by hygrometers directly measuring this variable.

The most common and easy-to-use instruments are based on electrical capacitive hygrometers. They provide a reasonable level of accuracy in the measurement range with continuous time series and have a shorter response time, which makes them suitable for flux tower requirements. As for temperature sensors, the conditioning of hygrometric sensors is critical, and they must be shielded and ventilated the same way. WMO (2014) recommends that measurement uncertainty should be in the order of magnitude of 2% in the range of 5–100% relative humidity and the maximum response time in the range 20–40 s. Details concerning air temperature measurements in flux tower may be found in Rebmann et al. (2018).

Soil water content is measured with dielectric sensors, such as time, frequency, or amplitude domain reflectometers. Selected dielectric sensors must have a measurement accuracy of $\pm 0.05 \, m^3 \, m^{-3}$ or better under factory calibration, valid for the type of soil and over the entire soil water content range expected at the site where the sensors are installed. Like for soil temperature, soil moisture may be subject to large variability, implying the installation of measurement profiles. For example, in the ICOS network, it is recommended to establish the soil humidity profile at the same depths as the temperature profile (Op de Beeck et al., 2018). The problem of horizontal heterogeneity and of representativeness in managed crops is the same as for soil temperature profiles.

4.6.4   Precipitation

Rain and snow precipitations may be measured using precipitation gauges. Two common systems are the weighing-recording gauge and the tipping-bucket gauge. In a weighing-recording gauge, water is collected in a container whose weight is continuously recorded. In the tipping bucket rain gauge, a small container is divided into two compartments of known volume (buckets) and balanced in unstable equilibrium about a horizontal axis. Rain fills one bucket and when filled, the system tips, the bucket drains, and rain fills the second one. Precipitation is estimated by the number of switching.

Measurement errors may be notably due to wind effects, wetting and evaporation losses, sampling errors due to weighing and tipping mechanisms, and in- and out-splashing effects due to device location, as well as random observational and instrumental errors. In automatic systems, power failures due to thunderstorm may induce large uncertainties, as precipitation during a storm may be lost.

Recommendations for the ICOS network are to locate gauges in places as open as possible, not further than 1 km from the EC tower, to record measurements every 60 s and to calibrate the gauge twice a year. More detail can be found in Dengel et al. (2018).

4.7 Summary

This chapter presented the fundamentals of some surface flux measurement methods with a special focus on eddy covariance. It discussed the many considerations necessary when taking measurements in the boundary layer and the way most of the major atmospheric variables can be quantified in time and space. The goal of this discussion is to familiarize the reader with the most common practices encountered in the boundary layer. It is not a comprehensive list of the variables or the techniques that can be used. We do not cover, for example, remote sensing-based instruments such as lidar, radar, sodar, or scintillometers. Some of these measurements are presented in other chapters. Measurement techniques and technology are always changing and improving, much like the boundary layer itself.

References

Acevedo, O., Moraes, O., Degrazia, G., Fitzjarrald, D., Manzi, A., Campos, J., 2009. Is friction velocity the most appropriate scale for correcting nocturnal carbon dioxide fluxes? Agric. For. Meteorol. 149 (1), 1–10.

Ammann, C., Spirig, C., Neftel, A., Steinbacher, M., Komenda, M., Schaub, A., 2004. Application of PTR-MS for measurements of biogenic VOC in a deciduous forest. Int. J. Mass Spectrom. 239, 87–101.

Arrouays, D., Saby, N.P., Boukir, H., Jolivet, C., Ratié, C., Schrumpf, M., Merbold, L., Gielen, B., Gogo, S., Delpierre, N., Vincent, G., Klumpp, K., Loustau, D., 2018. Soil sampling and preparation for monitoring soil carbon. Int. Agrophys. 32 (4), 633–643.

Aubinet, M., Grelle, A., Ibrom, A., Rannik, Ü., Moncrieff, J., Foken, T., Kowalski, A., Martin, P.H., Berbigier, P., Bernhofer, C., Clement, R., Elbers, J., Granier, A., Grünwald, T., Morgenstern, K., Pilegaard, K., Rebmann, C., Snijders, W., Valentini, R., Vesala, T., 2000. Estimates of the annual net carbon and water exchange of forests: the EUROFLUX methodology. Adv. Ecol. Res. 30, 113–175.

Aubinet, M., Feigenwinter, C., Bernhofer, C., Canepa, E., Heinesch, B., Lindroth, A., Montagnani, L., Rebmann, C., Sedlak, P., Van Gorsel, E., 2010. Advection is not the solution to the nighttime CO_2 closure problem—evidence from three inherently different forests. Agric. For. Meteorol. 150 (5), 655–664.

Aubinet, M., Vesala, T., Papale, D., 2012a. Eddy Covariance: A Practical Guide to Measurement and Data Analysis. Springer, Dordrecht.

Aubinet, M., Feigenwinter, C., Heinesch, B., Laffineur, Q., Papale, D., Reichstein, M., Rinne, J., Van Gorsel, E., Papale, D., 2012b. Nighttime flux correction. In: Aubinet, M., Vesala, T. (Eds.), Eddy Covariance: A Practical Guide to Measurement and Data Analysis. Springer, Berlin, Heidelberg, pp. 133–157.

Aubinet, M., Hurdebise, Q., Chopin, H., Debacq, A., De Ligne, A., Heinesch, B., Manise, T., Vincke, C., 2018. Inter-annual variability of Net Ecosystem Productivity for a temperate mixed forest: a predominance of carry-over effects? Agric. For. Meteorol. 262, 340–353.

Bai, M., Suter, H., Lam, S.K., Flesch, T.K., Chen, D., 2019. Comparison of slant open-path flux gradient and static closed chamber techniques to measure soil N_2O emissions. Atmos. Meas. Tech. 12, 1095–1102.

Baldocchi, D., 2014. Measuring fluxes of trace gases and energy between ecosystems and the atmosphere – the state and future of the eddy covariance method. Glob. Change Biol. 20, 3600–3609.

Baldocchi, D., 2020. How eddy covariance flux measurements have contributed to our understanding of Global Change Biology. Glob. Change Biol. 26, 242–260.

Baldocchi, D., Falge, E., Gu, L., Olson, R., Hollinger, D., Running, S., Anthoni, P., Bernhofer, C., Davis, K., Evans, R., Fuentes, J., Goldstein, A., Katul, G., Law, B., Lee, X., Malhi, Y., Meyers, T., Munger, W., Oechel, W., Paw U, K.T., Pilegaard, K., Schmid, H.P., Valentini, R., Verma, S., Vesala, T., Wilson, K., Wofsy, S., 2001. FLUXNET: a new tool to study the temporal and spatial variability of ecosystem-scale carbon dioxide, water vapor, and energy flux densities. Bull. Am. Meteorol. Soc. 82 (11), 2415–2434.

Baldocchi, D., Chu, H., Reichstein, M., 2018. Inter-annual variability of net and gross ecosystem carbon fluxes: a review. Agric. For. Meteorol. 249, 520–533.

Barr, A.G., Richardson, A.D., Hollinger, D.Y., Papale, D., Arain, M.A., Black, T.A., Bohrer, G., Dragoni, D., Fischer, M.L., Gu, L., Law, B.E., Margolis, H.A., McCaughey, J.H., Munger, J.W., Oechel, W., Schaeffer, K., 2013. Use of change-point detection for friction–velocity threshold evaluation in eddy-covariance studies. Agric. For. Meteorol. 171–172, 31–45.

Berger, B.W., Davis, K.J., Yi, C., Bakwin, P.S., Zhao, C.L., 2001. Long-term carbon dioxide fluxes from a very tall tower in a northern forest: flux measurement methodology. J. Atmos. Ocean Technol. 18 (4), 529–542.

Bowling, D.R., Turnipseed, A.A., Delany, A.C., Baldocchi, D.D., Greenberg, J.P., Monson, R. K., 1998. The use of relaxed eddy accumulation to measure biosphere-atmosphere exchange of iso-prene and of her biological trace gases. Oecologia 116, 306–315.

Brock, F.V., Richardson, S.J., 2001. Meteorological Measurement Systems. Oxford University Press. 290 p.

Burba, G., 2013. Eddy Covariance Method for Scientific, Industrial, Agricultural and Regulatory Applications: A Field Book on Measuring Ecosystem Gas Exchange and Areal Emission Rates. LI-CCOR Biosciences.

Burba, G., 2021. Atmospheric flux measurements. In: Chen, W., Venables, D.S., Sigrist, M.W. (Eds.), Advances in Spectroscopic Monitoring of the Atmosphere. Elsevier.

Burba, G., McDermitt, D.K., Grelle, A., Anderson, D.J., Xu, L., 2008. Addressing the influence of instrument surface heat exchange on the measurements of CO_2 flux from open-path gas analyzers. Glob. Change Biol. 14, 1854–1876.

Carrara, A., Kolari, P., Op de Beeck, M., Arriga, N., Berveiller, D., Dengel, S., Ibrom, A., Merbold, L., Rebmann, C., Sabbatini, S., Serrano-Ortiz, P., Biraud, S.C., 2018. Radiation measurements at ICOS ecosystem stations. Int. Agrophys. 32 (4), 589–605.

Crawford, T.L., Dobosy, R.J., McMillen, R.T., Vogel, C.A., Hicks, B.B., 1996. Air-surface exchange measurement in heterogeneous regions: extending tower observations with spatial structure observed from small aircraft. Glob. Change Biol. 2, 275–285.

de Gouw, J., Warneke, C., 2007. Measurements of volatile organic compounds in the earth's atmosphere using proton-transfer-reaction mass spectrometry. Mass Spectrom. Rev. 26, 223–257.

Dengel, S., Graf, A., Grünwald, T., Hehn, M., Kolari, P., Ottosson Löfvenius, M., Merbold, L., Nicolini, G., Pavelka, M., 2018. Standardized precipitation measurements within ICOS: rain, snowfall and snow depth: a review. Int. Agrophys. 32 (4), 607–617.

Denmead, O.T., Bradley, E.F., 1987. On scalar transport in plant canopies. Irrig. Sci. 8 (2), 131–149.

Denmead, O.T., Harper, L.A., Freney, J.R., Griffith, D.W.T., Leuning, R., Sharpe, R.R., 1998. A mass balance method for non-intrusive measurements of surface–air trace gas exchange. Atmos. Environ. 32, 3679–3688.

Desjardins, R.L., Denmead, O.T., Harper, L.A., McBain, M.C., Masse, D., Kaharabata, S., 2004. Evaluation of a micrometeorological mass balance method employing an open-path laser for measuring methane emissions. Atmos. Environ. 38, 6855–6866.

Detto, M., Verfaillie, J., Anderson, F., Xu, L., Baldocchi, D., 2011. Comparing laser-based open- and closed-path gas analyzers to measure methane fluxes using the eddy covariance method. Agric. For. Meteorol. 151 (10), 1312–1324.

Dolman, A.J., Noilhan, J., Durand, P., Sarrat, C., Brut, A., Piguet, B., et al., 2006. The CarboEurope regional experiment strategy. Bull. Am. Meteorol. Soc. 87 (10), 1367–1379.

Erisman, J.W., Otjes, R., Hensen, A., Jongejan, P., van den Bulk, P., Khlystov, A., Möls, H., Slanina, S., 2002. Instrument development and application in studies and monitoring of ambient ammonia. Atmos. Environ. 35 (11), 1913–1922.

Finnigan, J.J., 2004. A re-evaluation of long-term flux measurement techniques – Part II: coordinate systems. Bound.-Layer Meteorol. 113 (1), 1–41.

Foken, T., Oncley, S.P., 1995. Results of the workshop 'Instrumental and methodical problems of land surface flux measurements'. Bull. Am. Meteorol. Soc. 76, 1191–1193.

Foken, T., Wichura, B., 1996. Tools for quality assessment of surface-based flux measurements. Agric. For. Meteorol. 78, 83–105.

Foken, T., Göckede, M., Mauder, M., Mahrt, L., Amiro, B.D., Munger, J., W., 2004. Post-field data quality control. In: Lee, X., Massman, W., Law, B. (Eds.), Handbook of Micrometeorology: A Guide for Surface Flux Measurement and Analysis. Kluwer, Dordrecht, pp. 181–208.

Foken, T., Aubinet, M., Leuning, R., 2012a. The eddy covariance method. In: Aubinet, M., Vesala, T., Papale, D. (Eds.), Eddy Covariance: A Practical Guide to Measurement and Data Analysis. Springer, Berlin, Heidelberg, pp. 1–19.

Foken, T., Leuning, R., Oncley, S.R., Mauder, M., Aubinet, M., 2012b. Corrections and data quality control. In: Aubinet, M., Vesala, T., Papale, D. (Eds.), Eddy Covariance: A Practical Guide to Measurement and Data Analysis. Springer, Berlin, Heidelberg, pp. 85–131.

Franz, D., Acosta, M., Altimir, N., Arriga, N., Arrouays, D., Aubinet, M., Aurela, M., Ayres, E., López-Ballesteros, A., Barbaste, M., Berveiller, D., Biraud, S., Boukir, H., Brown, T., Brümmer, C., Buchmann, N., Burba, G., Carrara, A., Cescatti, A., Ceschia, E., Clement, R., Cremonese, E., Crill, P., Darenova, E., Dengel, S., D'Odorico, P., Gianluca, F., Fleck, S., Fratini, G., Fuß, R., Gielen, B., Gogo, S., Grace, J., Graf, A., Grelle, A., Gross, P., Grünwald, T., Haapanala, S., Hehn, M., Heinesch, B., Heiskanen, J., Herbst, M., Herschlein, C., Hörtnagl, L., Hufkens, K., Ibrom, A., Jolivet, C., Joly, L., Jones, M., Kiese, R., Klemedtsson, L., Kljun, N., Klumpp, K., Kolari, P., Kolle, O., Kowalski, A., Kutsch, W., Laurila, T., De Ligne, A., Linder, S., Lindroth, A., Lohila, A., Longdoz, B., Mammarella, I., Manise, T., Marañon-Jimenez, S., Matteucci, G., Mauder, M., Meier, P., Merbold, L., Mereu, S., Metzger, S., Migliavacca, M., Mölder, M., Montagnani, L.,

Moureaux, C., Nelson, D., Nemitz, E., Nicolini, G., Nilsson, M.B., Op de Beeck, M., Osborne, B., Ottosson Löfvenius, M., Pavelka, M., Peichl, M., Peltola, O., Pihlatie, M., Pitacco, A., Pokorny, R., Pumpanen, J., Ratié, C., Schrumpf, M., Sedlák, P., Serrano Ortiz, P., Siebicke, L., Šigut, L., Silvennoinen, H., Simioni, G., Skiba, U., Sonnentag, O., Soudani, K., Soulé, P., Steinbrecher, R., Tallec, T., Thimonier, A., Tuittila, E., Tuovinen, J., Vestin, P., Vincent, G., Vincke, C., Vitale, D., Waldner, P., Weslien, P., Wingate, L., Wohlfahrt, G., Zahniser, M., Vesala, T., 2018. Towards long-term standardised carbon and greenhouse gas observations for monitoring Europe's terrestrial ecosystems: a review. Int. Agrophys. 32 (4), 439–455.

Fratini, G., Ibrom, A., Arriga, N., Burba, G., Papale, D., 2012. Relative humidity effects on water vapour fluxes measured with closed-path eddy-covariance systems with short sampling lines. Agric. For. Meteorol. 165, 53–63.

Gash, J.H.C., Dolman, A.J., 2003. Sonic anemometer (co)sine response and flux measurement I. The potential for (co)sine error to affect sonic anemometer-based flux measurements. Agric. For. Meteorol. 119 (3–4), 195–207.

Gielen, B., Acosta, M., Altimir, N., Buchmann, N., Cescatti, A., Ceschia, E., Fleck, S., Hörtnagl, L., Klumpp, K., Kolari, P., Lohila, A., Loustau, D., Marañon-Jimenez, S., Manise, T., Matteucci, G., Merbold, L., Metzger, C., Moureaux, C., Montagnani, L., Nilsson, M.B., Osborne, B., Papale, D., Pavelka, M., Saunders, M., Simioni, G., Soudani, K., Sonnentag, O., Tallec, T., Tuittila, E., Peichl, M., Pokorny, R., Vincke, C., Wohlfahrt, G., 2018. Ancillary vegetation measurements at ICOS ecosystem stations. Int. Agrophys. 32 (4), 645–664.

Göckede, M., Markkanen, T., Hasager, C., Foken, T., 2006. Update of a footprint-based approach for the characterisation of complex measurement sites. Bound.-Layer Meteorol. 118, 635–655.

Göckede, M., Foken, T., Aubinet, M., Aurela, M., Banza, J., Bernhofer, C., Bonnefond, J.M., Brunet, Y., Carrara, A., Clement, R., Dellwik, E., Elbers, J., Eugster, W., Fuhrer, J., Granier, A., Grünwald, T., Heinesch, B., Janssens, I.A., Knohl, A., Koeble, R., Laurila, T., Longdoz, B., Manca, G., Marek, M., Markkanen, T., Mateus, J., Matteucci, G., Mauder, M., Migliavacca, M., Minerbi, S., Moncrieff, J., Montagnani, L., Moors, E., Ourcival, J.-M., Papale, D., Pereira, J., Pilegaard, K., Pita, G., Rambal, S., Rebmann, C., Rodrigues, A., Rotenberg, E., Sanz, M.J., Sedlak, P., Seufert, G., Siebicke, L., Soussana, J.F., Valentini, R., Vesala, T., Verbeeck, H., Yakir, D., 2008. Quality control of CarboEurope flux data – Part 1: COUPLING footprint analyses with flux data quality assessment to evaluate sites in forest ecosystems. Biogeosciences 5, 433–450.

Goulden, M.L., Munger, J.W., Fan, S.-M., Daube, B.C., Wofsy, S.C., 1996. Measurements of carbon sequestration by long-term eddy covariance: methods and a critical evaluation of accuracy. Glob. Change Biol. 2, 169–182.

Griffis, T.J., Hu, C., Baker, J.M., Wood, J.D., Millet, D.B., Erickson, M., Yu, Z., Deventer, M.J., Winker, C., Chen, Z., 2019. Tall tower ammonia observations and emission estimates in the U.S. Midwest. J. Geophys. Res. Biogeosci. 124, 3432–3447.

Griffith, D.W.T., Bryant, G.R., Hsu, D., Reisinger, A.R., 2008. Methane emissions from free-ranging cattle: comparison of tracer and Integrated horizontal flux techniques. J. Environ. Qual. 37, 582–591.

Gu, L., Falge, E., Boden, T., Baldocchi, D.D., Black, T.A., Saleska, S.R., Suni, T., Vesala, T., Wofsy, S., Xu, L., 2005. Observing threshold determination for nighttime eddy flux filtering. Agric. For. Meteorol. 128, 179–197.

Harper, L.A., Denmead, O.T., Freney, J.R., Byers, F.M., 1999. Direct measurements of methane emissions from grazing and feed lot cattle. J. Anim. Sci. 77, 1392–1401.

Harper, L.A., Byers, F.M., Sharpe, R.R., 2002. Treatment effects on methane emissions from grazing cattle. In: Takahashi, J., Young, B.A. (Eds.), Greenhouse Gases and Animal Agriculture. Elsevier Science B.V, Amsterdam, The Netherlands, pp. 167–170.

Harper, L.A., Denmead, O.T., Flesch, T.K., 2011. Micrometeorological techniques for measurement of enteric greenhouse gas emissions. Anim. Feed Sci. Technol. 166–167, 227–239.

Helbig, M., Wischnewski, K., Gosselin, G.H., Biraud, S.C., Bogoev, I., Chan, W.S., Euskirchen, E.S., Glenn, A.J., Marsh, P.M., Quinton, W.L., Sonnentag, O., 2016. Addressing a systematic bias in carbon dioxide flux measurements with the EC150 and the IRGASON open-path gas analyzers. Agric. For. Meteorol. 228–229, 349–359.

Hufkens, K., Gianluca, F., Cremonese, E., Migliavacca, M., D'Odorico, P., Peichl, M., Gielen, B., Hörtnagl, L., Soudani, K., Papale, D., Rebmann, C., Brown, T., Wingate, L., 2018. Assimilating phenology datasets automatically across ICOS ecosystem stations. Int. Agrophys. 32 (4), 677–687.

Hunt, J.E., Laubach, J., Barthel, M., Fraser, A., Phillips, R.L., 2016. Carbon budgets for an irrigated intensively grazed dairy pasture and an unirrigated winter-grazed pasture. Biogeosciences 13, 2927–2944.

Jordan, A., Haidacher, S., Hanel, G., Hartungen, E., Märk, L., Seehauser, H., Schottkowsky, R., Sulzer, P., Märk, T.D., 2009. A high resolution and high sensitivity proton-transfer-reaction time-of-flight mass spectrometer (PTR-TOF-MS). Int. J. Mass Spectrometry 286 (2–3), 122–128.

Judd, M.J., Kelliher, F.M., Ulyatt, M.J., Lassey, K.R., Tate, K.R., Shelton, I.D., Harvey, M.J., Walker, C.F., 1999. Net methane emissions from grazing sheep. Glob. Change Biol. 5, 647–657.

Kaimal, J.C., Businger, J.A., 1963. A continuous wave sonic anemometer-thermometer. J. Clim. Appl. Meteorol. 2, 156–164.

Kaimal, J.C., Finnigan, J.J., 1994. Atmospheric Boundary Layer Flows. Their Structure and Measurement. Oxford University Press.

Kamp, J.N., Häni, C., Nyord, T., Feilberg, A., Sørensen, L.L., 2020. The aerodynamic gradient method: implications of non-simultaneous measurements at alternating heights. Atmosphere 11 (10), 1067.

Kljun, N., Calanca, P., Rotach, M.W., Schmid, H.P., 2004. A simple parameterisation for flux footprint predictions. Bound.-Layer Meteorol. 112, 503–523.

Kormann, R., Meixner, F.X., 2001. An analytic footprint model for neutral stratification. Bound.-Layer Meteorol. 99, 207–224.

Kroon, P.S., Hensen, A., Jonker, H.J.J., Ouwersloot, H.G., Vermeulen, A.T., Bosveld, F.C., 2010. Uncertainties in eddy covariance flux measurements assessed from CH_4 and N_2O observations. Agric. For. Meteorol. 150, 806–816.

Langford, B., Acton, W., Ammann, C., Valach, A., Nemitz, E., 2015. Eddy-covariance data with low signal-to-noise ratio: time-lag determination, uncertainties and limit of detection. Atmos. Meas. Tech. 8, 4197–4213.

Laubach, J., Kelliher, F.M., 2004. Measuring methane emission rates of a dairy cow herd by two micrometeorological techniques. Agric. For. Meteorol. 125, 279–303.

Laubach, J., Kelliher, F.M., 2005. Methane emissions from dairy cows: comparing open-path laser measurements to profile-base techniques. Agric. For. Meteorol. 135, 340–345.

Laubach, J., Kelliher, F.M., Knight, T.W., Clark, H., Molano, G., Cavanagh, A., 2008. Methane emissions from beef cattle – a comparison of paddock- and animal-scale measurements. Aust. J. Exp. Agric. 48, 132–137.

Leclerc, M.Y., Thurtell, G.W., 1990. Footprint prediction of scalar fluxes using a Markovian analysis. Bound.-Layer Meteorol. 52, 247–258.

Lee, X., Black, T.A., 1994. Relating eddy correlation sensible heat flux to horizontal sensor separation in the unstable atmospheric surface layer. J. Geophys. Res. Atmos. 99 (D9), 18545–18553.

Lees, K.J., Quaife, T., Artz, R.R.E., Khomik, M., Clark, J.M., 2018. Potential for using remote sensing to estimate carbon fluxes across northern peatlands – a review. Sci. Total Environ. 615, 857–874.

Leuning, R., 2007. The correct form of the Webb, Pearman and Leuning equation for eddy fluxes of trace gases in steady and non-steady state, horizontally homogeneous flows. Bound.-Layer Meteorol. 123, 263–267.

Leuning, R., Judd, M.J., 1996. The relative merits of open- and closed path analysers for measurements of eddy fluxes. Glob. Change Biol. 2, 241–254.

Leuning, R., King, K.M., 1992. Comparison of eddy-covariance measurements of CO_2 fluxes by open- and closed-path CO_2 analysers. Bound.-Layer Meteorol. 59, 297–311.

Leuning, R.L., Moncrieff, J.B., 1990. Eddy covariance CO_2 flux measurements using open and closed path CO_2 analysers: correction for analyser water vapour sensitivity and damping of fluctuations in air sampling tubes. Bound.-Layer Meteorol. 53, 63–76.

Leuning, R., Baker, S.K., Jamie, I.M., Hsu, C.H., Klein, L., Denmead, O.T., Griffith, D.W.T., 1999. Methane emission from free-ranging sheep: a comparison of two measurement methods. Atmos. Environ. 33, 1357–1365.

Lindinger, W., Hansel, A., Jordan, A., 1998. On-line monitoring of volatile organic compounds at pptv levels by means of proton-transfer-reaction mass spectrometry (PTR-MS) medical applications, food control and environmental research. Int. J. Mass Spectrom. Ion Processes 173, 191–241.

Liu, H., Peters, G., Foken, T., 2001. New equations for sonic temperature variance and buoyancy heat flux with an omnidirectional sonic anemometer. Bound.-Layer Meteorol. 100, 459–468.

Long, S.P., Bernacchi, C.J., 2003. Gas exchange measurements, what can they tell us about the underlying limitations to photosynthesis? Procedures and sources of error. J. Exp. Bot. 54 (392), 2393–2401.

Loustau, D., Altimir, N., Barbaste, M., Gielen, B., Marañon-Jimenez, S., Klumpp, K., Linder, S., Matteucci, G., Merbold, L., Op de Beeck, M., Soulé, P., Thimonier, A., Vincke, C., Waldner, P., 2018. Sampling and collecting foliage elements for the determination of the foliar nutrients in ICOS ecosystem stations. Int. Agrophys. 32 (4), 665–676.

Mammarella, I., Kolari, P., Vesala, T., Rinne, J., 2007. Determining the contribution of vertical advection to the net ecosystem exchange at Hyytiälä forest, Finland. Tellus 59B, 900–909.

Massman, W.J., Ibrom, A., 2008. Attenuation of concentration fluctuations of water vapor and other trace gases in turbulent tube flow. Atmos. Chem. Phys. 8, 6245–6259.

Mauder, M., Foken, T., Aubinet, M., Ibrom, A., 2021. Eddy-covariance measurements. In: Foken, T. (Ed.), Springer Handbook of Atmospheric Measurements. Springer.

McInnes, K.J., Heilman, J.L., 2005. Relaxed eddy accumulation. In: Hatfield, J.L., Baker, J.M. (Eds.), Micrometeorology in Agricultural Systems. Agronomy Monograph, 47. American Society of Agronomy, Madison, WI, USA, pp. 437–453.

Metzger, S., Burba, G., Burns, S.P., Blanken, P.D., Li, J., Luo, H., Zulueta, R.C., 2016. Optimization of an enclosed gas analyzer sampling system for measuring eddy covariance fluxes of H_2O and CO_2. Atmos. Meas. Tech. 9, 1341–1359.

Moncrieff, J.B., Malhi, Y., Leuning, R., 1996. The propagation of errors in long-term measurements of land atmosphere fluxes of carbon and water. Glob. Change Biol. 2, 231–240.

Moncrieff, J.B., Massheder, J.M., de Bruin, H., Elbers, J., Friborg, T., Heusinkveld, B., Kabat, P., Scott, S., Soegaard, H., Verhoef, A., 1997. A system to measure surface fluxes of momentum, sensible heat, water vapour and carbon dioxide. J. Hydrol. 188–189, 589–611.

Montgomery, R.B., 1948. Vertical eddy flux of heat in the atmosphere. J. Meteorol. 5, 265–274.

Moore, C.J., 1986. Frequency response corrections for eddy correlation systems. Bound.-Layer Meteorol. 37, 17–35.

Munger, J.W., Loescher, H.W., Luo, H., 2012. Measurement, tower, and site design considerations. In: Aubinet, M., Vesala, T., Papale, D. (Eds.), Eddy Covariance – A Practical Guide to Measurement and Data Analysis. Springer, Dordrecht.

Neftel, A., Ammann, C., Fischer, C., Spirig, C., Conen, F., Emmenegger, L., Tuzson, B., Wahlen, S., 2010. N_2O exchange over managed grassland: application of a quantum cascade laser spectrometer for micrometeorological flux measurements. Agric. For. Meteorol. 150 (6), 775–785.

Nelson, A.J., Lichiheb, N., Koloutsou-Vakakis, S., Rood, M.J., Heuer, M., Myles, L.T., Joo, E., Miller, J., Bernacchi, C., 2019. Ammonia flux measurements above a corn canopy using relaxed eddy accumulation and a flux gradient system. Agric. For. Meteorol. 264, 104–113.

Nemitz, E., Mammarella, I., Ibrom, A., Aurela, M., Burba, G.G., Dengel, S., Gielen, B., Grelle, A., Heinesch, B., Herbst, M., Hörtnagl, L., Klemedtsson, L., Lindroth, A., Lohila, A., McDermitt, D.K., Meier, P., Merbold, L., Nelson, D., Nicolini, G., Nilsson, M.B., Peltola, O., Rinne, J., Zahniser, M., 2018. Standardisation of eddy-covariance flux measurements of methane and nitrous oxide. Int. Agrophys. 32 (4), 517–549.

Nicolini, G., Aubinet, M., Feigenwinter, C., Heinesch, B., Lindroth, A., Mamadou, O., Moderow, U., Mölder, M., Montagnani, L., Rebmann, C., Papale, D., 2018. Impact of CO_2 storage flux sampling uncertainty on net ecosystem exchange measured by eddy covariance. Agric. For. Meteorol. 248, 228–239.

Obukhov, A.M., 1951. Charakteristiki mikrostruktury vetra v prizemnom sloje atmosfery (Characteristics of the micro-structure of the wind in the surface layer of the atmosphere). Izv AN SSSR, ser Geofiz 3, 49–68.

Op de Beeck, M., Gielen, B., Merbold, L., Ayres, E., Serrano-Ortiz, P., Acosta, M., Pavelka, M., Montagnani, L., Nilsson, M., Klemedtsson, L., Vincke, C., De Ligne, A., Moureaux, C., Marañon-Jimenez, S., Saunders, M., Mereu, S., Hörtnagl, L., 2018. Soil-meteorological measurements at ICOS monitoring stations in terrestrial ecosystems. Int. Agrophys. 32 (4), 619–631.

Pattey, E., Desjardins, R.L., Boudreau, F., Rochette, P., 1992. Impact of density fluctuations on flux measurements of trace gases: implications for the relaxed eddy accumulation technique. Bound.-Layer Meteorol. 59 (1–2), 195–203.

Pattey, E., Desjardins, R.L., Rochette, P., 1993. Accuracy of the relaxed eddy-accumulation technique, evaluated using CO_2 flux measurements. Bound.-Layer Meteorol. 66 (4), 341–355.

Pattey, E., Edwards, G., Strachan, I.B., Desjardins, R.L., Kaharabata, S., Wagner Riddle, C., 2006. Towards standards for measuring greenhouse gas fluxes from agricultural fields using instrumented towers. Can. J. Soil Sci. 86 (3), 373–400.

Pavelka, M., Acosta, M., Kiese, R., Altimir, N., Brümmer, C., Crill, P., Darenova, E., Fuß, R., Gielen, B., Graf, A., Klemedtsson, L., Lohila, A., Longdoz, B., Lindroth, A., Nilsson, M., Marañon-Jimenez, S., Merbold, L., Montagnani, L., Peichl, M., Pihlatie, M., Pumpanen, J., Serrano Ortiz, P., Silvennoinen, H., Skiba, U., Vestin, P., Weslien, P., Janouš, D., Kutsch, W., 2018. Standardisation of chamber technique for CO_2, N_2O and CH_4 fluxes measurements from terrestrial ecosystems. Int. Agrophys. 32 (4), 569–587.

Peltola, O., Aslan, T., Ibrom, A., Nemitz, E., Rannik, Ü., Mammarella, I., 2021. The high frequency response correction of eddy covariance fluxes. Part 1: an experimental approach
and its interdependence with the time-lag estimation. Atmos. Meas. Tech. 14, 5071–
5088. https://doi.org/10.5194/amt-14-5071-2021,2021.

Prueger, J.H., Kustas, W.P., 2005. Aerodynamic methods for estimating turbulent fluxes. In:
Hatfield, V., Baker, J.M., Viney, M.K. (Eds.), Micrometeorology in Agricultural Systems.
Agronomy Monograph No. 47, ASA-CSSA-SSSA, Madison, WI, pp. 407–436.

Pumpanen, J., Kolari, P., Ilvesniemi, H., Minkkinen, K., Vesala, T., Niinistö, S., Lohila, A.,
Larmola, T., Morero, M., Pihlatie, M., Janssens, I., Curiel Yuste, J., Grünzweig, J., Reth,
S., Subke, J.-A., Savage, K., Kutsch, W., Østreng, G., Ziegler, W., Anthoni, P., Lindroth,
A., Hari, P., 2004. Comparison of different chamber techniques for measuring soil
CO_2 efflux. Agric. For. Meteorol. 123 (3–4), 159–176.

Rannik, Ü., 1997. On the damping of temperature fluctuations in a circular tube relevant to the
eddy covariance measurement technique. J. Geophys. Res. Atmos. 102 (D11), 12789–
12794. Article 97JD00362.

Rannik, Ü., Sogachev, A., Foken, T., Göckede, M., Kljun, N., Leclerc, M., Vesala, T., 2012.
Footprint analysis. In: Aubinet, M., Vesala, T., Papale, D. (Eds.), Eddy Covariance: A
Practical Guide to Measurement and Data Analysis. Springer, Berlin, Heidelberg,
pp. 211–262.

Rebmann, C., Kolle, O., Heinesch, B., Queck, R., Ibrom, A., Aubinet, M., 2012. Data acquisition and flux calculations. In: Aubinet, M., Vesala, T., Papale, D. (Eds.), Eddy
Covariance: A Practical Guide to Measurement and Data Analysis. Springer, Berlin, Heidelberg, pp. 59–84.

Rebmann, C., Aubinet, M., Schmid, H.P., Arriga, N., Aurela, M., Burba, G., Clement, R., De
Ligne, A., Fratini, G., Gielen, B., Grace, J., Graf, A., Gross, P., Haapanala, S., Herbst, M.,
Hörtnagl, L., Ibrom, A., Joly, L., Kljun, N., Kolle, O., Kowalski, A., Lindroth, A., Loustau,
D., Mammarella, I., Mauder, M., Merbold, L., Metzger, S., Mölder, M., Montagnani, L.,
Papale, D., Pavelka, M., Peichl, M., Roland, M., Serrano-Ortiz, P., Siebicke, L.,
Steinbrecher, R., Tuovinen, J.-P., Vesala, T., Wohlfahrt, G., Franz, D., 2018. ICOS eddy
covariance flux-station site setup: a review. Int. Agrophys. 32 (4), 471–494. https://doi.org/
10.1515/intag-2017-0044.

Ren, X., Sanders, J.E., Rajendran, A., Weber, R.J., Goldstein, A.H., Pusede, S.E., Browne, E.C.,
Min, K.-E., Cohen, R.C., 2011. A relaxed eddy accumulation system for measuring vertical
fluxes of nitrous acid. Atmos. Meas. Tech. 4, 2093–2103.

Rhew, R.C., Deventer, M.J., Turnipseed, A.A., Warneke, C., Ortega, J., Shen, S., Martinez, L.,
Koss, A., Lerner, B.M., Gilman, J.B., Smith, J.N., Guenther, A.B., de Gouw, J.A., 2017.
Ethene, propene, butene and isoprene emissions from a ponderosa pine forest measured by
relaxed eddy accumulation. Atmos. Chem. Phys. 17, 13417–13438.

Richardson, A.D., Hollinger, D.Y., Burba, G.G., Davis, K.J., Flanagan, L.B., Katul, G.G.,
Munger, J.W., Ricciuto, D.M., Stoy, P.C., Suyker, A.E., Verma, S.B., Wofsy, S.C.,
2006. A multi-site analysis of random error in tower-based measurements of carbon
and energy fluxes. Agric. For. Meteorol. 136, 1–18.

Richardson, A.D., Aubinet, M., Barr, A.G., Hollinger, D.Y., Ibrom, A., Lasslop, G., Reichstein,
M., 2012. Uncertainty quantification. In: Aubinet, M., Vesala, T., Papale, D. (Eds.), Eddy
Covariance: A Practical Guide to Measurement and Data Analysis. Springer, Berlin, Hei-
delberg, pp. 173–209.

Rinne, J., Ammann, C., 2012. Disjunct eddy covariance method. In: Aubinet, M., Vesala, T.,
Papale, D. (Eds.), Eddy Covariance: A Practical Guide to Measurement and Data Analysis.
Springer, Berlin, Heidelberg, pp. 291–307.

Rinne, J., Karl, T., Guenther, A., 2016. Simple, stable, and affordable: towards long-term eco-system scale flux measurements of VOCs. Atmos. Environ. 131, 225–227.

Ruppert, J., Thomas, C., Foken, T., 2006. Scalar similarity for relaxed eddy accumulation methods. Bound.-Layer Meteorol. 120, 39–63.

Sabbatini, S., Mammarella, M., Arriga, N., Fratini, G., Graf, A., Hörtnagl, L., Ibrom, A., Long-doz, B., Mauder, M., Merbold, L., Metzger, S., Montagnani, L., Pitacco, A., Rebmann, C., Sedlák, P., Šigut, L., Vitale, D., Papale, D., 2018. Eddy covariance raw data processing for CO_2 and energy fluxes calculation at ICOS ecosystem stations. Int. Agrophys. 32 (4), 495–515.

Saleska, S.R., Shorter, J., Herndon, S., Jimenéz, R., McManus, B., Nelson, D., Zahniser, M., 2006. What are the instrumentation requirements for measuring the isotopic composition of net ecosystem exchange of CO_2 using eddy covariance methods? Isotopes Environ. Health Stud. 42 (1), 117.

Schmid, H.P., 1994. Source areas for scalar and scalar fluxes. Bound.-Layer Meteorol. 67, 293–318.

Schmid, H.P., 2002. Footprint modeling for vegetation atmosphere exchange studies: a review and perspective. Agric. For. Meteorol. 113, 159–183.

Schotanus, P., Nieuwstadt, F.T.M., DeBruin, H.A.R., 1983. Temperature measurement with a sonic anemometer and its application to heat and moisture fluctuations. Bound.-Layer Meteorol. 26, 81–93.

Schuepp, P.H., Leclerc, M.Y., MacPherson, J.I., Desjardins, R.L., 1990. Footprint prediction of scalar fluxes from analytical solutions of the diffusion equation. Bound.-Layer Meteorol. 50, 355–373.

Sintermann, J., Spirig, C., Jordan, A., Kuhn, U., Ammann, C., Neftel, A., 2011. Eddy covariance flux measurements of ammonia by high temperature chemical ionisation mass spectrom-etry. Atmos. Meas. Tech. 4, 599–616.

Skov, H., Brooks, S.B., Goodsite, M.E., Lindberg, S.E., Meyers, T.P., Landis, M.S., Larsen, M. R.B., Jensen, B., McConville, G., Christensen, J., 2006. Fluxes of reactive gaseous mer-cury measured with a newly developed method using relaxed eddy accumulation. Atmos. Environ. 40 (29), 5452–5463.

Spirig, C., Neftel, A., Ammann, C., Dommen, J., Grabmer, W., Thielmann, A., Schaub, A., Beauchamp, J., Wisthaler, A., Hansel, A., 2005. Eddy covariance flux measurements of biogenic VOCs during ECHO 2003 using proton transfer reaction mass spectrometry. Atmos. Chem. Phys. 5, 465–481.

Swinbank, W.C., 1951. The measurement of vertical transfer of heat and water vapor by eddies in the lower atmosphere. J. Meteorol. 8, 135–145.

Tanimoto, H., Kameyama, S., Iwata, T., Inomata, S., Omori, Y., 2014. Measurement of air-sea exchange of dimethyl sulfide and acetone by PTR-MS coupled with gradient flux tech-nique. Environ. Sci. Technol. 48, 526–533.

Taylor, J.R., 1997. Introduction to Error Analysis: The Study of Uncertainties in Physical Mea-surements. University Science Books. 327 p.

Thomas, C., Martin, J., Law, B., Davis, K., 2013. Toward biologically meaningful net carbon exchange estimates for tall, dense canopies: multi-level eddy covariance observations and canopy coupling regimes in a mature Douglas-fir forest in Oregon. Agric. For. Meteorol. 173, 14–27.

van Gorsel, E., Delpierre, N., Leuning, R., Black, A., Munger, J.W., Wofsy, S., Aubinet, M., Feigenwinter, C., Beringer, J., Bonal, D., Chen, B., Chen, J., Clement, R.R., Davis, K. J., Desai, A.R., Dragoni, D., Etzold, S., Grünwald, T., Gu, L., Heinesch, B., Hutyra, L. R., Jans, W.W., Kutsch, W., Law, B.E., Leclerc, M.Y., Mammarella, I., Montagnani,

L., Noormets, A., Rebmann, C., Wharton, S., 2009. Estimating nocturnal ecosystem respiration from the vertical turbulent flux and change in storage of CO_2. Agric. For. Meteorol. 149, 1919–1930.

Vickers, D., Mahrt, L., 1997. Quality control and flux sampling problems for tower and aircraft data. J. Atmos. Ocean. Technol. 14, 512–526.

Webb, E.K., Pearman, G.I., Leuning, R., 1980. Correction of the flux measurements for density effects due to heat and water vapour transfer. Q. J. R. Meteorol. Soc. 106, 85–100.

Whitehead, J.D., Twigg, M., Famulari, D., Nemitz, E., Sutton, M.A., Gallagher, M.W., Fowler, D., 2008. Evaluation of laser absorption spectroscopic techniques for eddy covariance flux measurements of ammonia. Environ. Sci. Technol. 42 (6), 2041–2046.

Wilczak, J.M., Oncley, S.P., Stage, S.A., 2001. Sonic anemometer tilt correction algorithms. Bound.-Layer Meteorol. 99 (1), 127–150.

World Meteorological Organization (WMO), 2014. Guide to Meteorological Instruments and Methods of Observation. WMO-No. 8, World Meteorological Organization (updated in 2017).

Yamamoto, S., Saigusa, N., Gamo, M., Fujinuma, Y., Inoue, G., Hirano, T., 2005. Findings through the AsiaFlux network and a view toward the future. J. Geogr. Sci. 15 (2), 142–148.

Yu, G.-.R., Wen, X.-.F., Sun, X.-.M., Tanner, B.D., Lee, X., Chen, J.-.Y., 2006. Overview of ChinaFLUX and evaluation of its eddy covariance measurement. Agric. For. Meteorol. 137 (3–4), 125–137.

Zahniser, M.S., Nelson, D.D., McManus, J.B., Kebabian, P.L., 1995. Measurement of trace gas fluxes using tunable diode laser spectroscopy. Phil. Trans. R. Soc. Lond. A 351 (1696), 371–382.

Zhang, S.F., Wyngaard, J.C., Businger, J.A., Oncley, S.P., 1986. Response characteristics of the U.W. sonic anemometer. J. Atmos. Ocean. Technol. 2, 548–558.

Zhu, T., Pattey, E., Desjardins, R.L., 2000. Relaxed eddy-accumulation technique for measuring ammonia volatilization. Environ. Sci. Technol. 34 (1), 199–203.

What's next: Boundary layer prediction methods

Robert S. Arthur[a] and Wayne M. Angevine[b]
[a]Lawrence Livermore National Laboratory, Livermore, CA, United States, [b]Cooperative Institute for Research in Environmental Sciences, University of Colorado, and NOAA Chemical Sciences Laboratory, Boulder, CO, United States

5.1 Introduction: Why do we need models?

Simply put, boundary layer models are computer programs that solve equations to predict boundary layer dynamics. The wide variety of models available, the specifics of each model, and their suitability for different applications make for a rich and exciting field of study. This chapter will delve into the details of boundary layer models, providing an overview of how they work, how they're used, and what's next for boundary layer prediction.

Much of what we know about the atmospheric boundary layer comes from models. By taking advantage of the processing power of computers, we can predict atmospheric dynamics over a wide range of spatial and time scales, supplementing what we learn from observations. Although there are exceptions, observations are often limited to a small spatial extent, such as a point measurement or a vertical profile. On the other hand, models provide a four-dimensional picture of the boundary layer (three spatial dimensions plus time) and thus a more complete understanding of its dynamics.

With this in mind, we need models to predict how the boundary layer will affect our day-to-day lives. The weather forecast we check each morning comes from a model that predicts quantities like temperature, wind velocity, and precipitation. We also need models to predict boundary layer dynamics for safety or planning purposes. For example, where will the pollution from a new factory go? Or how much power will a proposed wind farm generate?

From a scientific perspective, models are also needed to further improve our understanding of how the boundary layer works. Since we can't control the real atmosphere, models can be used to run experiments in a controlled environment. Such experiments can test cause-and-effect relationships or other theories. Imagine being able to change the land cover or incoming solar radiation and observe how atmospheric dynamics change – models give us that ability.

Ultimately, the power of models is limited by two important caveats. First, model predictions are only as good as the model setup and input data. Choices made by the modeler can significantly affect the quality of the model output. Second, since no

Conceptual Boundary Layer Meteorology. https://doi.org/10.1016/B978-0-12-817092-2.00013-8

model can fully capture the dynamics of the real atmosphere, all models have inherent assumptions that limit their applicability. These caveats must be understood if models are to be used effectively and will therefore guide the discussions in this chapter.

5.2   How do we model the boundary layer?

5.2.1   Modeling basics

Computer models of the boundary layer are built on five main components: governing equations, computational grids, parameterizations for unresolved effects, boundary conditions, and input data.

(1) **Governing equations** are mathematical relationships that describe the dynamics of the boundary layer. Most importantly, these include conservation of mass and momentum for the air in the boundary layer. Equations describing the evolution of temperature and moisture, as well as an equation of state to calculate the pressure, are also included. Optionally, evolution equations for other scalar quantities (such as chemical constituents or pollutants) can be used for specific applications.

(2) **Computational grids** represent the locations in space at which model quantities are calculated, known as "grid points" or "grid cells." Grid shapes vary from simple rectangles to more complex shapes that conform to the terrain or other features. One of the most important aspects of all boundary layer models is the "grid spacing," or the distance between each grid point. This quantity is related to the model's "resolution" and determines what physical quantities can be captured explicitly (that is, "resolved") by the model.

(3) **Parameterizations** are simplified mathematical relationships that help models account for processes that are not explicitly resolved on the model grid. For example, a turbulent eddy or a cloud that is smaller in scale than the model grid spacing is "unresolved" (see Figure 5.1) and must be accounted for using a parameterization. A large portion of boundary layer modeling research is directed toward improving model parameterizations for anything from turbulent mixing to cloud processes to surface fluxes.

(4) **Boundary conditions** are mathematical definitions of what is happening at the edges of the computational domain. At the surface, this includes fluxes of momentum, heat, and moisture, which are usually parameterized based on Monin–Obukhov Similarity Theory (MOST, as covered in Chapter 3) in combination with a model of land surface effects. At the horizontal (or "lateral") edges of the domain, the boundary condition is often based on the results of a larger scale model or forecast. At the top of the domain, fluxes may also be specified, or "damping layers" may be used to prevent reflections or other undesired boundary effects.

(5) **Input data** is required to provide a starting point for model evolution. Similar to the lateral boundary conditions, input data is often derived from a larger scale model or forecast. Alternatively, observed atmospheric profiles or idealized conditions can be used to initialize the model for controlled experiments of boundary layer dynamics.

These building blocks are summarized in Figure 5.1, which shows a typical model grid with boundary conditions as well as resolved and unresolved processes. Knowing the building blocks, we can classify boundary layer models into two broad categories, called "prognostic" and "diagnostic" models (see Lundquist and Chow, 2012 for an extended discussion), which are introduced below.

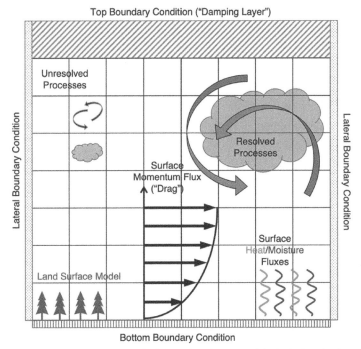

Figure 5.1 An example grid for a typical boundary layer model, depicting the boundary conditions as well as resolved and unresolved processes.
Credit: R.S. Arthur.

5.2.2 Prognostic models

Prognostic models solve the complete set of governing equations described in (1) above and rely on all five building blocks to model boundary layer dynamics. They are used in weather modeling and forecasting, as well as high-resolution computational fluid dynamics (CFD) studies. Due to the computational expense involved in solving prognostic equations, such models usually require the use of supercomputers in order to achieve reasonable run times.

Prognostic models numerically solve the governing equations in order to capture the evolution of each model variable. To do so, the equations must be discretized such that they can be represented on the model's computational grid. In the most basic form of discretization used by boundary layer models, called "finite difference," gradients are approximated as simple differences, for example:

$$\frac{\partial u}{\partial x} \approx \frac{u(x + \Delta x) - u(x)}{\Delta x}. \tag{5.1}$$

Here, u is the model variable (such as velocity) and Δx is the grid spacing.

In the field of boundary layer modeling, there is a wide variety of methods available to discretize model equations. This is also an active area of research. The field of numerical methods is devoted to developing and testing new discretizations that are both accurate and computationally efficient. Interested readers are referred to Moin (2010) and Ferziger et al. (2020) for more information on discretization methods and the numerical solution of equations.

5.2.3 Diagnostic models

Since solving the complete set of governing equations described above can be challenging and/or computationally expensive, diagnostic models with simplified governing equations are preferable to prognostic models for situations in which speed and ease-of-use are desired. Such models can usually run on personal computers in seconds to minutes. However, these benefits are balanced by a loss in accuracy relative to prognostic models. Diagnostic models are used for a variety of applications including pollutant transport and dispersion studies, wind resource characterization, and the calculation of wind loads on buildings.

There are two flavors of diagnostic models: mass consistent models and linearized models. Mass consistent models (Sherman, 1978; Goodin et al., 1980) extend the spatial extent of observations by first interpolating them over a given region and then enforcing conservation of mass. Linearized models (Jackson and Hunt, 1975; Hunt et al., 1988a, b) solve linear, steady-state equations for flow over terrain by dividing the atmosphere into near-surface and above-surface layers with different turbulence parameterizations. Both types of diagnostic models require terrain data and surface roughness information, but otherwise simple initial and boundary conditions. Although diagnostic models are useful in many applications, the main focus of this chapter is on prognostic models, which are used for most state-of-the art forecasting and research applications.

5.3 Boundary layer modeling paradigms

The most important parameter of any prognostic boundary layer model is the horizontal grid spacing. Smaller grid spacing, or finer resolution, allows more detailed atmospheric features to be captured on the computational grid. However, this results in additional computational expense as the number of grid points is increased. Due to the large range of scales of motion in the atmosphere – from global-scale circulations to fine-scale turbulence – the appropriate modeling approach must be chosen based on the scales of interest, balancing resolution with computational cost.

5.3.1 It's all about turbulence

Two overarching paradigms of prognostic boundary layer models – "mesoscale" and "microscale" models – have been developed over time, and their use depends primarily on the scales of boundary layer turbulence. Atmospheric boundary layer turbulence can range in scale from roughly the boundary layer depth ($\sim$1 km) to the dissipative scales of turbulence ($\sim$1 mm). These scales represent the sizes of turbulent eddies or coherent turbulent motions that are found in the boundary layer. However, because resolving turbulence down to the dissipative scale, generally thought to be the finest scale of air motion in the atmosphere, is not practical given today's computing resources; at least some of the effects of turbulence must be parameterized. The extent to which turbulence is parameterized, in turn, determines which modeling paradigm applies.

Mesoscale models, which have horizontal grid spacings of roughly 1 km or larger, can capture mesoscale weather events such as fronts and sea breezes, as well as larger "synoptic" scale features such as high/low pressure systems. However, all turbulent effects must be parameterized because the largest scale of boundary layer turbulence is smaller than the grid spacing. In mesoscale models, turbulent effects are parameterized using planetary boundary layer (PBL) schemes, which will be discussed further in Section 5.4.

Microscale models, often referred to as large-eddy simulation (LES) models, have horizontal grid spacings of roughly 100 m or smaller. Thus, LES models can resolve some portion of the scales of boundary layer turbulence, and indeed *must* capture the large or so-called energy-containing scales. In practice, this means that the grid scale must fall within the inertial subrange of turbulence discussed in Chapter 2. However, even high-resolution LES models generally do not capture turbulence all the way down to the dissipative scale and thus the effects of turbulence at scales smaller than the grid scale must still be parameterized. A wide variety of subgrid-scale models exist for this purpose and will be discussed further in Section 5.4.

As a comparison of the two paradigms, imagine that you are modeling the atmospheric boundary layer over the entire continental United States for a weather forecast. A mesoscale model would be most appropriate because it would allow your model to capture the weather features of interest with limited computational expense. It would be very computationally expensive to use a horizontal grid spacing smaller than a few kilometers in this case. Alternatively, imagine that you are now modeling pollutant dispersion in a city. In this case, turbulent effects such as building wakes would be quite important, and a microscale LES model would be most appropriate. If the area of the city is only several kilometers square, it would be computationally feasible to use a grid spacing of roughly 10–100 m, providing adequate resolution for LES.

The careful reader may have noticed a gap in the modeling paradigms, between the lower limit of mesoscale grid spacing ($\sim$1 km) and the upper limit of microscale grid spacing ($\sim$100 m). This gap, known as the *terra incognita* or "gray zone" (Wyngaard, 2004) is where model grid scales are nearly the same as the larger scales of boundary layer turbulence ($\sim$1 km). In the gray zone, neither PBL schemes nor LES turbulence models are technically appropriate. If a PBL scheme is used, some turbulent scales will be double counted (i.e., both resolved and parameterized). If an LES closure scheme is used, the energy containing scales of turbulence will generally be unresolved. In both cases, turbulent effects will be parameterized incorrectly. For these reasons, boundary layer modeling in the gray zone is generally avoided. However, improving turbulence treatment in the gray zone is also an active area of research (Chow et al., 2019).

5.3.2 How do we parameterize turbulence?

The Navier–Stokes equations describe conservation of momentum and mass for the air in the atmospheric boundary layer. More specifically, the primary governing equations in most boundary layer models are an averaged form of these equations known as the Reynolds Averaged Navier–Stokes (RANS) equations, which describe the mean flow. Due to the inherent nonlinearity of the Navier–Stokes equations, a term appears in the RANS equations containing unknown (that is, unresolved) turbulent fluctuations, referred to as the subgrid-scale (SGS) turbulent stress,

$$\tau_{ij,SGS} = -\rho \overline{u_i' u_j'}, \tag{5.2}$$

where ρ is the density of air, u is the velocity, the overbar denotes Reynolds averaging, the prime denotes departures from the Reynolds average, and index notation is used with $i, j = 1, 2, 3$ (see, for example, Pope, 2000; Kundu et al., 2012). The difficulty in calculating this term is known as the "turbulence closure problem," which was covered previously in Chapters 2 and 3.

In boundary layer models, the unknown subgrid-scale turbulent stress term must be parameterized using information that is known (that is, explicitly resolved) in the model. Since the subgrid-scale turbulent stress can be thought of as a turbulent flux of momentum, it is generally parameterized as a diffusive flux acting on the mean gradient, with an elevated diffusion coefficient known as the "eddy viscosity." Thus,

$$\tau_{ij,SGS} = K_M \left(\frac{\partial \overline{u}_i}{\partial x_j} + \frac{\partial \overline{u}_j}{\partial x_i} \right) = 2K_M \overline{S}_{ij}, \tag{5.3}$$

where K_M is the eddy viscosity and $S_{ij} = \frac{1}{2} \left(\frac{\partial u_i}{\partial x_j} + \frac{\partial u_j}{\partial x_i} \right)$ is the strain rate. This parameterization assumes that in an average sense, turbulent eddies create fluxes of momentum that can be modeled as diffusive fluxes. An analogous parameterization is used for turbulent fluxes of scalars such as heat and moisture, with the turbulent diffusion coefficient known as the eddy diffusivity K_H.

Ultimately, turbulence parameterizations depend on accurate approximation of the eddy viscosity and eddy diffusivity, and many techniques have been developed for this purpose. In fact, the appropriate use of mesoscale and microscale modeling paradigms depends on the appropriate calculation of these terms. While mesoscale models typically use PBL schemes, which parameterize vertical turbulent fluxes only, microscale LES models typically use three-dimensional closure schemes, which parameterize turbulent fluxes in all directions. These two approaches will be explored in greater detail in Sections 5.4 and 5.5.

5.3.3 Horizontal grid refinement

Because the horizontal grid spacing is such a critical factor in boundary layer model performance, it can be useful to refine the computational grid in the region(s) of interest. Refining the grid in particular areas, rather than over the entire domain, provides a

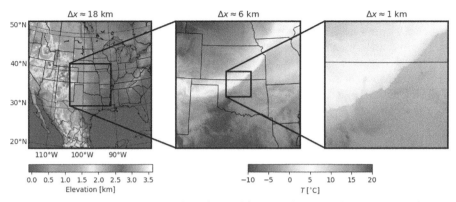

Figure 5.2 Mesoscale WRF simulation of a cold front passing through the central United States. Three nested domains with grid resolutions of roughly 18, 6, and 1 km are used. See Figure 5.3 in the next section for an extension of this example to the microscale.
Credit: Figure by R.S. Arthur, data from Arthur et al. (2020).

large savings in computational effort. Several techniques for controlled grid refinement are common in boundary layer modeling.

The first, known as grid nesting, involves placing successively smaller and finer "child" grids within larger and coarser "parent" grids (see Figure 5.2). The parent grid then supplies initial and boundary conditions to the child grid via interpolation. This technique has been adopted by many mesoscale weather models and forecast systems, including the commonly used Weather Research and Forecasting (WRF) model (Skamarock et al., 2019).

Grid nesting facilitates "multiscale" boundary layer modeling, in which a single model can transition seamlessly from the mesoscale to the microscale modeling paradigm as the grid is refined. Thus, depending on the grid scale, the appropriate turbulence treatment can be applied. Multiscale boundary layer modeling is an active area of research and will be used as an example below in Section 5.5.

Other techniques for grid refinement include unstructured grids, as well as adaptive mesh refinement (AMR). Unstructured grids use irregularly shaped grid cells that can be refined in the region(s) of interest. Such grids are becoming common in global scale weather and climate models such as the recently developed Model for Prediction Across Scales (MPAS; Skamarock et al., 2012). Furthermore, while most boundary layer models use static grids that do not change in time, AMR allows the grid to be refined dynamically based on given criteria (e.g., van Hooft et al., 2018). For example, the model could refine the grid in regions where sharp velocity or temperature gradients are detected, thus focusing computational expense in the areas where it is most needed.

5.4 Mesoscale models

In the recent past, mesoscale models were the primary tools used for numerical weather prediction (NWP), especially continental and regional scale weather forecasting. However, as computing power increases, global models with mesoscale grids are becoming more common. PBL schemes in mesoscale models (and global models on mesoscale grids) are usually responsible not only for vertical mixing within the PBL, but also for all vertical mixing. In fact, this separation of responsibility between the horizontal advection numerics and vertical PBL mixing can be thought of as a primary distinction between mesoscale and microscale models. Some schemes are also responsible for shallow clouds, which are intimately tied to the boundary layer.

The lower boundary condition for the PBL in a mesoscale model is provided by a land surface model (LSM). Although the details of LSMs are beyond the scope of this chapter, it is important to remember that they strongly constrain the PBL simulation and that they contain many physical parameters whose values are difficult to determine on large scales. Soil moisture is one such important variable.

5.4.1 Planetary boundary layer schemes

A useful distinction is made between stable and convective boundary layers when designing parameterization schemes. Stable boundary layers are usually handled by local eddy diffusion equations similar to Eq. (5.3) but in the vertical direction only. The flux of a variable is represented as

$$w'\psi' = K_{M,H}\frac{\partial\psi}{\partial z} \tag{5.4}$$

where $K_{M,H}$ are the eddy diffusivities for momentum and heat, and ψ represents any state variable. This reflects the physical reality that scales of motion in stable boundary layers are small. Performance in stable boundary layers depends on the details of the formulation, especially length scales and stability functions. The eddy diffusivities are commonly formulated as the product of a length scale l and a function of the local stability represented by the Richardson number Ri, as

$$K \simeq lf(Ri). \tag{5.5}$$

Most schemes behave well in weakly stable boundary layers with continuous turbulence. Large differences in scheme performance arise when the boundary layer is moderately or strongly stable and turbulence may be intermittent. This is a fundamental unsolved problem. The PBL scheme may also be distorted in order to compensate for errors in other parts of the model, particularly the land surface, or to avoid numerical instability and model crashes. Usually this distortion takes the form of increasing mixing beyond what would be observed in a similar real situation.

Convective boundary layers have much larger scales of motion and cannot be fully represented by local eddy diffusion. Several methods have been developed to represent the large-scale or non-local motions. One approach is to ignore them, which can be a reasonable approximation but results in a statically unstable (super-adiabatic) buoyancy profile throughout the boundary layer. The next level of improvement is to add an artificial non-local term to the eddy diffusion equation. A more physical approach is called Eddy Diffusivity Mass Flux (EDMF; Siebesma et al., 2007), which uses one or more explicit updrafts to carry the non-local fluxes,

$$\overline{w'\psi'} = K_{M,H}\frac{\partial \psi}{\partial z} + M(\psi_u - \psi), \tag{5.6}$$

where M is the updraft mass flux and ψ_u is the value of the state variable in the updraft. The updrafts are modeled as entraining plumes. Eddy diffusion is still used to model the small-scale fluxes. At the time of writing, most operational NWP models use some form of EDMF scheme, and several such schemes are available in WRF (Angevine et al., 2018).

The boundary layer is rarely in a purely stable or purely convective equilibrium state. Most of the time over most of the globe, both the depth of the boundary layer and the intensity of turbulence are changing. A common occurrence of such changes is between night and morning, known as the morning transition, and between midday and evening, known as the afternoon or evening transition (Angevine et al., 2020). Transitions also occur as air passes over coastlines, and internal boundary layers (multiple layers with different turbulence intensities and different histories) are common. Boundary layer schemes need to simulate these transitions and layers. For this reason, schemes that do not assume a specific profile of turbulence but can react to the vertical buoyancy distribution presented by the rest of the model are preferred.

5.4.2   Simulating a frontal passage

As an illustration of a mesoscale boundary layer modeling application, WRF is used to simulate a cold front passing through the central United States. The frontal passage can be considered a stability transition, which, as noted above, is an important test case for mesoscale modeling. This example is based on work by Arthur et al. (2020), and readers are referred to their paper for additional details of the study. The model is set up in a three-domain nested configuration with horizontal resolutions of roughly 18, 6, and 1 km (see Figure 5.2). Reanalysis data from the North American Mesoscale (NAM) Forecast System is used for initial and boundary conditions on the coarsest domain, and information is subsequently passed down to finer nests at their lateral boundaries. WRF has many PBL schemes available, and the Mellor–Yamada–Janjic (MYJ) scheme is used in this case. The WRF model also parameterizes atmospheric processes such as solar radiation, land surface dynamics, and surface fluxes.

As seen in Figure 5.2, the front comes in from the northwest with an accompanying change in temperature of more than 10 °C. Although it is not shown in the figure, a large increase in the near-surface velocity is also associated with the front. Because this region of the United States has experienced a large amount of wind energy development, the cold front interacts with wind farms as it passes through, modulating their power output. At the end of Section 5.5, this example will be continued with LES of the cold front moving through an actual wind farm.

5.5   Microscale models

LES is used for fine-resolution modeling of the boundary layer, focusing on microscale, turbulent dynamics. Due to its relatively large computational demands, LES is not typically used for forecasting, although this is an emerging field of research (Bauweraerts and Meyers, 2019). This section will explore the fundamentals of subgrid-scale turbulence schemes, followed by additional details about grid resolution and surface turbulence treatment for LES models.

5.5.1   Subgrid-scale turbulence schemes

The original and most basic subgrid-scale scheme was developed by Smagorinsky (1963) and is still used today. Conceptually, the Smagorinsky scheme calculates the eddy viscosity K_M, which has units of length squared per time, by multiplying a length scale by a velocity scale. Because the computational grid determines the scale of the resolved flow in the model, the length scale is chosen as the grid scale $\Delta = (\Delta x \Delta y \Delta z)^{1/3}$ multiplied by a constant scaling coefficient c_S, which is approximately 0.2. The velocity scale is given by this length scale, $c_S \Delta$, multiplied by the magnitude of the resolved strain rate $\overline{S}_{ij}$. Thus, the eddy viscosity is calculated as

$$K_M = (c_S \Delta)^2 |\overline{S}_{ij}|. \tag{5.7}$$

Once the eddy viscosity is calculated, it can be used to estimate the unresolved turbulent stress $\tau_{ij,\,SGS}$ (Eq. 5.2).

A similar turbulence scheme based on subgrid-scale turbulence kinetic energy (TKE) E_{SGS} was also developed by Deardorff (1980). In this scheme, the length scale Δ remains the same, although a different constant coefficient $c_E = 0.1$ is used. The velocity scale is then taken as the square root of the subgrid-scale TKE such that the eddy viscosity is calculated as

$$K_M = c_E \Delta \sqrt{E_{SGS}}. \tag{5.8}$$

When this scheme is used, a separate prognostic equation for the subgrid-scale TKE must be integrated by the model, adding some computational expense.

In LES, the grid can be conceptualized as a mathematical filter because it "filters out" motions that are smaller in scale than the grid spacing. Indeed, the computational grid in LES is sometimes referred to as an implicit filter. Building on this idea, explicit filtering can also be leveraged to gain more information about the resolved turbulence in a boundary layer model. By explicitly filtering the resolved velocity at a resolution that is coarser than the grid spacing, the turbulent stresses that occur in the range of scales between this "test filter" scale and the grid filter scale can be quantified. This is the basis for many advanced subgrid-scale turbulence schemes. The mathematical details of these schemes are beyond the scope of this chapter; however, interested readers are referred to Kirkil et al. (2012) for continued discussion.

While the discussion of grid resolution thus far has been focused on horizontal grid spacing (that is, Δx and Δy), vertical grid spacing Δz is also important, especially in LES. This is because turbulent eddies in the atmospheric boundary layer are often constrained in the vertical direction, either by the presence of the ground surface or by vertical temperature gradients. To illustrate this idea, imagine a convectively unstable boundary layer wherein the only vertical constraint on turbulent eddies is the depth of the boundary layer itself. In this case, a relatively large vertical grid spacing could still resolve the energy-containing scales of turbulence. Alternatively, imagine a stable boundary layer wherein vertical motions are constrained by a temperature inversion. In this case, a much smaller vertical grid spacing would be necessary to resolve the largest turbulent eddies. Due to this more stringent resolution requirement, the modeling of stable boundary layers is generally more computationally expensive than the modeling of unstable boundary layers (see, e.g., Wurps et al., 2020).

A similar restriction in the vertical length scales of turbulent eddies occurs near the ground surface, where turbulent motions are restricted by the ground itself, as well as by roughness features such as trees and buildings. This results in increased resolution requirements near the surface for accurate LES. A large portion of boundary layer modeling research is devoted to improving surface turbulence treatments, generally referred to as "surface layer models." An extended discussion of surface layer modeling can be found in Brasseur and Wei (2010).

The most common surface layer model is based on MOST, which is used in nearly all boundary layer modeng applications. It should be noted, however, that MOST includes assumptions of spatial homogeneity and steady state conditions, which are not necessarily present in fine-resolution microscale boundary layer models. If these assumptions are violated, or if near-surface vertical resolution is not adequate, errors can be introduced into the model via the subgrid-scale turbulence scheme, which will predict incorrect turbulent flux values.

Additional surface-layer treatments can alleviate these errors. For example, the near-surface eddy viscosity can be constrained to follow similarity theory, therefore overriding the value calculated by the subgrid-scale model near the surface (e.g., Mason and Thomson, 1992). The eddy viscosity can then be blended with the LES solution aloft. Furthermore, the effects of vegetation on turbulence can be captured using "canopy models," which introduce an additional drag term to the momentum equations and/or additional dynamics in the subgrid-scale turbulence scheme (see Patton and Finnigan, 2012).

5.5.2 Putting it all together: Multiscale modeling

As a culmination of the topics covered in this chapter, we return to our example simulation of a cold front moving through the central United States. Now, using LES, we are able to simulate the interaction of the cold front with an operational wind farm (see Figure 5.3). This microscale simulation uses 10 m grid spacing with a TKE-based subgrid turbulence scheme and represents flow interaction with wind turbines using an actuator disk model. The simulation is run in WRF with a multiscale setup: using grid nesting, it is embedded within the mesoscale simulation presented previously. Thus,

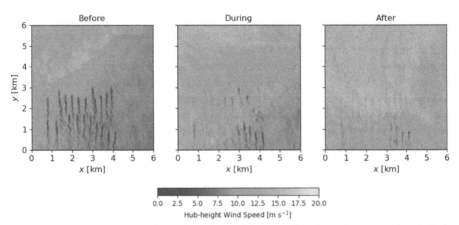

Figure 5.3 Microscale WRF simulation of a cold front passing through an operational wind farm. The simulation has 10 m grid spacing and is forced by the mesoscale simulation presented in Figure 5.2 using a multiscale WRF setup with grid nesting. Results are shown before, during, and after the frontal passage.
Credit: Figure by R.S. Arthur, data from Arthur et al. (2020).

the boundary layer dynamics are seamlessly downscaled to the region of interest and turbulence is parameterized using the appropriate paradigm for each nested grid.

While the mesoscale simulation captured the regional-scale features of the cold front, this microscale simulation resolves fine-scale turbulent structures, including individual turbine wakes (deep purple colors in Figure 5.3). Before the frontal passage, the flow is relatively quiescent with wind speeds less than 5 m/s at turbine hub height. However, the passage of the cold front is associated with a nearly 10 m/s increase in hub-height wind speeds and much higher turbulence intensity. The increase in wind speed brings the turbines from a relatively low power production state to near their rated (that is, maximum) power production in a matter of minutes. The wakes can also affect power output by reducing the wind speed felt by downstream turbines. If wind plant operators can predict such abrupt changes in wind speed and wake dynamics, they can more effectively manage the wind farm's performance.

5.6 Summary and future outlook

This chapter has covered the basics of boundary layer modeling and prediction, including governing equations, computational grids, parameterizations, boundary conditions, and input data. As a major takeaway, consider the distinction between the mesoscale and microscale modeling paradigms. While mesoscale models resolve synoptic and regional scale atmospheric dynamics, they must parameterize turbulent effects using PBL schemes. On the other hand, microscale models resolve the energy-containing scales of turbulence using LES, but must parameterize smaller scale dissipative effects. The appropriate use of boundary layer models depends on the appropriate turbulence treatment. In summary, it's all about turbulence!

So, what's next for boundary layer modeling and prediction? Because models depend on computers, model performance will always be linked to computing technology. As computers get more powerful and efficient, models will run faster and at finer resolution. This will enable exciting advances in global-scale modeling, multi-scale modeling, LES forecasting, and improved understanding of boundary layer phenomena. However, computing power is only part of the story. As this chapter has emphasized, boundary layer models can only be as powerful as their underlying assumptions allow. In particular, model performance depends on the use of appropriate governing equations, boundary conditions, and model inputs. Furthermore, accurate parameterizations must be used for unresolved processes, such as turbulence, clouds, and land–surface interactions. The success of future models will depend on improved theories for how these processes affect the boundary layer at different scales. In this way, boundary layer modeling is intrinsically tied to other aspects of boundary layer science.

Acknowledgments

Prepared by LLNL under Contract DE-AC52-07NA27344. W.M.A. was supported by the National Oceanic and Atmospheric Administration Chemical Sciences Laboratory. R.S.A. gratefully acknowledges Katie Lundquist and Lee Glascoe at LLNL for the opportunity to work on this chapter.

References

Angevine, W.M., et al., 2018. Shallow cumulus in WRF parameterizations evaluated against LASSO large-eddy simulations. Mon. Weather Rev. 146, 4303–4322.

Angevine, W.M., et al., 2020. Transition periods in the diurnally-varying atmospheric boundary layer over land. Bound. Lay. Meteorol. https://doi.org/10.1007/s10546-020-00515-y.

Arthur, R.S., et al., 2020. Multi-scale simulation of wind farm performance during a frontal passage. Atmosphere 11 (3), 245.

Bauweraerts, P., Meyers, J., 2019. On the feasibility of using large-eddy simulations for real-time turbulent-flow forecasting in the atmospheric boundary layer. Bound. Lay. Meteorol. 171 (2), 213–235.

Brasseur, J.G., Wei, T., 2010. Designing large-eddy simulation of the turbulent boundary layer to capture law-of-the-wall scaling. Phys. Fluids 22 (2), 021303.

Chow, F.K., et al., 2019. Crossing multiple gray zones in the transition from mesoscale to microscale simulation over complex terrain. Atmosphere 10 (5), 274.

Deardorff, J.W., 1980. Stratocumulus-capped mixed layers derived from a three-dimensional model. Bound. Lay. Meteorol. 18 (4), 495–527.

Ferziger, J.H., Peric, M., Street, R.L., 2020. Computational Methods for Fluid Dynamics. Springer.

Goodin, W.R., McRae, G.J., Seinfeld, J.H., 1980. An objective analysis technique for constructing three-dimensional urban scale wind fields. J. Appl. Meteorol. 19, 98–108.

Hunt, J.C.R., Leibovich, S., Richards, K.J., 1988a. Turbulent shear flow over low hills. Q. J. R. Meteorol. Soc. 114, 1435–1470.

Hunt, J.C.R., Richards, K.J., Brighton, P.W.M., 1988b. Stably stratified shear-flow over low hills. Q. J. R. Meteorol. Soc. 114, 859–886.

Jackson, P.S., Hunt, J.C.R., 1975. Turbulent wind flow over a low hill. Q. J. R. Meteorol. Soc. 101, 929–955.

Kirkil, G., et al., 2012. Implementation and evaluation of dynamic subfilter-scale stress models for large-eddy simulation using WRF. Mon. Weather Rev. 140, 266–284.

Kundu, P.K., Cohen, I.M., Dowling, D.R., 2012. Fluid Mechanics, fifth ed. Elsevier, Academic Press.

Lundquist, K.A., Chow, F.K., 2012. Flow over complex terrain, numerical modeling of. In: El-Shaarawi, A.H., Piegorsch, W.W. (Eds.), Encyclopedia of Environmetrics, second ed. John Wiley and Sons, Ltd., Chichester, UK, pp. 1054–1063.

Mason, P.J., Thomson, D.J., 1992. Stochastic backscatter in large-eddy simulations of boundary layers. J. Fluid Mech. 242, 51–78.

Moin, P., 2010. Fundamentals of Engineering Numerical Analysis. Cambridge University Press.

Patton, E.G., Finnigan, J.J., 2012. Canopy turbulence. In: Fernando, H.J.S. (Ed.), Handbook of Environmental Fluid Dynamics. CRC Press, pp. 311–328.

Pope, S.B., 2000. Turbulent Flows. Cambridge University Press.

Sherman, C.A., 1978. A mass-consistent model for wind fields over complex terrain. J. Appl. Meteorol. 17, 312–319.

Siebesma, A.P., Soares, P.M.M., Teixeira, J., 2007. A combined eddy-diffusivity mass-flux approach for the convective boundary layer. J. Atmos. Sci. 64, 1230–1248.

Skamarock, W.C., et al., 2012. A multiscale nonhydrostatic atmospheric model using centroidal voronoi tesselations and C-grid staggering. Mon. Weather Rev. 140, 3090–3105.

Skamarock, W.C., et al., 2019. A Description of the Advanced Research WRF Model Version 4. NCAR Tech. Note NCAR/TN-556+STR.

Smagorinsky, J., 1963. General circulation experiments with the primitive equations: I. The basic experiment. Mon. Weather Rev. 91 (3), 99–164.

van Hooft, J.A., et al., 2018. Towards adaptive grids for atmospheric boundary-layer simulations. Bound. Lay. Meteorol. 167, 421–443.

Wurps, H., Steinfeld, G., Heinz, S., 2020. Grid-resolution requirements for large-eddy simulations of the atmospheric boundary layer. Bound. Lay. Meteorol. 175, 179–201.

Wyngaard, J.C., 2004. Toward numerical modeling in the "Terra Incognita". J. Atmos. Sci. 61 (14), 1816–1826.

Who's afraid of the dark: The not so stable stable boundary layer

Carmen J. Nappo
CJN Research Meteorology, Oak Ridge, TN, United States

A semi-quantitative discussion of what happens at night within the first mile of the atmosphere.

6.1 Introduction

Of the various layers comprising the atmosphere, none is more challenging than the nocturnal boundary layer (NBL) over land. Indeed, a thorough discussion of the mechanisms involved in the formation and maintenance of the NBL would require a textbook of its own; see for example Nappo and Johansson (1999) and Steeneveld (2007). In general, the NBL is a nonlinear combination of several variables and processes that can vary from night to night and from hour to hour. The NBL is composed of layers, each with its own characteristics. These layers are not always present, and sometimes a few, but not all, the layers are present.

During the day, rising plumes of warm air generated near the ground surface mix the air throughout the daytime convective boundary layer (CBL). This stirring is such that the local effects of terrain and land use are generally obliterated when time averages of an hour or so are made. However, at night in the absence of this stirring, local topographic and secondary flows can dominate the meteorology within a few hundred meters above the ground surface. Examples of these secondary effects include surface inhomogeneity, density currents, sea and lake breezes, flow blocking by hills and mountains, elevated levels of shear-generated turbulence, atmospheric gravity waves, sporadic episodes of turbulence, bursts, and sudden warmings. On any night, one can expect to see one or more of these various processes to some degree. Because these processes generally interact, the combined effects can mask the individual processes. These effects greatly complicate the analyses of nocturnal meteorological observations, especially near the ground surface. This in turn impedes the development of a unified theory and accurate predictions of the exchanges of heat, momentum, water vapor, and pollutants between the lower near-surface regions and the upper free atmosphere are quite difficult to make. In other words, just about everything this book discusses gets more difficult at night. The SBL scientist must first make detailed measurements of the atmospheric variables throughout a given height above the ground surface, analyze these data in hopes that processes can be identified, seek to parameterize the processes in terms of observable variables, and then test these results

Conceptual Boundary Layer Meteorology. https://doi.org/10.1016/B978-0-12-817092-2.00012-6

in numerical models. It must be noted, as with all theoretical thermodynamic processes, the variables must be *steady-state* values. In the atmosphere, and especially in the NBL, all variables are in continuous change and a steady-state value is represented by an average value.

In this chapter, we provide a description of selected features of the NBL over land. More details of the idealized NBL can be found, for example, in Arya (1988), Stull (1988), Sorbjan (1989), Garratt (1992), Nappo and Johansson (1999), Steeneveld (2007), and Foken (2016). We begin in Section 6.2 with an idealized view of the NBL. In Section 6.3, we present a more realistic description of the SBL. In Section 6.4, we discuss the structure of the SBL. In Section 6.5, we discuss SBL classification schemes. In Section 6.6, the low-level jet (LLJ) is described in terms of observations and analyses. In Section 6.7, we describe non-stationary turbulence.

6.1.1   What do we mean by "stable stratification"?

A system in a state of equilibrium is said to be *stable* if after a small perturbation it returns to its initial state. Imagine an air parcel in equilibrium with its environment at height z_0 where the air temperature is $T(z_0)$. Now imagine the parcel is raised a small distance, Δz, where the air temperature is $T(z_0 + \Delta z)$. If the lifting process is relatively rapid, the air in the parcel will not mix with the ambient air and the temperature in the parcel will remain constant. The difference in temperature between the air parcel and the local air is $T(z_0 + \Delta z) - T(z_0) = T'$. The density of air is inversely proportional to its temperature; thus, a parcel of cold air will be denser than a parcel of warm air, and we all know that warm air rises. If $T' < 0$, then the air parcel is heavier than the air at height $z_0 + \Delta z$ and the parcel will sink to its initial height. This is an example of negative buoyancy. However, if $T' > 0$, the weight of the air parcel will be less than the surrounding air and the parcel will be pushed upward and will continue to rise. This is the principle of hot air balloons. Let's express the *stratification*, that is, the rate of change of air temperature with height, as $\partial T/\partial z$. Then, if $\partial T/\partial z > 0$, the air temperature increases with height. In this case, an air parcel lifted upward will be negatively buoyant and will return to its initial position. This is a *stable* stratification and is most often seen during nighttime conditions. Note that an *inversion* describes the case when air temperature increases with height. If $\partial T/\partial z < 0$, the air parcel will be positively buoyant. This is an *unstable* stratification and is generally the case during daytime conditions. If $\partial T/\partial z = 0$, the stratification is *neutral*, and wherever we place the air parcel, it will remain there.

A stably stratified fluid is one in which the density decreases with height. This type of flow is called *laminar* flow. If the fluid is also stably stratified, we have a *stably stratified flow*. A stably stratified flow is non-turbulent. Indeed, it must be because that is the way we defined it (density decreasing with height). However, this type of flow is seldom observed in the real world. It is safe to say that turbulence can be found in almost all geophysical flows. We must point out that while we do not have a consistent theory of turbulence, we have a good idea of what turbulence is, how it is produced, how it dissipates, and its effect on fluid flows (see Chapter 2). One important characteristic of turbulence is its transience. Turbulence exists in a continuous state of dissipation by friction and can be observed only if it is continuously being produced. When the production is shut down, turbulence immediately begins to decay exponentially. Time scales for this decay range from a few seconds to several minutes on the boundary-layer scale.

It might seem that "stably stratified turbulence" is an oxymoron or a contradiction in terms. To clarify, we must first describe another characteristic of turbulence: *it mixes things*. This mixing is due to the action of three-dimensional eddies within the fluid. The eddies cascade to smaller and smaller eddies that are eventually dissipated into frictional heat. Do you put cream in your coffee or tea? If you put in a spoon of cream into your drink and did nothing, it would take a long time for the cream to diffuse throughout the cup. That is because the process of diffusion is molecular. But we all know that the coffee or tea must be stirred. The stirring generates turbulence and the turbulent eddies rapidly disperse the cream throughout the cup. When the stirring stops, you have taken away the generation mechanism and the turbulence quickly decays.

To put it all together, we have seen that in a stably stratified, laminar flow, sheets of fluid flow over one another. In a laminar flow, "friction" between layers is produced by fine-scale sheets of turbulence. In this case, friction takes the form of a *stress*, the so-called *shear stress*. We have seen that turbulence must be continually generated, and this is the case as long as the fluid motion continues. Now because the flow is stably stratified, if a turbulent eddy transports a fluid parcel upward from one level into the next level, buoyancy will cause the parcel to return to its original layer. When the effects of buoyancy balance the effects of shear stress, the flow is in equilibrium, and we refer to this as a stably stratified flow.

6.2   The idealized stable boundary layer

In the idealized stable boundary layer, the ground is flat, smooth, and homogeneous. The skies are clear and the atmosphere is incompressible and barotropic, that is, secondary flows such as sea breezes and drainage flows do not occur. The main features of the idealized SBL include:

1. a surface-based temperature inversion
2. mechanically generated turbulence
3. clear-air radiative cooling
4. an elevated wind speed maximum or jet
5. a capping residual layer

We will now explain each of these features, but first it is helpful to see all the parts together. Figure 6.1 shows the three main layers of the SBL.

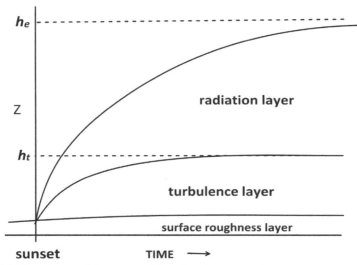

Figure 6.1 Three layers of the nocturnal boundary layer, note these curves are not drawn to scale.

6.2.1 The surface-based temperature inversion

Sometime before sunset, the shortwave solar radiation reaching the ground surface becomes less than the sum of the energy leaving the surface through a combination of longwave radiation, the upward flux of heat flux by convection, and the downward conduction of heat into the soil. Consequently, the ground surface begins to cool. Figure 6.2 shows a diagram of the energy flow. Soon, the upward sensible heat flux

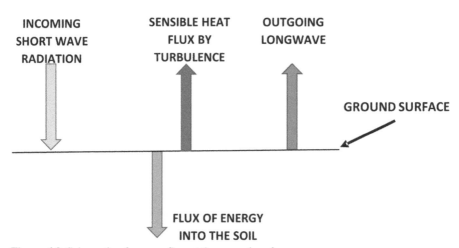

Figure 6.2 Schematic of energy flux at the ground surface.

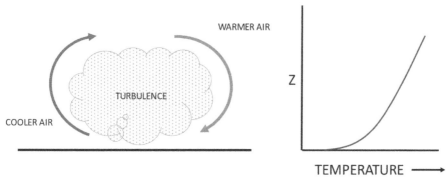

Figure 6.3 Schematic of overturning motion and resultant temperature profile.

near the ground surface reverses direction and the air within the first meter or so above the ground also begins to cool. Mechanical turbulence generated at the ground surface brings warm air downward from above and cool air upward from below, as illustrated in Figure 6.3. Quickly, a temperature inversion, $\partial T/\partial z > 0$, develops, beginning at the ground surface and extending to some height, h. As cooling of the near-surface air continues, the depth of the ground-based inversion increases.

This is illustrated in Figure 6.4, which shows idealized profiles of potential temperature as functions of time. Note that the potential temperature at height h_i is taken to be constant, that is, $\partial\theta/\partial z = 0$. (While this assumption simplifies SBL models, it may not always be accurate.) This layer constitutes the SBL.

Because the ground surface is the only source of turbulence in the SBL, h_i could also be considered as the depth of the effective turbulence.

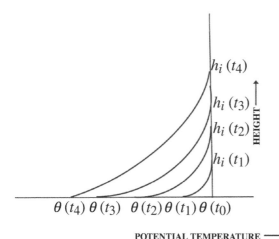

Figure 6.4 Ideal profiles of potential temperature throughout the development of the stable boundary layer.

6.2.2 Mechanically generated turbulence

Mechanically generated turbulence is a constant feature of the planetary boundary layer (PBL) regardless of the stability (see Chapter 2). Because the atmospheric flow must be zero at the ground surface, strong velocity shears develop immediately above the ground surface. Strong velocity shear leads to turbulence; however, turbulence is also generated by wakes produced by surface obstacles such as rocks, shrubs, trees, buildings, and heterogeneous topography. Collectively, these obstacles are referred to as *roughness* or *surface roughness*. This turbulence extends upward to depths of a few meters during stable conditions. This layer of turbulence is known as the *surface roughness layer* (SRL) and is the only source of turbulence in the SBL. Typically, the depth of the SRL is less than a few meters; however, the turbulence can extend much higher than the roughness elements. The turbulence fluxes in the SRL are relatively constant and the SRL is often considered to be a layer of constant flux. The assumption of a *constant flux layer* (CFL) is often used as a lower boundary condition in atmospheric numerical forecast models.

6.2.3 Clear-air radiative cooling

Although longwave radiational cooling continues throughout the SRL and up to h_i, the downward transport of heat by turbulence dominates. However, above the SRL turbulence weakens and gives way to radiational cooling. Above h_i, radiational cooling dominates in a layer extending much higher than the turbulence layer, as illustrated in Figure 6.1. This region presents a problem to PBL modelers. If one assumes the observed cooling is due to turbulence, then the downward transport of pollutants will be overestimated and the air concentrations will be underpredicted. This can have dangerous consequences if toxic substances are being considered. The radiation layer height asymptotically approaches an equilibrium level height, h_e, where the temperature change, integrated across the boundary layer, is completely balanced by the radiative and turbulent processes. A succinct discussion of radiative cooling in the SBL can be found in Duynkerke (1999).

6.2.4 The low-level jet

Shortly after sunset, a wind maximum or jet develops at a height ranging from 100 to 500 m above the ground surface. Blackadar (1957) proposed that LLJ develops when the eddy viscosity caused by the daytime thermal mixing is removed. The air is then free to respond to the larger-scale pressure gradients. The LLJ is frequently observed in the central United States, especially in the Great Plains. The jet usually flows from the south to southwest. The jet is not always present, and whether it is an integral part of the SBL is debatable. Holton (1967) proposed that the LLJ develops in response to differential heating of the sloping surface of the Great Plains in the Midwestern United States. While the LLJ is seen primarily at night, it is considered not to be a part of the SBL. The LLJ wind maximum is sometimes found near the top or within the SBL; then the LLJ can generate turbulence in the upper parts of the SBL. More will be said about the LLJ later.

6.2.5 The residual layer

When daytime convection is cut off by the development of the ground-based inversion, turbulent motions above the SBL do not immediately stop. Energetic eddies remain up to the top of the mixed layer and decay within a few over turnings, about 20 min. However, turbulent motions can persist throughout the night and new turbulence can be generated by breaking gravity waves. This layer extends from the top of the SBL to the previous day's mixing depth and is referred to as the *residual layer* and the turbulence there is referred to as *fossil* turbulence. While not a part of the SBL dynamics, it is a persistent feature of the nighttime atmospheric physics and connects the free atmosphere with the SBL. The various layers of the SBL are illustrated in Figure 6.1.

6.3 The observed stable boundary layer

This section could have been titled, the "real" SBL; however, the concept of reality is more philosophical than scientific. The objective of science is to understand nature, and understanding can come only by observation. This being said, it is not always easy to observe nature, especially when different "actions" occur simultaneously. In Section 6.2, we described these actions individually. In this section, we take a more general, real-world view.

6.3.1 How it starts. The afternoon-to-evening transition

In Section 6.2.1, we described how "sometime before sunset" the ground surface starts to cool. However, for modeling purposes to say "sometime" is useless. Thus, it is more accurate to say, "during the boundary layer afternoon-to-evening transition (AET) over land, changes occur thermodynamically and kinematically near the ground surface." These changes include a decrease in surface temperature, a decrease in wind speed, an increase in water vapor, and transitions of heat flux from positive (up) to negative (down). Sudden changes do not exist in nature and so the AET is seamless. However, because a general theory of turbulence does not exist, atmospheric modelers are forced to use available parameterizations for either the CBL or the SBL. Thus, in atmospheric models the change from convective to stable parameterization schemes are usually more like flipping a switch than the actual more gradual transition. There has been much research into the AET so that modelers can determine accurately when and how to switch from unstable to SBL regimes. To date, a satisfactory resolution of this problem is yet to be made. The main difficulty is establishing a consistent definition of the AET, especially the start time. It is interesting to note that many models of the time-dependent growth of the SBL begin with an established initial height, which begs the question of the origin of this initial height.

To define the afternoon-to-evening transition period, some criteria must be used to demark when the transition period begins. There are several ways to identify the start and end of the AET: level of turbulence, direction of heat flux, mesoscale temperature fluctuations, or from profiler measurements. Each of these is explained here.

6.3.1.1 Turbulence

The general assumption applied to parameterizations is that the SBL is defined by turbulence alone. Monin-Obukhov similarity theory (MOST) (Monin, 1970) provides a reasonable description of the effects of turbulence for a broad range of conditions (Sorbjan, 2010; Grachev et al., 2013) once the SBL is established. The decaying turbulence itself, before the onset of the SBL, can also exhibit MOST behavior, at least for strongly convective situations with weak synoptic forcing (Nadeau et al., 2011; van Heerwaarden and Mellado, 2016). Thus, our understanding of both the (quasi) steady CBL and the (quasi) steady SBL has progressed significantly, despite their own challenges (Holtslag et al., 2013; Lothon et al., 2014). However, this understanding and resulting parameterizations may be of limited use during the sunset transition itself (Sun et al., 2003), since the onset of the SBL typically occurs a few hours before the net radiation becomes negative (Van der Linden et al., 2017).

6.3.1.2 Heat flux

The most widely used definition of the AET is the onset of negative surface heat flux (Caughey et al., 1979; Grant, 1997; Beare et al., 2006). According to this definition, the AET is an almost instantaneous local event that occurs when the heat flux near the ground surface switches sign. This is a convenient definition because surface flux measurements are easily made and can be calculated numerically. Edwards et al. (2006) expand on this definition by adding that the AET starts when shortwave heating begins to decrease even in the presence of a positive surface heat flux. This modification extends the AET to earlier in the day but still does not account for any changes that occur after the heat flux changes sign. Nieuwstadt and Brost (1986) assumed that the AET occurs when upward surface sensible heat flux ceases, but this definition does not encompass all physical processes of the transition that occur over a deeper layer and an extended period. Mahrt (1981) used a kinematic approach by choosing the start times of the AET as when the winds below 50–100 m begin to decrease, and the ending times when the flow at all levels in the SBL has rotated toward high pressure. This definition extends the time of AET to 4 or 5 h before sunset; however, while it accounts for changes away from the ground surface, it does not consider changes occurring near the ground surface. Because the SBL starts near the ground surface, changes occurring there could be missed using this definition.

6.3.1.3 Mesoscale variability of surface temperature

Nappo (1977) demonstrated that the mesoscale flow over the complex terrain of the Tennessee River Valley is more variable during stable conditions than during unstable conditions. During daytime CBL conditions, the stirring actions of large-scale eddies masks the effects of terrain irregularities (i.e., ridges, valleys, and hills); however, these irregularities have a strong effect on the mesoscale flow during stable conditions. Acevedo and Fitzjarrald (2001) used this terrain-stability effect to examine the AET over the moderately complex terrain near Albany, New York. They use data taken from a dense network of 26 surface stations over an area the size of a typical mesoscale grid cell, about 25×25 km. The analysis is based on the area-averaged temporal standard deviation of temperature, $\sigma_{T,\text{temp}}$, calculated at each station over 30-min periods, and the area-average spatial standard deviation of temperature, $\sigma_{T,\text{space}}$, calculated each minute. Figure 6.5, taken from Acevedo and Fitzjarrald (2001), is a plot of these variables as functions of time. In the CBL, $\sigma_{T,\text{space}}$ will be small, but the values of $\sigma_{T,\text{temp}}$ will be typical for the CBL. When daytime mixing is absent, terrain effects will dominate and $\sigma_{T,\text{space}}$ will be large. In the afternoon after maximum insolation, convective eddy sizes and temperatures begin to decrease and $\sigma_{T,\text{temp}}$ decreases. Acevedo and Fitzjarrald (2001) define the beginning of the AET as the time when $\sigma_{T,\text{temp}}$ reaches a minimum value, and the end of the AET as the time when $\sigma_{T,\text{space}}$ is maximum. This definition is convenient because it depends only on temperature measurements that are made easily; however, over simple (smooth) topography this method may not be reliable.

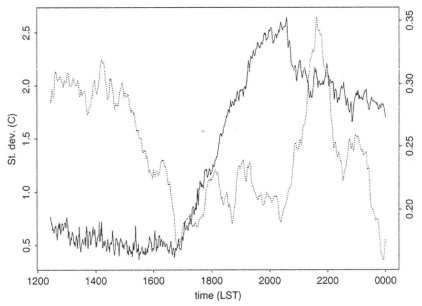

Figure 6.5 *Solid line*: spatial standard deviation of temperature, calculated at each minute from 26 station values (scale on left axis). *Dotted line*: mean temporal standard deviation. It is calculated at each station, for 30-min periods, after trend removal. Average results for the 26 stations are presented (scale on right axis).
Taken from Acevedo and Fitzjarrald (2001).

6.3.1.4   Profiler measurements

Another definition of the AET is based on measurements from a wind profiler (Grimsdell and Angevine, 2002). Their definition states that the AET occurs when the CBL height, as defined by 915-MHz profiler measurements of return power, begins to decrease and there is a sharp decrease in spectral width of the turbulence in a given layer. Busse and Knupp (2012) examine the AET using sodar data, a 915-MHz wind profiler, as well as surface temperature, dewpoint, and surface wind data to provide a comprehensive definition of the early evening boundary layer transition. Sodar backscatter is sensitive to temperature fluctuations and therefore as the heat flux decreases, the sodar return power exhibits changes from a time-varying convective structure to a more-stratified and steady structure. A relative minimum in intensity and height of the sodar backscatter is one indication that the transition is occurring. As the boundary layer evolves from the unstable convective afternoon conditions to the more stable nocturnal conditions, the fine-scale temporal variations in many parameters, including temperature, the 10- to 2-m temperature difference, dewpoint, and wind speed decrease.

As an illustration of their results, Figure 6.6, taken from Busse and Knupp (2012), shows time-height cross sections of sodar back scatter and sodar derived vertical velocity. The sodar data illustrates the development of the AET. While the afternoon CBL is present, surface-layer plumes, as represented by columns of enhanced back scatter, are indicated in the sodar signal-to-noise ratio (SNR). As the transition occurs, the fluctuations in SNR decrease and a more continuous, horizontally stratified layer

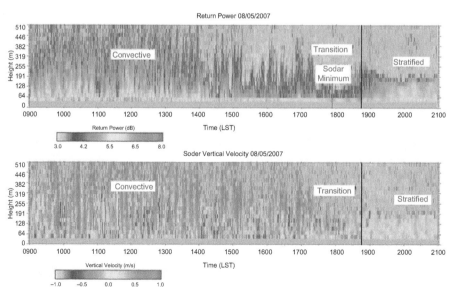

Figure 6.6 Sodar (top) backscatter and (bottom) vertical velocity from August 5, 2007 (*vertical black line* indicates time of sunset).
Taken from Busse and Knupp (2012).

develops. The higher frequency fluctuations in sodar SNR decrease by 1800 local standard time (LST) at the time of minimum sodar backscatter and the associated transition 45 min before sunset. The stratified layer associated with the incipient SBL can be seen by 1840 LST (5 min before sunset). Following this change, the stratified layer deepens and strengthens as the SBL develops. The sodar vertical velocity profiles reveal updrafts and downdrafts associated with the fluctuations in sodar return power during convective conditions. The AET is closely associated with the dissipation of these significant vertical motion perturbations. This change occurred 45 min after the transition at approximately 1845 LST (at sunset) and corresponds to the formation of the SBL.

Busse and Knupp (2012) conclude that the AET has a relatively consistent pattern regardless of season. The AET often starts with a decrease in wind speed variance followed by a temperature decrease, a mixing ratio increase, and a wind speed decrease all measured at 10 m, with increases in the wind speeds aloft. The final stage of the AET is typically represented by a minimum in the sodar return and a coincident development of a stably stratified layer that is visible as a low-level maximum in sodar return. Based on their dataset, the AET duration over the entire boundary layer depth is about 160 min for the summer cases and is 85 min for the autumn cases. This study shows that the use of sodar and 915-MHz profiler backscatter and vertical velocity data can provide information about the AET from the CBL to the SBL that would not otherwise be available from only surface observations. The definition of the AET developed by Busse and Knupp (2012) should be considered as valid for cases with low 10-m wind speed and scattered cumulus clouds or clear conditions.

6.3.1.5   AET research

It should be clear from the preceding discussion that the AET is very challenging from both modeling and observational perspectives; it is transitory, most of the forcings are small or null and the turbulence regime changes from fully convective, close to homogeneous and isotropic, toward a more heterogeneous and intermittent state (Lothon et al., 2014). Moreover, large-eddy simulation (LES) is complicated owing to the changing resolution and domain requirements during the transition (Basu et al., 2008) and the importance of other processes, such as radiative transfer (Edwards, 2009). Therefore, it is very important to gain observationally based insight into the period just after the onset of the SBL. The most comprehensive exploration of the AET took place in the Boundary-Layer Late Afternoon and Sunset Turbulence (BLLAST; Lothon et al., 2014) field campaign that was conducted in 2011 in Southern France, in an area of complex and heterogeneous terrain.

Using BLLAST data, Román-Cascón et al. (2015) examined the interactions among several SBL processes, including drainage flows, gravity waves, and turbulence, occurring just after the AET of July 2, 2011. Hooijdonk et al. (2017) used BLLAST data to study the rate of growth of the ground-based temperature inversion during the early stages of the AET. Their results show that, on average, the growth rate of the temperature inversion (normalized by the maximum inversion during the night) weakly declines with increasing wind speed. The observed growth rate is quantitatively consistent among observations at other sites (i.e., Karlsruhe station, Germany, and the Dome C observatory, Antarctica) and appears insensitive to various other parameters. The results were also insensitive to the afternoon decay rate of the net radiation except when this decay rate was very weak.

6.3.2 SBL depth

The depth of the SBL, h, is important for several reasons. For example, Gopalakrishan et al. (1998) point out that the value of h is of major concern in air pollution meteorology, and that most industrial stacks are located within this layer. Hence, consideration of the dispersion of hazardous materials is highly dependent on the state of the SBL.

Unlike the convective boundary layer, where there is a well-defined lid, what constitutes the top of the SBL can be ambiguous. Ideally, the top of the SBL can be defined as the height at which turbulence disappears and above which shear stress and sensible heat flux becomes negligible (Yu, 1978). This height is often vaguely defined, especially when the turbulence is not steady. An alternative sometimes used is the height corresponding to the maximum wind speed in the LLJ (Clarke, 1970); however, as we previously noted, the LLJ isn't always present. Models have been used to estimate the SBL depth, but many of these have proven unreliable. Arya (1981) suggests that an upward looking sodar and a lidar provide a reliable means of measuring the height of the shallow and weakly mixed boundary layer that commonly occurs during nighttime stable conditions over land. The High-Resolution Doppler Lidar (Grund et al., 2001; Banta et al., 2003) and the Cooperative Institute for Research in Environmental Sciences (CIRES) tethered lifting system (TLS; Balsley et al., 2003) can provide high-resolution measurements of turbulence through the SBL and thus SBL depth measurements (Pichugina and Banta, 2010). However, these are complex and expensive systems and not suitable for operational use.

6.3.2.1 Estimating the SBL depth

When direct measurements of the turbulence structure of the SBL are not available, the height of the SBL is estimated using parametric relations between h and other boundary-layer parameters. Several parameterizations have been proposed in the literature. In those studies, detailed, high-quality measurements of the SBL structure are necessary to provide the parametric values required by the model as well as the estimated values of h with which to test the model. Thus, a given parameterization of a model predicts a value of h and the researcher compares this value with one or several definitions of the SBL height. For example, André and Mahrt (1982) made use of various SBL heights, such as, h_i, the height where the temperature gradient, vanishes (Yamada, 1979); h_θ, the height of maximum SBL cooling rate (Melgarejo and Deardorff, 1974); h_s, the height where the potential temperature gradient exceeds 3.5×10^{-3}K m^{-1} (the low-level value from the Standard Atmosphere); and, h_R, the depth of the SBL defined as the height where the Richardson number exceeds a threshold value of 0.5. These are all convenient scales for estimating the temperature structure. The predicted values of h are then compared with these estimates. It should be noted that both diagnostic and prognostic equations for the SBL depth are still an active area of research. The text that follows summarizes the highlights, but readers are encouraged to use the flowchart provided by Seibert (2000) and the work of Zilitinkevich and Baklanov (2002) for practical considerations.

A problem with assuming that SBL depth is determined solely by the surface-layer turbulence is that while the CBL structure is governed primarily by surface fluxes, the SBL is affected as well by radiative cooling, synoptic scale subsidence, elevated shear layers, the LLJ, complex terrain, and intermittent turbulence. Often, it is difficult to unambiguously calculate temperature gradients from profiles and so the height, h_i, where the temperature gradient vanishes can be uncertain, and calculating the height, h_θ, of maximum cooling rate requires a time series of measurements. Thus, Clarke (1970) used the height corresponding to the first maximum of the LLJ, as the SBL top, h_u.

Yu (1978) tested diagnostic and prognostic equations for parameterizing the SBL depth. Based on similarity theory, the diagnostic equations tested included (Monin, 1970; Clarke, 1970)

$$h \propto k u_* / f \tag{6.1}$$

where k is the van Karman constant, u_* is the surface friction velocity, and f is the Coriolis parameter (Deardorff, 1972).

$$h = \left(\frac{1}{30L} + \frac{1}{0.35\, u_*} + \frac{1}{H_T} \right)^{-1} \tag{6.2}$$

where L is the Monin–Obukhov length and H_T is the height of the tropopause (Businger and Arya, 1974).

$$h \propto \left(\frac{u_* L}{f} \right)^{1/2} \tag{6.3}$$

These definitions use MOST, which requires stationary turbulence and constant SBL height, neither of which seldom exist or may be present only for a short time. Generally, the similarity conditions should last at least an hour. However, observations indicate that h typically increases with time during the night (Blackadar, 1957; Izumi and Barad, 1963). For example, the height of the surface inversion, h_i, has been found to grow from a value of less than 100 m shortly after sunset to a height typically 200–500 m by early next morning. Zilitinkevich and Deardorff (1974) suggest that the tendency for h_i to increase, despite increasing stability, may only reflect the tendency of a nocturnal jet to develop near the top of the SBL.

Other formulations have been proposed to account for the time-dependent behavior of h_i, and have been tested in various environments. In most cases, predications are best for the slightly stable stability class (Yu, 1978). This is expected because all the diagnostic formulae are based on similarity theory, which holds well during neutral or slightly stable conditions. Yu (1978) suggests that under extreme stability, the

SBL heights are very shallow with less scatter in the data that are seen for very stable or moderately stable conditions. It is concluded that the diagnostic equations are appropriate for slight-to-moderate and extreme stability. During other stability conditions, the diagnostic equations may be useful. In all stability classes, the prognostic equations are unsatisfactory. This is similar to the conclusion of Zilitinkevich (2012), who calls for more validation for the SBL prognostic equations.

The SBL height has also been defined as the height, h_t, at which shear stress and sensible heat flux disappear and above which shear stress and sensible heat flux becomes negligible or below a threshold value (Businger and Arya, 1974; Yu, 1978; Brost and Wyngaard, 1978; Zeman, 1979). For studies of stratified turbulence, consideration of the distribution of fluxes through the SBL is necessary. However, h_i and h_t are not the same thing. Mahrt et al. (1979) argue that the temperature profile is not solely influenced by turbulence but also by radiative cooling and external disturbances. Generally, these "external" disturbances and their origins are unknown and lack mathematical treatments. However, the theory of longwave radiation is well developed; see, for example, Wallace and Hobbs (2006), Stull (1988), Steeneveld (2007), and Foken (2016). Thus, to first order we can consider turbulence and longwave radiation as the most influential processes in the SBL. Nieuwstadt (1984) examined the relative importance to the SBL of turbulence fluxes and radiative cooling. He distinguishes two cases: (1) one with radiation but no turbulent heat flux, and (2) one with turbulent heat flux but no radiation. Observations by Nieuwstadt and Driedonks (1979) and Mahrt et al. (1979) show quasi-stationary SBL depths as well as the models proposed by Zeman (1979) and Brost and Wyngaard (1978), which do not contain radiation effects.

Arya (1981) points out that a good prognostic model of h incorporating the essential physics of the evolutionary processes in the NBL should do better than the best diagnostic relations based on the questionable assumptions of steady state and equilibrium during nighttime conditions. Unfortunately, in many applications the required information, such as the SBL height at the initial time, t_0, after the evening transition and the potential temperature, $\overline{\theta}_h$, at $z = h$ may not be available. In such cases, the diagnostic height relations may provide the only means of estimating h. For a succinct review of many of the equations available to estimate SBL depth readers are referred to Baklanov (2005).

6.3.3   Radiation and turbulence in the SBL structure

The structure of the SBL is a function of turbulence, longwave radiation, and environment. Environmental forcing includes terrain slope, surface roughness, near-surface winds and stability, cloud cover, intermittent turbulence, gravity waves, and baroclinic flows. These forcing elements are discussed in Monahan et al. (2015). However, the dominate control mechanisms are radiation, stability, and turbulence. As the late afternoon air cools, its density increases and it becomes increasingly difficult for the turbulence to lift the heavier air. The more stratified a fluid is, the more it resists upward displacements. Thus, the stable stratification within an inversion acts against the turbulence. Businger and Arya (1974) show that for the stable stratification to completely suppress the SBL turbulence, the stratification must be greater than about $25\,°C\ m^{-1} \approx 82\,°F(ft)^{-1}$. This is an unrealistic value and suggests that there will always be some turbulence in the SBL.

The ideal SBL has three layers: a surface roughness layer, which may extend vertically from a few to several meters above the ground surface; an inversion layer, which may extend from the top of the roughness layer to a few hundreds of meters; and a residual layer, which extends from the top of the inversion layer to the top of the previous day's mixing layer (i.e., several hundreds of meters). As we have seen previously, we must augment this structure by including a "turbulence layer" (see Figure 6.1). It is important to remember that the SBL is never solely turbulent or radiative. At any time, the SBL will be a combination of the two. Note that in Figure 6.1 the labels "turbulence layer" and "radiation layer" mean that the primary cooling in the layer is done by either turbulence or radiation.

Previous studies have attempted to explain the development of the SBL in terms of turbulent heat fluxes (e.g., Delage, 1974; Wyngaard, 1975; Brost and Wyngaard, 1978; Zeman, 1979; Nieuwstadt and Tennekes, 1981; Garratt, 1982) or have examined the surface inversion in terms of only clear-air radiative cooling (e.g., Brunt, 1934; Anfossi et al., 1976; Klöppel et al., 1978; Garratt and Brost, 1981). A few studies considered both turbulence and radiation (e.g., André et al., 1978; Yamada, 1979; Nieuwstadt, 1980; Tjemkes and Duynkerke, 1989; Gopalakrishan et al., 1998).

Garratt and Brost (1981) using a numerical model of the NBL (Brost and Wyngaard, 1978) and *in-situ* field observations make a comprehensive study of turbulence and radiation in the SBL. The SBL height is defined according to the so-called flux criterion, that is, that height where turbulent fluxes decrease to 5% of their surface values. Shortly after sunset, the cooling rate decreases rapidly with height above the ground generating large temperature gradients in the first few hours. As the inversion develops, the total cooling rate in the lower half of the SBL approaches a constant value with height. The tendency toward a constant cooling rate throughout the SBL is a natural requirement for the SBL to approach a steady state. Since the radiative cooling rate near the ground surface is much larger than the total cooling rate within the bulk of the boundary layer, a constant SBL cooling rate may require turbulent warming near the ground surface, while at greater levels the heat flux divergence provides most of the cooling. The result is a low-level maximum in the downward heat flux near the ground surface.

Garratt and Brost (1981) partition the SBL into three layers according to the relative strengths of turbulence and radiation. In the first layer, where $z \ll 0.10h$, radiative cooling dominates and at $z \approx 0.1h$ the cooling rate due to turbulence is equal to that due to radiation. Throughout the bulk of the SBL, $0.1h \leq z \leq 0.8h$, turbulent cooling dominates and radiative cooling is small and decreases slowly with time. At longer timescales, the radiative cooling might even change to warming in the central region. In the third or uppermost layer, $0.8h \leq z \leq h$, radiative cooling again dominates.

André and Mahrt (1982) analyze the interaction of turbulence and radiation using data from the Wangara experiment (Clarke et al., 1971) and the Voves experiment (e.g., André and Lacarrére, 1980). Their results included the following:

1. Turbulent heat flux divergence and clear-air radiative cooling, on average, contribute
   equally to the development of the SBL.
2. Turbulence contributions to the SBL are confined to the lower part of the surface inversion,
   while clear-air radiative cooling extends the surface inversion to levels several times higher
   than the turbulent layer.
3. Clear-air radiative cooling is maximum near the top of the inversion, which acts to not only
   increase the depth of the layer but also to reduce the stratification. Consequently, the temperature inversion height increases throughout the night while the depth of the turbulence
   layer remains nearly constant.
4. The crucial role of radiative cooling precludes the successful relation between the inversion
   depth and surface fluxes.

Figure 6.7, taken from André and Mahrt (1982), shows temperature profiles from the
Wangara and Voves data on selected periods. Indicated are the heights of h_i, h_s, and h_u.
Note that the depth of the SBL turbulence as indicated by h_R is greater than the height
of the LLJ for the Voves curve but not for the Wangara curve. This illustrates the variability of the SBL structure and the difficulty in generalizing it. In their conclusions,
André and Mahrt suggest the development of a two-layer model of SBL. In the lower
layer, turbulence dominates, while in the upper layer clear-air radiative cooling
dominates.

In the absence of clouds or fog, the air in the SBL will cool throughout the night by
longwave radiation. It is important to note that the turbulence generated near the gro-
und surface is the only continuous source of turbulence in the SBL. The strength or
intensity of this turbulence is determined by the wind speed near the ground and the

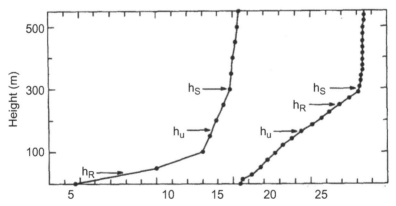

Figure 6.7 Potential temperature (C) 6. Potential temperature profiles for Wangara Day 8 at
0020 LST (left) and for Voves I I July at 0226 LST (right). *Arrows* indicate the respective
heights of the surface inversion (h_s), low-level wind maximum (h_u), and turbulent boundary
layer (h_R).
Taken from André and Mahrt (1982).

surface roughness. The assumption of a flat surface rules out drainage flows. The wind speed or LLJ is more a consequence than a product of the SBL.

6.4 SBL classification

The stable atmospheric boundary layer has been divided into two categories: the weakly stable boundary layer and the very stable boundary layer (Malhi, 1995; Mahrt et al., 1998; Ohya et al., 1997). Businger (1973) notes that boundary layer meteorology has long distinguished between cases of SBLs with continuous turbulence and those with stronger stability and intermittent turbulence. The weakly stable boundary layer is the usual NBL in which turbulence is more or less continuous; it has been examined in terms of observations (Lenschow et al., 1987; Van Ulden and Wieringa, 1996), scaling arguments (Derbyshire, 1990), similarity theory (Zilitinkevich and Mironov, 1996), and laboratory studies (Ohya et al., 1997). Sun et al. (2012) identified three turbulence regimes: regime 1 with weak but continuous turbulence, regime 2 with strong turbulence, and regime 3 with moderate sporadic turbulence. Monahan et al. (2015) discuss several regimes based on wind speed, stratification, and turbulence.

Mahrt (1998) identifies and discusses three qualitative turbulence regimes based on the behavior of the heat flux alone:

1. Weakly stable ($0 < z/L < \in \ll 1$): the turbulent heat flux increases with z/L due to increasing amplitude of the temperature fluctuations with increasing stability, where $\in$ is the maximum stability for the weakly stable regime, equal to 0.06 for $z = 10$ m.
2. Transition stability regime ($\in < z/L < O(1)$): the turbulent heat flux decreases rapidly with increasing z/L due to decreasing amplitude of the vertical velocity fluctuations (i.e., turbulence).
3. Very stable regime ($z/L > O(1)$): the heat flux is small but continues to decrease slowly with increasing stability.

Note that $-z/L$ is a dimensionless parameter expressing the ratio of convective to mechanical production of turbulence in near-neutral conditions, and that for stable conditions $L < 0$. Thus, for example, for the weakly stable regime with $z/L = 0.06$, the turbulence heat flux at 10 m is about 170 times greater than the negative buoyancy flux due to radiative cooling.

For weak stratification (regime 1), when the potential temperature gradient is small, the magnitude of the downward heat flux is limited by the smallness of the temperature fluctuations. The heat flux vanishes when the temperature gradient vanishes at neutral stratification. When the stratification is strong (regime 3), negative buoyancy limits the size of the turbulence eddies and the heat flux decreases with increasing stratification, theoretically vanishing when the stratification becomes sufficiently large. Between these stability extremes, the downward heat flux reaches a maximum. Malhi (1995) finds that the maximum heat flux occurs when $z/L = 0.2$ at $z = 9$ m. Mahrt (1998), using data from the Microfronts project (Howell and Sun, 1999) finds the maximum heat flux at $z/L = 0.06$ at 10 m and $z/L = 0.02$ at 3 m. Mahrt (1998) comments that "there is no evidence that the value of $-z/L$ corresponding to maximum downward heat flux is universal."

6.4.1 SBL vertical structure

Mahrt (1999) describes what he labels as the "classical vertical structure" of the SBL based partly on the concepts of Nieuwstadt (1984) and Holtslag and Nieuwstadt (1986). These idealized layers are based mostly on the turbulence structure of the of the SBL rather than the complication of considering both a turbulence plus a radiative boundary layer. However, it should be noted that while the turbulent SBL is essential for numerical models, the inversion boundary layer is essential for air pollution models.

Figure 6.8, taken from Mahrt (1999), illustrates the disposition of the layers. The relative stability is expressed as z/L.

1. The layer in immediate contact with the ground is the roughness sublayer. The time-averaged flow within this layer varies spatially on the scale of the roughness elements and a universal flux-gradient relationship seems unobtainable. A way to visualize this is to imagine calculating velocity gradients within the rapids of a swiftly running river. Then, no single value is representative of the overall velocity gradient if indeed one even exists.

2. The surface layer resides above the roughness sublayer. The fluxes in the surface layer are numerically close to the surface value and the flux-gradient relation depends only on z/L. In the surface layer, the turbulent fluxes are approximately constant, varying about 10% through its depth.

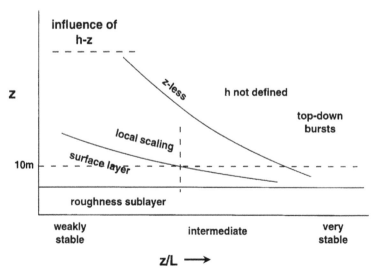

Figure 6.8 Schematic illustration of layers in the stable boundary layer as determined by stability.
Taken from Mahrt (1999).

3. Above the surface layer, the approximation of height-independent flux is no longer valid. However, if the Obukhov length is redefined in terms of local fluxes, then similarity theory can apply in the form of local scaling where z/L is replaced by z/Λ, where Λ is the Obukhov length based on local fluxes at height z.
4. If z/Λ becomes sufficiently large, then various quantities are predicted to become independent of z/Λ corresponding to "z-less" stratification (Wyngaard, 1973; Hicks, 1976; Nieuwstadt, 1984; Dias et al., 1995).
5. Near the boundary layer top, when definable, the distance from the boundary layer top may become a relevant length scale in which case $(z - h)/\Lambda$ becomes a relevant stability parameter (Holtslag and Nieuwstadt, 1986).

6.4.2   The very stable SBL

In the very stable boundary layer (VSBL), which is characterized by weak winds, clear skies, and strong net radiative cooling at the surface, the turbulence is weak or even intermittent near the surface and is perhaps layered (Mahrt et al., 1998, 2014). The VSBL also admits boundary-layer breakdowns, bursting events, and gravity waves and their interactions with turbulence. Various aspects related to the VSBL are described by Mahrt (1985), Derbyshire (1990), and Mahrt (1999). Derbyshire (1990) points out that in the VSBL, the weak flow near the surface may not be able to maintain turbulence by itself because the turbulence cannot support the downward heat flux demanded by surface cooling. Then regions of turbulence can develop away from the surface, leading to the "upside-down boundary layer" as described later. The VSBL is of considerable practical importance. The absence of significant mixing allows buildup of high concentrations of contaminates. For example, Sharan et al. (1995) show that the Bhopal gas leak on December 3, 1984, which resulted in 3000 fatalities, occurred during strongly stable, weak-wind conditions.

The VSBL has been discussed in several scientific papers including, for example, Kondo et al. (1978), Mahrt (1985), Smedman (1988), Mahrt et al. 1998), Gopalakrishan et al. (1998), Mahrt (1999), Mahrt and Vickers (2003), and Sun et al. (2012). Banta et al. (2007) provide an excellent review of VSBL studies and a detailed analysis of five nights with weak winds and VSBLs, that is, characteristics of the VSBL, observed during the Cooperative Atmosphere–Surface Exchange Study (CASES-99; Poulos et al., 2002). These characteristics included a shallow traditional boundary layer only 10–30 m deep with weak intermittent turbulence within the strong surface-based radiation inversion. Above this shallow surface layer, sits a layer of very weak turbulence and negligible turbulent mixing. Banta et al. (2007) focused on the effects of a quiescent layer just above the surface layer and the impacts of this quiescent layer on turbulent transport and numerical modeling. The presence of this quiescent layer indicates that the atmosphere above the shallow surface layer was isolated and detached from the surface layer (i.e., the turbulence structure was layered).

Boundary layers are difficult to define when the principal source of turbulence is shear generation detached from the ground surface (Mahrt and Vickers, 2003; Smedman et al., 1993), sometimes occurring at the top of the surface inversion layer. Such shear generation could be associated with enhancement of shear by the nocturnal

LLJ (e.g., Blackadar, 1957; Ostdiek and Blumen, 1997), modulation of shear by internal gravity waves (Chimonas, 2002), acceleration associated with decoupling (Derbyshire, 1999), or generation of turbulence associated with unstable waves and density currents (Sun et al., 2003). In these cases, shear generated turbulence may intermittently burst downward toward the surface (Nappo, 1991; Cuxart et al., 2000; Ohya, 2001). This situation has been referred to as the upside-down boundary layer by Mahrt (1999) and others. The VSBL may be extremely thin, less than 10 m deep (Smedman, 1988). Models generally do not employ sufficient vertical resolution to resolve very thin SBLs. In the worst scenario, the boundary layer is so shallow that a surface layer does not exist, that is, the roughness sublayer (Raupach, 1994; additional references in Mahrt, 1999) occupies more than the lowest 10% or 20% of the boundary layer. In this case, Monin-Obukhov theory does not apply at any level.

Mahrt et al. (2014) point out that for very stable conditions, the turbulence intensity defined as σ_w/V, where σ_w is standard deviation of the vertical velocity and V is the mean wind speed, is roughly independent of the temperature stratification. That is, for a given wind speed, the impact of suppression of the turbulence by increasing stratification is not observed, as also found in Sun et al. (2012). The reason for this unexpected near independence from the stratification is complex but is partly associated with development of strong directional shear with strong stratification. Strong directional shear often occurs with cold air drainage.

As a final note, from a purely applied perspective, stability classes developed by Pasquill (1961) describe the magnitude of turbulence as it relates to plume dispersion. This classification scheme is discussed in further detail in Chapter 11.

6.5 Nocturnal LLJ

It is of interest to note that a search of the American Meteorological Society's (AMS) online library returned 3,523 references for "nocturnal boundary layer" extending from 1894 to 2019, while a search for "low-level jet" returned 16,690 references from 1882 to 2019. From 1970 to 2019, the AMS journals contain 3,319 references for the NBL and 15,523 for the LLJ. This disparity suggests that the LLJ has been of greater scientific interest and for a longer time than the NBL. Parish and Oolman (2010) state, "The summertime Great Plains low-level jet (LLJ) of the central United States is one of the most intensely studied mesoscale features of the past 50 years (e.g., Lettau and Davidson, 1957; Hoecker, 1963; Bonner, 1968)." Considering the vast number of AMS papers, and not counting those in other scientific journals, this section can be only introductory, and readers interested in the LLJ should consult the references herein.

6.5.1 LLJ observations

There have been many observational studies of the LLJ, for example, Bonner (1968), Garratt (1985), Kraus et al. (1985), Whiteman et al. (1997), Andreas et al. (2000), Banta et al. (2002), Song et al. (2005), Seefeldt and Cassano (2008), and Baas et al. (2009). Stensrud (1996) in a review article on the importance of the LLJ to climate points out that although LLJs were first described in the 1930s over Africa (Goualt, 1938; Farquharson, 1939), it was not until the 1950s that interest in the LLJ blossomed. In one of the earlier studies, Bonner (1968) compiled one of the early climatologies of the LLJ over the Great Plains of the United States. The southerly LLJ determined to be an isolated phenomenon occurring at about 800 m above the ground in the south-central United States. The greatest frequency of LLJ occurrences is in south-central region of the Great Plains. Whiteman et al. (1997) examine the LLJ in the Great Plains using enhanced radiosonde observations at a site located in north-central Oklahoma and determined that LLJs are present in 47% of the warm season soundings and 45% of the cold season soundings. More than 50% of the LLJs have wind maxima less than 500 m above ground level. The southerly jets show a pronounced diurnal variation in frequency, with the nighttime frequency almost double the daytime frequency. This nocturnal nature of the southerly LLJs has been attributed mainly to diurnal processes occurring in the atmospheric boundary layer, including the inertial oscillation of boundary layer winds after they become decoupled from surface friction at night (Blackadar, 1957) and the diurnal oscillation of the horizontal pressure gradient over the sloping Great Plains (Holton, 1967). The frequency of occurrence of the northerly jets, however, shows only a small diurnal variation, indicating that diurnal boundary layer processes play a less important role in the development of these jets, perhaps because of the relatively more important role of advection in the cold air outbreaks associated with northerly jets, the damped diurnal cycle in the cold air mass, or the role of postfrontal cloudiness.

Baas et al. (2009) gives a climatology of nocturnal LLJs for the topographically flat measurement site at Cabauw, the Netherlands. They find that LLJs at Cabauw originate from an inertial oscillation, which develops after sunset in a layer decoupled from the surface by stable stratification. It is found that LLJs occur in about 20% of the nights, are typically situated at 140–260 m above ground level, and have a speed of 6–10 m s^{-1}. Moderate geostrophic forcing and high radiative cooling (no clouds) are the most favorable circumstances for the development of a substantial LLJ. For stronger nocturnal cooling and lower geostrophic forcing, the LLJs form at a lower altitude. These results suggest that frictional decoupling after sunset because of stable stratification is the main mechanism for LLJ formation at Cabauw. Indeed, about 60% of the detected LLJs show features of a well-developed inertial oscillation. However, the characteristics of these jets (in terms of height and turning compared to the 10-m wind) do not differ from those of the remaining 40% of the LLJs.

6.5.2 LLJ analysis

Several theories have been used to explain the occurrence of the LLJ. The absence of strong daytime vertical mixing allows the SBL to respond to other influences such as gravity-driven flows, terrain-induced perturbations, and synoptic scale pressure gradients and fronts. Kraus et al. (1985) list some of the causes of the LLJ as follows: synoptic scale baroclinity associated with weather patterns; baroclinity associated with sloping terrain (i.e., drainage winds, fronts, advective accelerations); splitting, ducting, and confluence around mountain barriers; land and sea breezes; mountain and valley winds; and inertial oscillations.

The most used explanations of the LLJ are given by Blackadar (1957) and Holton (1967). Blackadar proposed that the inertial oscillation could explain the LLJ. When the boundary layer transitions from convective to stable conditions, the eddy diffusivity can decrease from 1 to $10^{-3} \, m^2 \, s^{-1}$. When the turbulent drag force is removed, the boundary layer flow becomes unbalanced and accelerates toward the geostrophic wind direction, leading to the development of a super-geostrophic wind speed maximum, known as the nocturnal jet, at the top of the SBL. Holton (1967), reviving an original suggestion by Bleeker and Andre (1951), shows that alternate heating and cooling of the slopes of the Rocky Mountains can produce diurnal oscillations of the wind in a boundary layer with constant eddy viscosity.

Over sloping terrain, the along-slope component of the buoyancy force associated with daytime heating and nighttime cooling of the surface can become an important secondary forcing mechanism. Fedorovich et al. (2017)[1] performed a numerical study LLJ over gently sloping terrain. Using scale analysis, they show that the slope angle needs only to be on the order of 0.0–0.02 degrees, which is typical of the slope of the Great Plains, in order for the component of buoyancy in the along-slope equation of motion to be of the same order of magnitude as a typical LLJ acceleration required to attain, say, $5 \, m \, s^{-1}$ wind over 6 h. Fedorovich et al. (2017) point out that although the Blackadar (1957) description of an LLJ with a wind vector that veers in time has been generally confirmed qualitatively, more detailed analyses of observed jets suggest that the theory may be incomplete. For example, Blackadar's theory cannot explain how peak wind speeds in some observed LLJs can reach twice the free-atmospheric geostrophic speed. It also fails to explain the higher frequency of stronger LLJs formed over the gently sloping terrain of the Great Plains. Their results document that the along-slope advection of environmental potential temperature throughout the boundary layer during the night has a major impact on the structure and evolution of the LLJ. This advection alters the buoyancy field and can reignite static instability in the weakly turbulent LLJ flow. The turbulence that reemerges as result of the instability leads to a complete or partial remix of the lower portion of the boundary layer flow and drastically changes the appearance of the LLJ in terms of its shape and vertical position. Additionally, they confirm that a pronounced nighttime jet-like flow develops from the daytime tilted CBL in the absence of any free-atmospheric geostrophic forcing. These results support the so-called Holton mechanism.

Shapiro et al. (2016) seeks a unified theory of the Great Plains LLJ in which the jet emerges in the sloping atmospheric boundary layer as the nocturnal phase of an oscillation arising from diurnal variations in turbulent diffusivity (Blackadar mechanism) and surface buoyancy (Holton mechanism). Although neither the Holton's slope theory nor the Blackader's inertial oscillation theory is generally sufficient to explain observations of Great Plains LLJs, the physical mechanisms underlying these theories are plausible, and it has long been speculated that both can be important in the development of Great Plains LLJs. They present a unified theory for the Great Plains LLJ in

---

[1] This paper contains a comprehensive review of previous LLJ studies.

which the jet appears in the nighttime phase of oscillations arising from diurnal cycles of turbulent mixing (Blackadar mechanism) and heating/cooling of the slope (Holton mechanism). As in Holton (1967), the equations of motion are supplemented with a thermal energy equation. The buoyancy evolves in accord with the coupled governing equations, unlike the thermal field proxy in previous studies, which is specified. A reference experiment in which both mechanisms are operating provides a baseline description of the fair-weather warm-season diurnal cycle over the sloping portion of the southern Great Plains, including the emergence of a strong LLJ in the nocturnal phase of the diurnal cycle. The strength, timing, and vertical structure of the analytical LLJ are in good qualitative agreement with typical LLJ observations over the southern Great Plains.

Lundquist (2003) attempted to verify the inertial oscillation theory of the LLJ over the Great Plains using data from CASES-99. Lundquist notes that, "Despite their theoretical importance to nocturnal boundary layer dynamics, inertial oscillations have been rarely observed in the atmosphere." Discussions of inertial oscillations in the literature are few and mostly associated with fronts rather than the evening transition. Notable exceptions are the work of Mori (1990), which suggests a correlation between high-amplitude inertial oscillations and frontal passages, and the work of Ostdiek and Blumen (1997), which focuses on one particular frontal passage to document the presence of inertial oscillations at several sites in the vicinity of the frontal zone at levels from the surface up through 1 km. The theoretical work of Blumen (1997) shows that frontogenesis would modify the amplitudes of inertial oscillations, thus enhancing the difficulty of quantifying their occurrences with statistical significance. It is possible that inertial oscillations are ubiquitous in the atmosphere, only with very small amplitudes. Intermittent events, such as the evening transition or frontal passages, could increase their amplitudes so that large-amplitude (or even observable) inertial oscillations could be considered intermittent events. Therefore, an analysis method suited to apparently intermittent and varying-amplitude events would be required to identify inertial oscillations.

From the analyses described earlier, it seems that the Holton mechanism is more applicable to the Great Plains LLJ, while the Blackadar mechanism is more applicable to the Netherlands LLJ. Of course, there will be many exceptions to this generalization especially when consideration is given to the many processes that can and often occur in the NBL.

6.6   Non-stationary turbulence

Turbulence theories are generally based on the assumptions of stationarity, homogeneity, and isotropy (i.e., turbulence statistics are constant both in time and in space). Such conditions may be created in the laboratory, but in the field and especially at night these conditions might be approximated but only over a limited period of time and at a certain location. The term *intermittency* is often used in describing the unsteady nature of turbulence. Sun et al. (2012) quote from the *Glossary of Meteorology* Glickman, 2000, p. 410), "the property of turbulence within one air mass that occurs at some times and some places and does not occur at intervening times and places." This seems to imply that between each episode of turbulence, the air flow is calm. Sun et al. (2012) modifies the definition of intermittent turbulence by noting that turbulence in the SBL never completely disappears but can become extremely weak. Thus, they use the word *intermittency* "to describe a temporal variation of turbulence strength observed at a fixed location." Yet, this definition can fail when considering singular events such as atmospheric bores or solitary waves (see, e.g., Christie et al., 1978; Rottman and Einaudi, 1993; Koch et al., 2008). These types of large-amplitude disturbances are not rare and may even be missed in time averages; however, they can lead to a rapid destruction of the SBL stratification, which identify as bursts or break-downs. Thus, perhaps the term non-stationary turbulence may be sufficiently general to include these isolated events.

Several mechanisms can account for non-stationary events including wave insta-bilities (e.g., Blumen et al., 2001; Balsley et al., 2002; Sun et al., 2004; Meillier et al., 2008; Nappo, 2012, Sun et al., 2015); density currents (e.g., Sun et al., 2002 Cuxart, 2008; Seefeldt and Cassano, 2008); wind gusts (e.g., Acevedo and Fitzjarrald, 2003; Doviak and Ge, 1984; White, 2009), and bursting (e.g., Durst, 1932; Gifford, 1952; Mahrt, 1985; Nappo, 1991; Coulter and Doran, 2002; Costa et al., 2011). Intermittent turbulent events are most common during strongly stable conditions; however, they are likely to occur on most nights regardless of stability, and patches of turbulence in various states of decay can almost always be found throughout the NBL.

6.6.1 Intermittency classification

Mahrt (1985) describes two possible forms of turbulence intermittency: "local" or "global." Local intermittency is identified with the fine-scale structure internal to the main eddies and is associated with the dissipation of the turbulence kinetic energy. The sharp edges of the main eddies contribute to the intermittency due to the gener-ation of smaller-scale turbulence by the eddy-scale shear. In some cases, the small-scale intermittency occurs as numerous narrow zones of shear with uncertain relation to the main-current coherent structures. Global intermittency is identified with patches of turbulence with large intervening areas with little turbulence. Typically, the dis-tance between these patches is large compared with the scale of the patches. One of the biggest problems in measuring surface layer fluxes is the question of sampling time in the presence of global intermittency. Sufficiently long records may include significant nonstationarity (Wyngaard, 1973), partly due to diurnal trends, and may include advection as well. Thus, in some cases it may not be possible to accurately measure fluxes at a fixed point.

Van de Wiel et al. (2003) examine global intermittency near the ground surface (first 10 m) using data from the CASES-99 and the results from a one-dimensional bulk model of three coupled nonlinear differential equations (Van de Wiel et al., 2002). The model focuses on an intermittency generating mechanism that arises from a direct interaction of the lower atmosphere with the vegetation surface, without inter-action with the air aloft. Model results show that intermittent turbulence is most likely to occur over land surfaces with low vegetation during clear-sky conditions in the presence of a moderate to low synoptic pressure gradient. The existence of a vegeta-tion layer has a strong influence on intermittency dynamics. Due to its small heat capacity, the vegetation temperature can respond quickly to rapidly changing conditions. This in turn affects the stability of the lower atmosphere causing an important feedback mechanism. Their proposed mechanism closely follows Businger (1973) who gives a qualitative description of the SBL breakdown process leading to global intermittency:

> As the ground cools at night, the establishment of a stable thermal stratification near the ground causes the Richardson number, Ri, to approach and finally exceed its critical value, Ri_c, at some height above the ground. When this happens, turbulence is suppressed, a laminar layer develops, and the downward transfer of heat and momentum from higher layers is impeded. Hence, the wind near the ground surface diminishes and a near-surface calm develops. Above the laminar layer, momentum is still transferred downward, but little heat is transferred because the stratification at this level is nearly neutral. The winds aloft are essentially disconnected from the surface, and the air is free to accelerate with little resistance from surface friction. A strong wind shear then builds up and because there is not a similar increase in the heat flux, Ri must decrease and eventually drop below the Ri_c value. When this happens, the laminar layer is destroyed from above by turbulence. The turbulence then diffuses downward, eventually reaching the ground as a "burst" of heat and momentum.

Van de Wiel et al. (2003) show that intermittent turbulence is most likely to occur in clear-sky conditions with a moderately weak effective pressure gradient (i.e., low wind speed). However, these results pertain to the lower SBL. Sun et al. (2012) use the CASES-99 data to show three possible turbulence regimes at each measurement height. Regime 1 shows weak turbulence when the wind speed is less than a threshold value, regime 2 shows strong turbulence when the wind speed exceeds a threshold value, and regime 3 shows moderate turbulence when top-down turbulence sporadically bursts into the otherwise weak turbulence. For regime 1, the strength of small turbulence eddies is correlated with local shear and weakly related to the stratification. For regime 2, the turbulence strength increases with wind speed as a result of turbulence generation by the bulk shear, which scales with observation height. Banta et al. (2003) use data from a high-resolution Doppler lidar to measure the bulk shear and turbulence generated in the layer between the maximum of the LLJ and the earth's surface. They show that a LLJ Richardson number defined by the SBL stability and bulk shear U_x/Z_x, where U_x is the maximum wind speed of the jet and Z_x is its height above the ground, relates with the turbulence in the layer below the jet maximum. The threshold wind speed separating regimes 1 and 2 marks the transition above which the boundary layer approaches near neutral conditions, that is, where turbulent mixing substantially reduces the stratification and temperature fluctuations.

Sun et al. (2012) identify three categories of intermittency associated with the three turbulence regimes. In category A, the wind speed is observed to oscillate across its threshold value. Turbulence is enhanced when the wind speed exceeds its threshold value and is reduced when the speed falls below it (i.e., turbulence oscillates between regimes 1 and 2). In category B, when the wind speed remains less than its threshold value, the disturbances intermittently reduce the local stability and slightly increase the local turbulence. In category C, top-down turbulent events suddenly intrude downward into a weak turbulent environment and turbulence regimes occurs.

6.6.2   Bursting and wave-turbulence interactions

In the proceeding section, it is given that turbulence is almost always present to some degree in the SBL and that episodic changes in the turbulence originate locally through shear instability; however, this does not address the causes of the instabilities. These causes often exist above the SBL and often originate large distances away from the observer. For example, upper-level "jet streaks" can result in tropospheric meso-scale gravity waves that can propagate at high speed for thousands of kilometers (Uccellini and Koch, 1987; Koch et al., 2005). These waves can trip SBL turbulence through wave-turbulence coupling. Thunderstorms can also initiate mesoscale gravity waves either from gust fronts or bow waves (Uccellini, 1975; Doviak and Ge, 1984; Chimonas and Nappo, 1987).

6.6.2.1   Bursting

Bursting events, sometimes referred to as "sudden warmings," were reported early on by Durst (1932) and later by, for example, Gifford (1952), Lyons et al. (1964), Bean et al. (1973), Kondo et al. (1978), Mahrt (1985), and Gossard et al. (1985). For long-term air quality considerations, the assumption that vertical transfer is small in the SBL is generally acceptable; however, in treating the problem of short-term exposure to hazardous materials, consideration of the non-stationary aspect of the SBL is crucial. For example, during breakdown events pollutants residing in the upper levels of the SBL or the residual layer can be brought downward to the surface layer with relatively little dilution. Harrison et al. (1978) and Winkler (1980) show that sudden increases of ozone concentration at the ground surface are correlated with disturbances in the SBL as indicated by sodar traces. They relate these concentration increases to turbulent transport from higher levels (RL), where ozone concentrations are greater than at the ground surface. Kondo et al. (1978) suggest that under stable conditions, there exists a distinguishable interface dividing the active turbulent ground-based layer from a quiet layer above. Undulations in this interface are produced by Kelvin-Helmholtz billows, which develop in response to speed and density differences across this interface. The movement of these undulations past an instrumented tower are marked by periods of enhanced turbulence and negative heat flux. Kondo's data indicate a time scale of about 10 min for these periods. However, they point out that it is not the propagation and undulations of these bellows, but rather their breaking and consequent mixing process that contribute to the long-term average heat flux.

Nappo (1991) analyzed burst events using surface data taken in an urban setting in St. Louis, MO, a rural site outside of St. Louis, and a complex terrain site near Oak Ridge, TN. Time series of 1-min averaged wind speeds and temperatures were first high-frequency filtered to form a slowly varying signal, effectively a 30-min running mean. This was subtracted from the same raw data but now filtered with a 5-min running mean. Then a wind-speed temperature covariance was calculated. The theory assumes that during a breakdown event warm high-speed air is brought downward and cool low speed air is brought upward. In these cases, a positive covariance is produced. The relation between breakdown events and turbulence is illustrated in Figure 6.9, which shows the time series of the covariance and the monostatic sodar trace for one night at the Oak Ridge site.

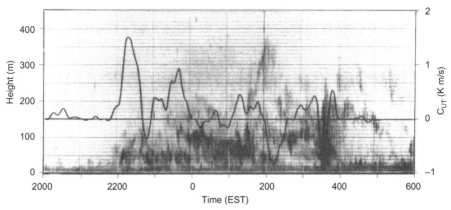

Figure 6.9 Sodar trace and wind speed-temperature covariance for a typical night at the Oak Ridge, TN site. Wind speed and temperature data taken at 40 m height of the sodar record. From Nappo (1991).

The elevation of the tower instruments is about 40 m above the sodar. The covariance is essentially zero until just before 2200 EST when a sharp rise occurs. This jump in covariance corresponds with the sudden appearance of turbulence as indicated by the sodar trace at about the 40 m height. After this initial period, the covariance and the turbulence vary with time. A correlation between these quantities exists if examined at the 40 m level of the sodar trace. At about 0400 EST, the turbulence and covariance both fall below zero and remain at that value. Breakdown events are counted by how often and for how long covariance is greater than a threshold value, which is taken to be 10% of the average nighttime covariance. The sampling time is from 2000 to 0600 LST. Seasonal variation is not considered. The number of nights of useful data are 165 for the complex terrain site, 103 for the urban site, and 76 for the rural site. The results are summarized in Table 6.1.[2]

Table 6.1 Average values of breakdown characteristics.

	RAMS #1 (urban)	RAMS (rural)	Walker Branch (complex terrain)
Breakdowns per night	13	15	18
Duration of breakdowns (min)		19 (10)[a]	16 (6)[a]
Interval between breakdowns (min)	21 (12)[a]	17 (6)[a]	14 (4)[a]
Covariance per breakdown (k-m s^{-1})	3.2×10^{-2}	5.5×10^{-2}	4.8×10^{-2}
Nighttime average covariance (k-m s^{-1})	0.6×10^{-3}	12×10^{-3}	-7.8×10^{-3}

[a] Numbers in parentheses are the modal values of the distributions.

[2] Note that Walker Branch refers to the Oak Ridge site.

The average values of the individual characteristics are not very different over the three sites, even though these sites are located in contrasting environments. This is significant because it suggests that breakdowns are not exceptional events limited to a certain terrain type, but rather a regular feature of the SBL. The differences between these values can be explained qualitatively. At the urban station, both the heat island effect (e.g., Bornstein, 1968; Foken, 2016) and the large surface roughness act to limit both the strength of the temperature inversion and the velocity shear near the ground surface (Arya, 1988). These effects will result in small values of the temperature and wind-speed perturbations caused by a breakdown. This effect also causes the large times between breakdowns at the urban station. At the rural station, temperature and wind-speed gradients are larger than at the urban station and this will result in stronger perturbations and greater values of C_{UT} than at the urban site. At the Oak Ridge site, the breakdowns are frequent and of short duration. This is consistent with terrain generated disturbances such as gravity waves, streamline deformation, vortex shedding, and so on.

6.6.2.2 Wave-turbulence interactions

Surface obstacles can cause streamline deformation sufficient to trigger bursts of turbulence in stable conditions, but turbulence episodes have been observed by DeBaas and Driedonks (1985) in the Netherlands, an area well away from and presumably unaffected by upwind hills and mountain ranges. In the SBL, it is expected that turbulence will decrease with distance from the surface roughness layer; however, (DeBaas and Driedonks, 1985) using turbulence data from the Cabauw tower observed turbulence intensities decrease up to about 80 m, and then increase to a maximum at about 160 m, and then decrease upward to where measurements ended. Using linear wave theory, they demonstrated that the increase in turbulence was due a Kelvin-Helmholtz instability. This is perhaps an example of wave oscillations being confused with low-frequency turbulence.

Large-amplitude mesoscale gravity waves typically have horizontal wavelengths of 50–500 km, vertical wavelengths of 1–4 km, intrinsic periods of 0.5–4 h, surface pressure amplitudes of 0.5–15 hPa, and phase velocities of 15–35 m s^{-1} (Wang and Zhang, 2007). The rapid destruction of the SBL due to internal gravity waves generated by distant thunderstorms has been documented by, for example, Curry and Murty (1974), Uccellini (1975), Balachandran (1980), Shreffler and Binkowski (1981), and Doviak and Ge (1984). These gravity waves are ducted or trapped between the ground surface and some upper-level reflecting level, which can be a critical level or the tropopause. Chimonas and Nappo (1987) demonstrated that the gust front described by Doviak and Ge (1984) matches the conditions for a ducted bow wave being dragged through the troposphere by the thunderstorm much like a bow wave of a moving boat.

Once produced by a wave, the turbulence is free to interact with the gravity wave and the mean flow. Analyses of these interactions has been carried out in a series of papers by Einaudi and Finnigan (1993) and references therein.

6.7 Summary

This chapter addressed the complexities of the SBL. Key points include:

- The SBL has a layered structure; layers are generally defined based on the existence of turbulence.
- The SBL occurs primarily at night when shortwave radiation stops driving convection.
- The SBL is statically stable with weaker turbulence occurring inconsistently.
- The height of the SBL can be defined in many ways.
- Parameterizations of the SBL are still evolving as new observations and theories become available.
- An LLJ impacts turbulence in the SBL, but the mechanisms for formation are still not clearly identified.

6.8 Back to the beginning

This chapter concludes the first part of this book. Now that we've covered all the mains parts of the boundary layer, we can see how many changes occur in the course of a 24-h period. In Figure 6.10 you can see all the main sublayers of the boundary layer: The convective boundary layer or mixed layer during the day where mixing is active, turbulence is driven by convection, and the layer grows by entrainment from above, heat transfer is from the surface to the atmosphere, the SBL at night with its own sublayers discussed above. The residual layer contains well-mixed air from the day before that is disconnected from the surface. Throughout the entire cycle we see the surface layer or the air in direct contact with the earth's surface, typically defined

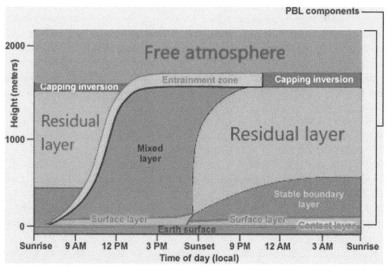

Figure 6.10 The diurnal structure of the boundary layer.
Image from Spiridonov and Ćurić (2021).

as 10% of the mixed layer. Capping it all is the free atmosphere above, where the surface no longer *directly* influences the atmosphere.

References

Acevedo, O.C., Fitzjarrald, D.R., 2001. The early evening surface-layer transition: temporal and spatial variability. J. Atmos. Sci. 58, 2650–2670.

André, J.C., Lacarrére, P., 1980. Simulation détaillée de la couche limite atmosphérique. Comparaison avec la situation des 2 et 3 juillet 1977 á Voves. La Météorlogie VI 22, 5–49.

André, J.C., Mahrt, L., 1982. The nocturnal surface inversion and influence of clean-air radiative cooling. J. Atmos. Sci. 39, 864–878.

Acevedo, O.C., Fitzjarrald, D.R., 2003. In the core of the night-effects of intermittent mixing on a horizontally heterogeneous surface. Bound.-Layer Meteorol. 106, 1–33. https://doi.org/10.1023/A:1020824109575.

André, J.C., De Moor, G., Lacarrére, P., Therry, G., du Vachat, R., 1978. Modeling the 24-hour evolution of the mean and turbulent structures of the planetary boundary layer. J. Atmos. Sci. 35, 1861–1883.

Andreas, E.L., Claffey, K.J., Makshtas, A.P., 2000. Low-level atmospheric jets and inversions over the Western Weddell Sea. Bound.-Layer Meteorol. 97, 459–486.

Anfossi, D., Bacci, P., Longhetto, A., 1976. Forecasting of vertical temperature profiles in the atmosphere during nocturnal inversions from air temperature trends at screen height. Q. J. R. Meteorol. Soc. 102, 173–180.

Arya, S.P.S., 1981. Parameterizing the height of the stable atmospheric boundary layer. J. Appl. Meteorol. 20, 1192–1202.

Arya, S.P.S., 1988. Introduction to Meteorology. Academic Press, New York. 303 pp.

Baas, P., Bosveld, F.C., Baltink, H.K., Holtslag, A.A.M., 2009. A climatology of nocturnal low-level jets at Cabauw. J. Appl. Meteorol. 48, 1627–1642.

Baklanov, A., 2005. Parameterisation of SBL height in atmospheric pollution models. In: Air Pollution Modeling and Its Application XV. Kluwer Academic Publishers, pp. 415–424.

Balachandran, N.K., 1980. Gravity waves from thunderstorms. Mon. Weather Rev. 108, 804–816.

Balsley, B.B., Fritts, D.C., Frehlich, R.G., Jones, R.M., Vadas, S.L., Coulter, R., 2002. Up-gully flow in the great plains region: a mechanism for perturbing the nighttime lower atmosphere? Geophys. Res. Lett. 29, 1931.

Balsley, B.B., Frehlich, R.G., Jensen, M.L., Meillier, L.Y., Muschinski, A., 2003. Extreme gradients in the nocturnal boundary layer: structure, evolution, and potential causes. J. Atmos. Sci. 60, 2496–2508.

Banta, R.M., Newsom, R.K., Lundquist, J.L., Pichugina, Y.L., Coulter, R.L., Mahrt, L., 2002. Nocturnal low-level jet characteristics over Kansas during CASES-99. Bound.-Layer Meteorol. 105, 221–252.

Banta, R.M., Pichugina, Y.L., Newsom, R.K., 2003. Relationship between low-level jet properties and turbulence kinetic energy in the nocturnal stable boundary layer. J. Atmos. Sci. 60, 2549–2555.

Banta, R., et al., 2007. The very stable boundary layer on nights with weak low-level jets. J. Atmos. Sci. 64, 3068–3090.

Basu, S.F., Holtslag, A.A.M., Van de Wiel, B.J.H., Moene, A.F., Steeneveld, G.-J., 2008. An inconvenient "truth" about using sensible heat flux as a surface boundary condition in models under stably stratified regimes. Acta Geophys. 56, 88–99.

Bean, B.R., Frisch, A.S., McAllister, L.G., Pollard, J.R., 1973. Planetary boundary-layer turbulence studies from acoustic echo sounder and in situ measurements. Bound.-Layer Meteorol. 4, 449–474.

Beare, R.J., Edwards, J.M., Lapworth, A.J., 2006. Simulation of the observed evening transition and nocturnal boundary layers: large-eddy simulation. Q. J. R. Meteorol. Soc. 132, 81–99.

Blackadar, A., 1957. Boundary layer wind maxima and their significance for the growth of nocturnal inversions. Bull. Am. Meteorol. Soc. 33, 373–379.

Bleeker, W., Andre, M.J., 1951. On the diurnal variation of precipitation, particularly over central U.S.A., and its relation to large-scale orographic circulation systems. Q. J. R. Meteorol. Soc. 77, 260–271.

Blumen, W., 1997. A model of inertial oscillations with deformation frontogenesis. J. Atmos. Sci. 54, 2681–2692.

Blumen, W., et al., 2001. Turbulence statistics of a Kelvin–Helmholtz billow event observed in the night-time boundary layer during the cooperative atmosphere–surface exchange study field program. Dyn. Atmos. Oceans 34, 189–204.

Bonner, W.D., 1968. Climatology of the low-level jet. Mon. Weather Rev. 96, 833–850.

Bornstein, R.D., 1968. Observations of the urban heat island effect in New York city. J. Appl. Meteorol. 7, 575–582.

Brost, R.A., Wyngaard, J.C., 1978. A model study of the stably stratified planetary boundary layer. J. Atmos. Sci. 35B, 1427–1440.

Brunt, D., 1934. Physical and Dynamical Meteorology. Cambridge University Press, pp. 124–146.

Businger, J.A., 1973. Turbulent transfer in the atmospheric surface layer. In: Haugen, D.H. (Ed.), Workshop on Micrometerology. American Meteorological Society, Boston, MA, pp. 67–100.

Businger, J.A., Arya, S.P.S., 1974. The height of the mixed layer in the stably stratified boundary layer. Adv. Geophys. 18A, 73–92.

Busse, J., Knupp, K., 2012. Observed characteristics of the afternoon evening boundary layer transition based on sodar and surface data. J. Appl. Meteorol. 51, 571–582.

Caughey, S.J., Wyngaard, J.C., Kaimal, J.C., 1979. Turbulence in the evolving stable boundary layer. J. Atmos. Sci. 36, 1041–1052.

Chimonas, G., 2002. On internal gravity waves associated with the stable boundary layer. Bound.-Layer Meteorol. 102, 139–155.

Chimonas, G., Nappo, C.J., 1987. A thunderstorm bow wave. J. Atmos. Sci. 44, 533–541.

Christie, D.R., Muirhead, K.J., Hales, A.L., 1978. On solitary waves in the atmosphere. J. Atmos. Sci. 35, 805–825.

Clarke, R.H., 1970. Observational studies in the atmospheric boundary layer. Q. J. R. Meteorol. Soc. 96, 91–114.

Clarke, R.H., Dyer, A.J., Brook, R.R., Reid, D.G., Troup, A.J., 1971. The Wangara experiment. Boundary-layer data. Pap. No 19, Div. Meteor. Phys., CSIRO, Australia.

Costa, F.D., Acevedo, O., Mombach, J.C.M., Degrazia, G.A., 2011. A simplified model for intermittent turbulence in the nocturnal boundary layer. J. Atmos. Sci. 68, 1714–1729.

Coulter, R.L., Doran, J., 2002. Spatial and temporal occurrences of intermittent turbulence during CASES-99. Bound.-Layer Meteorol. 105, 329–349.

Curry, M., Murty, R.C., 1974. Thunderstorm-generated gravity waves. J. Atmos. Sci. 31, 1402–1408.

Cuxart, J., 2008. Nocturnal basin low-level jets: an integrated study. Acta Geophys. 56, 100–113.

Cuxart, J., et al., 2000. Stable atmospheric boundary-layer experiment in Spain (SABLES 98): a report. Bound.-Layer Meteorol. 96, 337–370.

Deardorff, J.W., 1972. Rate of growth of the nocturnal boundary layer. In: Preprints Symp. Air Pollution, Turbulence and Diffusion, Las Cruces, Amer. Meteor. Soc, pp. 183–190.

DeBaas, A.F., Driedonks, A.G., 1985. Internal gravity waves in a stably stratified boundary layer. Bound.-Layer Meteorol. 31, 303–323.

Delage, Y., 1974. A numerical study of the nocturnal atmospheric boundary layer. Q. J. R. Meteorol. Soc. 100, 351–364.

Derbyshire, S.H., 1990. Nieuwstadt's stable boundary layer revisited. Q. J. R. Meteorol. Soc. 116, 127–158.

Derbyshire, S.H., 1999. Boundary-layer modelling: established approaches and beyond. Bound.-Layer Meteorol. 90, 423–446. https://doi.org/10.1023/A:1001749007836.

Dias, N.L., Brutsaert, W., Wesley, M.L., 1995. Z-less stratification under stable conditions. Bound.-Layer Meteorol. 75, 175–187.

Doviak, R.J., Ge, R., 1984. An atmospheric solitary gust observed with a Doppler radar, a tall tower, and a surface network. J. Atmos. Sci. 41, 2560–2573.

Durst, C.S., 1932. The breakdown of steep wind gradients in inversions. Q. J. R. Meteorol. Soc. 58, 165–168.

Duynkerke, P., 1999. Turbulence, radiation, and fig in Dutch stable boundary layers. Bound.-Layer Meteorol. 90, 447–477.

Edwards, J.M., 2009. Radiative processes in the stable boundary layer: part II. The development of the nocturnal boundary layer. Bound.-Layer Meteorol. 131, 127–146.

Edwards, J.M., Beare, R.J., Lapworth, A.J., 2006. Simulation of the observed evening transition and nocturnal boundary layers: single-column modeling. Q. J. R. Meteorol. Soc. 132, 61–80.

Einaudi, F., Finnigan, J.J., 1993. Wave-turbulence dynamics in the stably stratified boundary layer. J. Atmos. Sci. 50, 1841–1864.

Farquharson, S.J., 1939. The diurnal variation of wind over tropical Africa. Q. J. R. Meteorol. Soc. 65, 165–183.

Fedorovich, E., Gibbs, J.A., Shapiro, A., 2017. Numerical study of nocturnal low-level jets over gently sloping terrain. J. Atmos. Sci. 74, 2813–2834.

Foken, T., 2016. Micrometeorology. Springer Nature, Berlin. 362 pp.

Garratt, J.R., 1982. Observations in the nocturnal boundary layer. Bound.-Layer Meteorol. 22, 21–48.

Garratt, J.R., 1985. Inland boundary layer at low latitudes. Part 1, the nocturnal jet. Bound.-Layer Meteorol. 32, 307–327.

Garratt, J.R., 1992. The Atmospheric Boundary Layer. Cambridge University Press. 316 pp.

Garratt, J.R., Brost, R.A., 1981. Radiative cooling effects within and above the nocturnal boundary layer. J. Atmos. Sci. 38, 2730–2746.

Gifford, F., 1952. The breakdown of a low-level inversion studied by means of detailed soundings with a modified radiosonde. Bull. Am. Meteorol. Soc. 33, 373–379.

Glickman, T.S., 2000. Glossary of Meteorology, 2nd. American Meteorological Society. 855 pp.

Gopalakrishan, S.G., Sharan, M., McNider, R.T., Singh, M.P., 1998. Study of radiative and turbulent processes in the stable boundary layer under weak wind conditions. J. Atmos. Sci. 55, 954–960.

Gossard, E.E., Gaynor, J.E., Zamora, R.J., Neff, W.D., 1985. Fine structure of elevated stable layers observed by sounder and in situ tower sensors. J. Atmos. Sci. 42 (20), 2156–2169.

Goualt, J., 1938. Vents en altitude á Fort Lamy. Ann. Phys. 5, 70–91. Du Globe de la France d'Outre-Mer.

Grachev, A.A., Andreas, E.L., Fairall, C.W., Guest, P.S., Persson, P.O.G., 2013. The critical Richardson number and limits of applicability of local similarity theory in the stable boundary layer. Bound.-Layer Meteorol. 147, 51–82.

Grant, A.L.M., 1997. An observational study of the evening transition boundary-layer. Q. J. R. Meteorol. Soc. 123, 657–677.

Grimsdell, A.W., Angevine, W.M., 2002. Observations of the afternoon transition of the convective boundary layer. J. Appl. Meteorol. 41, 3–11.

Grund, C.J., et al., 2001. High-resolution Doppler lidar for boundary layer and cloud research. J. Atmos. Sci. 18, 376–393.

Harrison, R.M., Holman, C.D., McCortney, H.A., McIlvenn, J.F.R., 1978. Nocturnal depletion of photochemical ozone at a rural site. Atmos. Environ. 12, 2021–2026.

Hicks, B.B., 1976. Wind profile relationships from "Wangara" experiments. Q. J. R. Meteorol. Soc. 102, 535–551.

Hoecker, W.H., 1963. Three southerly low-level jet systems delineated by the Weather Bureau special pibal network of 1961. Mon. Weather Rev. 91, 573–582.

Holton, J.R., 1967. The diurnal boundary layer wind oscillation above sloping terrain. Tellus 19A, 199–205.

Holtslag, A.A.M., Nieuwstadt, F.T.M., 1986. Scaling the atmospheric boundary layer. Bound.-Layer Meteorol. 36, 201–209.

Holtslag, A.A.M., et al., 2013. Stable atmospheric boundary layers and diurnal cycles: challenges for weather and climate models. Bull. Am. Meteorol. Soc. 94, 1691–1706.

Hooijdonk, I.G.S., et al., 2017. Near-surface temperature inversion growth rate during the onset of the stable boundary layer. J. Atmos. Sci. 74, 3433–3449.

Howell, J., Sun, J., 1999. Surface-layer fluxes in stable conditions. Bound.-Layer Meteorol. 90, 495–520.

Izumi, Y., Barad, M., 1963. Wind and temperature variations during development of a low-level jet. J. Appl. Meteorol. 2, 28–33.

Klöppel, M., Stilke, G., Wamser, C., 1978. Experimental investigation into variations and comparisons with results of simple boundary-layer models. Bound.-Layer Meteorol. 15, 135–146.

Koch, S.E., et al., 2005. Turbulence and gravity waves within an upper-level front. J. Atmos. Sci. 62, 3885–3908.

Koch, S.E., et al., 2008. Turbulent mixing processes in atmospheric bores and solitary waves deduced from profiling systems and numerical simulation. Mon. Weather Rev. 136, 1373–1400.

Kondo, J., Kanechika, O., Yasuda, N., 1978. Heat and momentum transfers under strong stability in the atmospheric surface layer. J. Atmos. Sci. 35, 1012–1021.

Kraus, H., Malcher, J., Schaller, E., 1985. A nocturnal low-level jet during PUKK. Bound.-Layer Meteorol. 31, 187–195.

Lenschow, D.H., Li, X.S., Zhu, C.J., Stankov, B.B., 1987. The stably stratified boundary layer over the Great Plains' part I. Bound.-Layer Meteorol. 42, 95–121.

Lettau, H., Davidson, B., 1957. Exploring the Atmosphere's First Mile. vol. 2 Pergamon Press. 202 pp.

Lothon, M., et al., 2014. The BLLAST field experiment: boundary-layer late afternoon and sunset turbulence. Atmos. Chem. Phys. 14, 10931–10960.

Lundquist, J.K., 2003. Intermittent and elliptical inertial oscillations in the atmospheric boundary layer. J. Atmos. Sci. 60, 2661–2673.

Lyons, R., Panofsky, H.A., Wallaston, S., 1964. The critical Richardson number and its implications for forecast problems. J. Appl. Meteorol. 3, 136–142.

Mahrt, L., 1981. The early evening boundary layer transition. Q. J. R. Meteorol. Soc. 107, 329–343.

Mahrt, L., 1985. Vertical structure and turbulence in the very stable boundary layer. J. Atmos. Sci. 42, 2333–2349.

Mahrt, L., 1999. Stratified atmospheric boundary layers. Bound.-Layer Meteorol. 90, 375–396.

Mahrt, L., Vickers, D., 2003. Formulation of turbulent fluxes in the stable boundary layer. J. Atmos. Sci. 60, 2538–2548.

Mahrt, L., Heald, R.C., Troen, I., Lenschow, D., Stankov, B., 1979. An observational study of the structure of the nocturnal boundary layer. Bound.-Layer Meteorol. 17, 247–264.

Mahrt, L., Richardson, S., Stauffer, D., Seaman, N., 2014. Nocturnal wind-direction shear in complex terrain. Q. J. R. Meteorol. Soc. 140, 2393–2400.

Mahrt, L., Sun, J., Blumen, W., Delany, T., Oncley, S., 1998. Nocturnal boundary layers. Bound.-Layer Meteorol. 88, 255–278.

Malhi, Y.S., 1995. The significance of the dual solutions for heat fluxes measured by the temperature fluctuation method in stable conditions. Bound.-Layer Meteorol. 74, 389–396.

Meillier, Y.P., Frehlich, R.G., Jones, R.M., Balsley, B.B., 2008. Modulation of small-scale turbulence by ducted gravity waves in the nocturnal boundary layer. J. Atmos. Sci. 65, 1414–1427.

Melgarejo, J.W., Deardorff, J.W., 1974. Stability functions for the boundary-layer resistance laws based upon boundary-layer heights. J. Atmos. Sci. 31, 1324–1333.

Monahan, A.H., Rees, T., He, Y., McFarlane, N., 2015. Multiple regimes of wind, stratification, and turbulence in the stable boundary layer. J. Atmos. Sci. 72, 3178–3198.

Monin, A.S., 1970. The atmospheric boundary layer. Annu. Rev. Fluid Mech. 2, 225–250.

Mori, Y., 1990. Evidence of inertial oscillations of the surface wind at Marcus Island. J. Geophys. Res. 95, 11777–11783.

Nadeau, D.F., Pardyjak, E.R., Higgins, C.W., Fernando, H.J.S., Parlange, M.B., 2011. A simple model for the afternoon and early evening decay of convective turbulence over different land surfaces. Bound.-Layer Meteorol. 141, 301–324.

Nappo, C.J., 1977. Mesoscale flow over complex terrain during the eastern Tennessee trajectory experiment (ETTEX). J. Appl. Meteorol. 16, 1186–1196.

Nappo, C.J., 1991. Sporadic breakdowns of stability in the PBL over simple and complex terrain. Bound.-Layer Meteorol. 54, 69087.

Nappo, C.J., 2012. An Introduction to Atmospheric Gravity Waves, second ed. International Geophysics.

Nappo, C.J., Johansson, P.-E., 1999. Summary of the Lövånger internal workshop on turbulence and diffusion in the stable planetary boundary layer. Bound.-Layer Meteorol. 90, 345–374.

Nieuwstadt, F.T.M., 1980. A rate equation for the inversion height in a nocturnal boundary layer. J. Appl. Meteorol. 19, 1445–1447.

Nieuwstadt, F.T.M., 1984. The turbulent structure of the stable, nocturnal boundary layer. J. Atmos. Sci. 41, 2202–2216.

Nieuwstadt, F.T.M., Brost, R.A., 1986. The decay of convective turbulence. J. Atmos. Sci. 43, 532–546.

Nieuwstadt, F.T.M., Driedonks, A.G.M., 1979. The nocturnal boundary layer: a case study compared with model calculations. J. Appl. Meteorol. 18, 1398–1405.

Nieuwstadt, F.T.M., Tennekes, H., 1981. A rate equation for the nocturnal boundary-layer height. J. Atmos. Sci. 38, 1418–1428.

Ohya, Y., 2001. Wind-tunnel study of atmospheric stable boundary layers over a rough surface. Bound.-Layer Meteorol. 98, 57–82.

Ohya, Y., Neff, D.E., Meroney, R.N., 1997. Turbulence structure in a stratified boundary layer under stable conditions. Bound.-Layer Meteorol. 83, 139–161.

Ostdiek, V., Blumen, W., 1997. A dynamic trio: inertial oscillation, deformation frontogenesis, and the Ekman–Taylor boundary layer. J. Atmos. Sci. 54, 190–1502.

Parish, T.R., Oolman, L.D., 2010. On the role of sloping terrain in forcing the great plains low-level jet. J. Atmos. Sci. 67, 2690–2699.

Pasquill, F., 1961. The estimation of the dispersion of windborne. Material. Meteorol. Mag. 90, 33–49.

Pichugina, Y.L., Banta, R.M., 2010. Stable Boundary Layer Depth from High-Resolution Measurements of the Mean Wind Profile. Journal of Applied Meteorology and Climatology 49 (1), 20–35.

Poulos, G., et al., 2002. CASES-99: a comprehensive investigation of the stable nocturnal boundary layer. Bull. Am. Meteorol. Soc. 83, 555–581.

Raupach, M., 1994. Simplified expressions for vegetation roughness length and zero-plane displacement as functions of canopy height and area index. Bound.-Layer Meteorol. 71, 211–216.

Román-Cascón, C., et al., 2015. Interactions among drainage flows, gravity waves and turbulence: a BLLAST case study. Atmos. Chem. Phys. 15, 9031–9047.

Rottman, J.W., Einaudi, F., 1993. Solitary waves in the atmosphere. J. Atmos. Sci. 50, 2116–2136.

Seefeldt, M.W., Cassano, J.J., 2008. An analysis of low-level jets in the Greater Ross Ice Shelf region based on numerical simulation. Mon. Weather Rev. 136, 4188–4205.

Seibert, P., 2000. Review and intercomparison of operational methods for the determination of the mixing height. Atmos. Environ. 34, 1001–1027. https://doi.org/10.1016/S1352-2310 (99)00349-0.

Shapiro, A., Fedorovich, E., Rahimi, S., 2016. A unified theory for the Great Planes nocturnal low-level jet. J. Atmos. Sci. 73, 3037–3057.

Sharan, M., McNider, R.T., Gopalakrishnan, S.G., Singh, M.P., 1995. Bhopal gas leak: a numerical simulation of episodic dispersion. Atmos. Environ. 29, 2061–2074.

Shreffler, J.H., Binkowski, F.S., 1981. Observations of pressure jump lines in the Midwest, 10–12 August 1976. Mon. Weather Rev. 109, 1713–1725.

Smedman, A., 1988. Observations of a multi-level turbulence structure in a very stable atmospheric boundary layer. Bound.-Layer Meteorol. 44, 231–253.

Smedman, A., Tjernstro, H., Högström, U., 1993. Analysis of the turbulence structure of a marine low-level jet. Bound.-Layer Meteorol. 66, 105–126.

Song, J., Liao, K., Coulter, R.L., Lesht, B.M., 2005. Climatology of the low-level jet at the southern Great Plains atmospheric boundary layer experiment site. J. Appl. Meteorol. 44, 1593–1606.

Sorbjan, Z., 1989. Structure of the Atmospheric Boundary Layer. Prentice Hall, Englewood Cliffs, NJ. 317 pp.

Sorbjan, Z., 2010. Gradient-based scales and similarity laws in the stable boundary layer. Q. J. R. Meteorol. Soc. 136, 1243–1254.

Spiridonov, V., Ćurić, M., 2021. Atmospheric boundary layer (ABL). In: Fundamentals of Meteorology. Springer, Cham, https://doi.org/10.1007/978-3-030-52655-9_14.

Steeneveld, G.J., 2007. Understanding and Prediction of Stable Atmospheric Boundary Layers over Land. Wageningen University and Research ProQuest Dissertations Publishing, 28238503.

Stensrud, D.J., 1996. Importance of low-level jets to climate: a review. J. Clim. 9, 1698–1711.

Stull, R.B., 1988. An Introduction to Boundary Layer Meteorology. Kluwer Academic Publishers, Dordrecht, The Netherlands. 666 pp.

Sun, J., et al., 2002. Intermittent turbulence associated with a density current passage in the stable boundary layer. Bound.-Layer Meteorol. 105, 199–219.

Sun, J., et al., 2003. Heat balance in the nocturnal boundary layer during CASES-99. J. Atmos. Sci. 42, 1649–1666.

Sun, J., Lenschow, D.H., Burns, S.P., 2004. Atmospheric disturbances that generate intermittent turbulence in nocturnal boundary layers. Bound.-Layer Meteorol. 110, 255–279.

Sun, J., Mahrt, L., Banta, B., Pichugina, Y.L., 2012. Turbulence regimes and turbulence intermittency in the stable boundary layer during CASES-99. J. Atmos. Sci. 69, 338–351.

Sun, J., Mahrt, L., Nappo, C., Lenschow, D.H., 2015. Wind and temperature oscillations generated by wave-turbulence interactions in the stably stratified boundary layer. J. Atmos. Sci. 72, 1484–1503.

Tjemkes, S.A., Duynkerke, P.G., 1989. The nocturnal boundary layer: model calculations compared with observations. J. Appl. Meteorol. 28, 161–175.

Uccellini, L.W., 1975. A case study of apparent gravity wave initiation of severe convective storms. Mon. Weather Rev. 103, 497–513.

Uccellini, L.W., Koch, S.E., 1987. The synoptic setting and possible energy sources for mesoscale wave disturbances. Mon. Weather Rev. 115, 721–728.

Van de Wiel, B.J.H., Ronda, R.J., Moene, A.F., de Bruin, H.A.R., Holtslag, A.A.M., 2002. Intermittent turbulence and oscillations in the stable boundary layer over land. Part I: a bulk model. J. Atmos. Sci. 59, 942–958.

Van de Wiel, B.J.H., Moene, A.F., Hartogensis, O.K., de Bruin, H.A.R., Holtslag, A.A.M., 2003. Intermittent turbulence in the stable boundary layer over land. Part III: a classification for observations during CASES-99. J. Atmos. Sci. 60, 2509–2522.

Van der Linden, S.J.A., et al., 2017. Local characteristics of the nocturnal boundary layer in response to external pressure forcing. J. Appl. Meteorol. Climatol., 3035–3047.

van Heerwaarden, C.V., Mellado, J.P., 2016. Growth and decay of a convective boundary layer over a surface with a constant temperature. J. Atmos. Sci. 73, 2161–2177.

Van Ulden, A.P., Wieringa, J., 1996. Atmospheric boundary layer research at Cabauw. Bound.-Layer Meteorol. 78, 39–69.

Wallace, J.M., Hobbs, P.V., 2006. Atmospheric Science. 92. Academic Press. 483 pp.

Wang, S., Zhang, F., 2007. Sensitivity of mesoscale gravity waves to the baroclinicity of jet-front systems. Mon. Weather Rev. 135, 670–688.

White, L.D., 2009. Sudden nocturnal warming events in Mississippi. J. Appl. Meteorol. Climatol. 48, 738–775.

Whiteman, C.D., Bian, X., Zhong, S., 1997. Low-level jet climatology from enhanced rawinsonde observations at a site in the Southern Great Plains. J. Appl. Meteorol. 36, 1363–1376.

Winkler, K., 1980. Störungen der nachtlichen Grenzschicht. Meteorol. Rdsch. 33, 90–94.

Wyngaard, J.C., 1973. On surface-layer turbulence. In: Haugen, D.A. (Ed.), Workshop on Micrometeorology. American Meteorological Society, Boston.

Wyngaard, J., 1975. Modeling the planetary boundary layer—extension to the stable case. Bound.-Layer Meteorol. 9, 441–460.

Yamada, T., 1979. Prediction of the nocturnal surface inversion height. J. Atmos. Sci. 36, 792–804.

Yu, T.-W., 1978. Determining height of the nocturnal boundary layer. J. Appl. Meteorol. 17, 28–33.

Zeman, O., 1979. Parameterization of the dynamics of stably boundary layers and nocturnal jets. J. Atmos. Sci. 36, 792–804.

Zilitinkevich, S.S., 2012. The height of the atmospheric planetary boundary layer: state of the art and new development. In: Fernando, H.J.S., Klaić, Z.B., McCulley, J.L. (Eds.), National Security and Human Health Implications of Climate Change. NATO Science for Peace and Security Series C: Environmental Security, vol. 125. Springer, pp. 147–161.

Zilitinkevich, S., Baklanov, A., 2002. Calculation of the height of the stable boundary layer in practical applications. Bound.-Layer Meteorol. 105, 389–409. https://doi.org/10.1023/A:1020376832738.

Zilitinkevich, S.S., Deardorff, J.W., 1974. Similarity theory for the planetary boundary layer of time-dependent height. J. Atmos. Sci. 31, 1449–1452.

Zilitinkevich, S.S., Mironov, D., 1996. A multi-limit formulation for the equilibrium depth of a stably stratified boundary layer. Bound.-Layer Meteorol. 81, 325–351.

Further reading

Darby, L.S., Allwine, K.J., Banta, B., 2006. Nocturnal low-level jet in a mountain basin complex. Part II: transport and diffusion of tracer under stable conditions. J. Appl. Meteorol. Climatol. 45, 740–753.

Deardorff, J.W., 1971. Rate of growth of the nocturnal boundary layer. In: Paper Presented at Symposium on Air Pollution, Turbulence and Diffusion, Las Cruces, N. M., NCAR ms, pp. 71–246.

Izumi, Y., 1971. Kansas 1968 field program data report. Environmental Research Papers, No. 379, AFCRL-72-0041, 79 pp.

Spiridonov, V., Ćurić, M., 2021. Atmospheric boundary layer (ABL). In: Fundamentals of Meteorology. Springer, Cham, https://doi.org/10.1007/978-3-030-52655-9_14.

Steeneveld, G.J., Holtslag, A.A.M., Nappo, C.J., van De Wiel, B.J.H., Mahrt, L., 2008. Exploring the possible role of small-scale terrain drag on stable boundary layers over land. J. Atmos. Sci. 47, 2518–2529.

Wilczak, J.M., et al., 1995. Contamination of wind profiler data by migrating birds: characteristics of corrupted data and potential solutions. J. Atmos. Oceanic Technol. 12B, 449–467.

Yus-Diez, J., et al., 2019. Nocturnal boundary layer turbulence regimes analysis during the BLLAST campaign. Atmos. Chem. Phys. 19, 9495–9514.

Zilitinkevich, S.S., Monin, A.S., 1974. Similarity theory for the atmospheric boundary layer. Izv. Akad. Nauk SSSR, Fiz. Atmos. Okeana 10, 587–599.

We can't move mountains: Flow in complex environments

Chuixiang Yi and Eric Kutter

School of Earth and Environmental Sciences, Queens College of the City University of New York, New York, United States

7.1 Introduction

You've seen in previous chapters how the wind behaves in the absence of any obstruction, but what happens when a large object blocks the flow? This chapter focuses on the largest obstacles the wind will encounter: mountains. In addition to the joy of pursuing understanding for understanding's sake, there are a number of practical reasons for an in-depth analysis of wind dynamics over mountainous terrain. An airplane pilot may need to know where pockets of strong wind shear occur around a mountain. A power company official needs to know where to place a wind turbine for steady, strong winds. A farmer should be aware of the differences in moisture and temperature at various locations around a mountain slope. Fire control professionals plan their actions based on the predicted direction and speed of wind flows. Climate modelers must account for the impacts of any obstruction of the synoptic wind caused by complex terrain.

Using the fundamentals of turbulent and advective flows that you have learned in previous chapters, we will investigate the wind flow and the exchange of energy and matter between the terrestrial biosphere and the atmosphere around sloped terrain. Throughout the discussion, we focus on the physical causes of the observed phenomena, and the potential impacts on local and regional climate in mountainous biomes.

7.2 Windward wet and leeward dry

The highest mountain areas are located in the Himalaya-Tibet region, and the longest mountain range extends nearly continuously along the west coast of the Americas from Alaska in the north to Chile in the south (Figure 7.1). The other famous mountain areas include those in the European Alps, the Southern Alps of New Zealand, and so on, as shown in Figure 7.1. A common characteristic across the world's mountains is that the windward sides of mountains receive plenty of precipitation with lush green vegetation cover, while the leeward sides are dry with a barren land cover (Figure 7.2). Why are the windward sides of mountains wet and the leeward sides dry?

When prevailing winds move across mountain ranges, the moisture concentration of incoming air masses is modified by the mountains. How many water molecules a given parcel of air can hold is determined by the saturation vapor pressure, $e_s(T)$,

Conceptual Boundary Layer Meteorology. https://doi.org/10.1016/B978-0-12-817092-2.00003-5

Physical Map of the World. April 2004

Figure 7.1 The worldwide distribution of mountain lands.

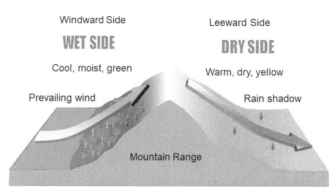

Figure 7.2 Two different climates between windward side and leeward side of a mountain. Credit: Chuixiang Yi.

which is a unique function of temperature T (Figure 7.3). $e_s(T)$ is independent of actual moisture content at the given temperature and represents an ability of the air to hold water molecules in suspension, like salt dissolved in water. $e_s(T)$ increases with rising temperature. As a result, hotter air can hold more water molecules than cooler air. This basic physical principle is one reason that a warmer climate will mean an increase in the frequency and intensity of extreme weather events. While climate change's impacts are not the primary focus of this chapter, the boundary layer is inextricably linked with climate (see Chapter 12) and interested readers can find more details about this topic in Yi et al. (2015).

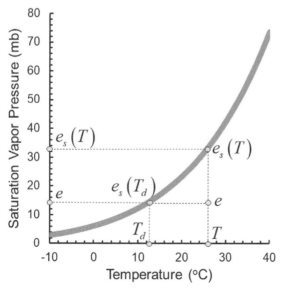

Figure 7.3 Saturation vapor pressure as a function of temperature. T is current air temperature and e is current vapor pressure for which the saturation vapor pressure is $e_s(T)$. T_d is dew point temperature at which the current vapor pressure e becomes saturated $e = e_s(T_d)$. Credit: Chuixiang Yi.

The actual amount of water molecules is denoted by vapor pressure, e, which is not a function of the temperature. The ratio of the actual vapor pressure, e, to its saturation value, $e_s(T)$, at a given temperature is called relative humidity, RH. In other words, the relative humidity is the amount of water the air is carrying, divided by the amount of water the air could possibly carry at that moment. RH is usually expressed as a percentage by multiplying the ratio by 100%. When the RH reaches 100%, condensation occurs, turning the excess water vapor to liquid water. In equation form, this happens when $e = e_s(T)$, $RH = 100\%$. This is the curve in Figure 7.3.

The difference between saturation vapor pressure and actual vapor pressure is called the vapor pressure deficit (VPD). VPD is a measure of how far a parcel of air is from saturation. A high VPD means that the air is far from being saturated, so it has low RH and the air is dry. There are two ways for an unsaturated parcel of air to become saturated: (1) by evaporation of water into the air without temperature change (i.e., e moves vertically toward $e_s(T)$ in Figure 7.3), and (2) by cooling the parcel to a temperature, T_d, without water vapor concentration change (i.e., e moves horizontally toward $e_s(T_d)$ in Figure 7.3).

The temperature, T_d, is called the dew point temperature. T_d is an indicator of the amount of water vapor in the air. The higher the dew point temperature, the greater the moisture content of the air at a given temperature. When the air parcel's temperature, T_p, is close to dew point temperature T_d, the parcel is nearly saturated. Therefore, the condensation condition can be also expressed as $T_p = T_d$, $RH = 100\%$.

VPD is an important control parameter for plants' photosynthesis and transpiration. When plants take CO_2 from the air through stomata on the leaves, they lose moisture through the stomata. Within stomata the air is saturated, with its saturation vapor pressure a function of leaf temperature, $e_s(T_{leaf})$. If the leaf temperature, T_{leaf}, is approximated as the air temperature, T, the VPD represents vapor pressure difference between the stomata and its surroundings. This VPD is a driving force for water vapor diffusion (transpiration) from plants to the atmosphere. Under high VPD conditions, plants open less or close their stomata completely to protect themselves from water loss, even though this means no incoming CO_2 for the plant. Mountain ranges generate low VPD (humid) on the windward side and high VPD (dry) on the leeward side. On the humid windward side of mountains, biological diversity is rich and ecological production is high, in contrast to the dry leeward side (Figure 7.2).

How do mountains change the moisture properties of prevailing wind? When prevailing winds meet a mountain range, the air is usually forced to rise over the terrain. The rising parcel cools and expands. How much air temperature drops per kilometer of altitude change is called a lapse rate, which depends on if the parcel is saturated or unsaturated. For an unsaturated parcel of air, its lapse rate is constant ($10\,°C\,km^{-1}$) everywhere in the troposphere. This constant is also called the dry adiabatic lapse rate and can be derived from the first law of thermodynamics with an adiabatic assumption, which means that a parcel air is rising or sinking so rapidly that there is not enough time to exchange heat with its surroundings. Figure 7.5 shows temperature profiles for both sides of the mountain.

As illustrated in Figure 7.4, the rising parcel is unsaturated below 1 km altitude because $T_p > T_d$ (Figure 7.5A) and hence parcel temperature T_p decreases by the

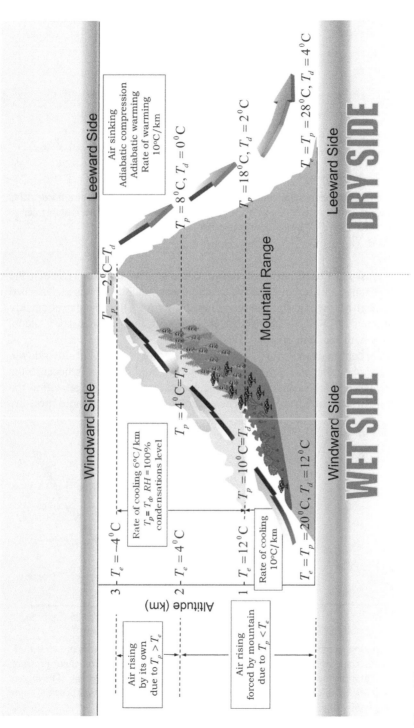

Figure 7.4 Properties of ascending air on windward side and descending air on leeward side. T_p is parcel temperature, T_d is dew point temperature, and T_e is environmental temperature. Data are taken from Figure 5.14 in Ahrens (2012). Credit: Chuixiang Yi.

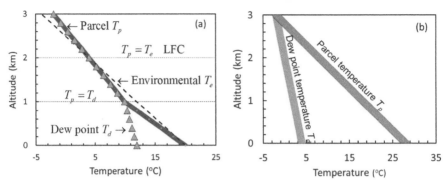

Figure 7.5 Vertical profiles of temperature on (A) windward side, and (B) leeward side. LFC stands for the level of free convection as $T_p = T_e$. Below the LFC, atmosphere is stable due to $T_p < T_e$, while above the LFC, atmosphere is unstable $T_p > T_e$.
Credit: Chuixiang Yi.

dry adiabatic lapse rate $(10\,^\circ\mathrm{C\,km}^{-1})$. Note that the moisture contents of a parcel in the rising process change a little, but its ability to hold water vapor decreases as described by the dry adiabatic lapse rate (*blue bubble* in Figure 7.6A). Thus, the parcel quickly becomes saturated $(T_p = T_d,\ RH = 100\%)$ at 1 km and then condensation starts, forming clouds with rain and/or snow (Figure 7.4).

As the parcel continues ascending, its cooling rate is not constant but slows down, following the moist adiabatic lapse rate (average $6\,^\circ\mathrm{C\,km}^{-1}$). This is because the latent heat released by the condensation process warms the parcel, offsetting the cooling rate from adiabatic expansion. The capacity of the parcel to hold moisture

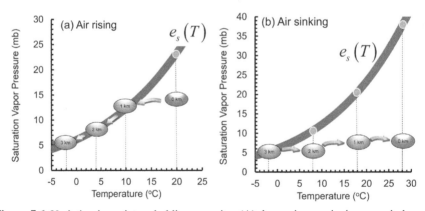

Figure 7.6 Variation in moisture holding capacity: (A) decreasing as air rises on windward side, and (B) increasing air sinks on leeward side. The *red curve* is saturation vapor pressure, which is an indicator of how much moisture air can hold for a given temperature. The *arrows* indicate the direction of air parcel motion in vertical, and the *blue bubble* indicates actual vapor pressure at the given altitude marked on the bubble.
Credit: Chuixiang Yi.

is reduced with increasing altitude up to the top of the mountain due to the parcel temperature T_p $(=T_d)$ decreasing $6\,°C\,km^{-1}$. As a result, most of parcel moisture is drawn out of it as precipitation, indicated by significant decreasing of saturation vapor pressure in Figure 7.6A.

Below 2 km in Figure 7.4, the air is forced to rise by the wind's momentum reaching the mountain because the parcel of air is more dense than its surroundings ($T_p < T_e$), and the air in this layer is stable. Above 2 km level, the air parcel becomes warmer than the air surrounding it ($T_p > T_e$) due to the release of latent heat, causing an unstable thermal condition, and the air rises by itself. The level ($T_p = T_e$) in the atmosphere where the parcel of air changes from stable ($T_p < T_e$) to unstable ($T_p > T_e$) is called the level of free convection.

We have talked about the air's journey uphill, but what happens as our intrepid parcel reaches the mountaintop? As the air passes over the mountain range, it moves downslope on the leeward side (Figure 7.4). The sinking parcel temperature increases $10\,°C\,km^{-1}$ due to adiabatic compression and warming (Figure 7.5B). Thus, the capacity of the sinking parcel to hold water vapor increases greatly. However, the actual amount of moisture in the parcel keeps relatively constant at the amount it had on the top of the mountain (Figure 7.6B). As a result, the relative humidity, RH, of sinking air rapidly decreases (Figure 7.6B). In addition, the difference between parcel temperature and dew point temperature reaches its greatest value at the bottom of the mountain and the air is far from being saturated (Figure 7.5B).

To make these processes easier to picture imagine the following highly technical example: two circus performers are so strong that they delight the audience by lifting a platform with many people standing on it. If the amount of people on the platform is less than the performers can lift (e is less than $e_s(T)$) then everyone stays on the platform. You may know from your own workout routine in the gym that the higher you have to lift a weight, the harder it is to lift it still further. As the performers lift the platform, they can carry fewer and fewer people ($e_s(T)$ is reduced with altitude as the temperature and pressure both drop), until the "relative humanity" is 100%, and then the circus crew has to shove some people off if the platform is going to rise any further (precipitation occurs). On the way down, with fewer people on the platform, it is much easier for the performers (e has been reduced by all that precipitation on the way up the windward side). They have no trouble with the platform because they can carry more and more people as the platform lowers ($e_s(T)$ is increasing as the parcel comes down the mountain). The audience will not see any people falling off the platform on the way down (e stays constant while $e_s(T)$ increases, meaning no precipitation on the leeward side).

The warm and dry region on the leeward side of a mountain is called a rain shadow. The process of a mountain changing the moisture properties of prevailing winds is called the orographic effect. Some deserts in the world resulted from orographic effects in rain shadow regions. For instance, the Gobi Desert in Mongolia and China is in a rain shadow of the Himalaya Mountains, the Atacama Desert in Chile is in a rain shadow of the Andes Mountains, and the Great Basin Desert in the United States of America is created by the rain shadow of the Sierra Nevada and Cascade Mountains.

7.3   Forces and slope flows

Winds blow from high-pressure areas to low-pressure areas. The greater the pressure difference, the stronger the winds. The pressure difference is usually generated by temperature difference. The daytime sea breeze and nighttime land breeze are good examples of the above principle's applications. They are generated by the temperature difference between land and water due to heat capacity: land heats and cools faster than water. Hills and mountains stand up in the vertical direction, which plays different roles in generating a temperature difference between the daytime and nighttime: the slope heats and cools faster than the air. During daytime, sun heats the slope causing the air adjacent to the slope to warm (low pressure), while the air far from the slope stays cool (high pressure) at the same altitude. This heat-induced pressure difference makes airflow move toward the slope and then rise upslope, which is called valley breeze. During nighttime, the slope becomes cold faster than the air because of radiation cooling. The cold air adjacent to the slope is denser and moves downslope driven by horizontal pressure gradient force and gravity force. The cold downslope flow is termed katabatic flow. Strong katabatic flow can be generated over certain terrain features, such as a high-altitude plateau covered with snow or ice (Figure 7.7). These strong katabatic winds are typical in Antarctica and Greenland, where the wind may be extremely strong and gusty near the coasts and less severe in mountain regions.

Katabatic flows occur usually with gentle speeds ($1-3\,\mathrm{m\,s}^{-1}$) in the early morning around mountains and hills, particularly in clear and calm nights. Nearly 70% of Earth's land surface is covered with hills and mountains. Katabatic flows are a persistent feature over the rugged land surfaces. These flows participate in the redistribution of moisture, greenhouse gases, and energy in the atmosphere. Consequently,

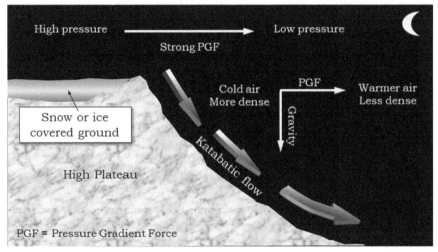

Figure 7.7 Strong katabatic flow can be generated over high-elevated plateau covered with snow or ice. The PGF stands for pressure gradients force.
Credit: Chuixiang Yi.

they affect local weather (Hodur, 1997; Helbig et al., 2021), including fog (Duynkerke, 1999), clouds (Bromwich et al., 2001), rain (Steiner et al., 2003), and snow (Massom et al., 2001). Particularly, the katabatic flows within canopy (drainage flows) persist even in cases where synoptic forcing is strong (Feigenwinter et al., 2010). These drainage flows result in many problems in measuring and modeling land-atmosphere interactions (Helbig et al., 2021), such as serious advection errors in nocturnal eddy flux measurements (Goulden et al., 1996; Yi et al., 2000; Massman and Lee, 2002; Aubinet et al., 2003; Staebler and Fitzjarrald, 2004; Feigenwinter et al., 2008; Yi et al., 2008; Montagnani et al., 2009). Therefore, studying the control mechanisms of katabatic flows within the canopy has many practical implications.

7.4 Governing equations of slope flows

Physically, katabatic flows are associated with many factors, such as slope cooling, adiabatic downslope compression warming, ambient stratification, slope angle, vegetation structure, and other factors (Yi et al., 2005; Haiden and Whiteman, 2005; Froelich and Schmid, 2007; Sun et al., 2007; Schaeffer et al., 2008; Burns et al., 2011). How these factors control katabatic flow developments is complicated and remains unclear. Some studies indicate that katabatic winds are stronger on steep slopes (Mahrt, 1982; Horst and Doran, 1986; Davies et al., 1995; Princevac et al., 2008), while others have demonstrated that katabatic winds are stronger on gentle slopes (Fleagle, 1950; Zhong and Whiteman, 2008; Axelsen and van Dop, 2009).

To better understand basic behaviors of slope flows as a function of slope angle, cooling and warming rates, ambient stratification, and canopy structure, we introduce a single layer model of slope flows. Fleagle (1950) initially conceived of the layer model that considers a parcel being cooled on a slope (Figure 7.8). As the parcel cools,

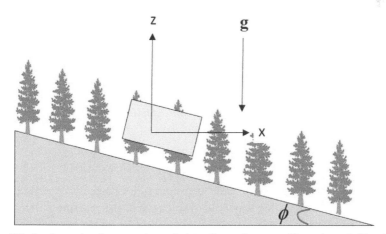

Figure 7.8 A schematic of canopy thermal-slope flows along a uniform slope ϕ. The slope is covered with uniform vegetation. The z direction of the coordinate is opposite to the gravity **g**. Credit: Chuixiang Yi.

it moves downslope due to gravity. As the parcel descends, it warms due to adiabatic compression. Furthermore, McNider (1982) incorporated ambient stratification and Yi (2009) formulated canopy drag force into the model. Hereafter, we call it a canopy thermal-slope model.

Starting from Newton's second law of motion, the parcel of air will be governed by external forces acting on it. We consider that the parcel is driven by three kinds of forces: pressure gradient, gravity, and drag force resulting from the canopy and ground surface, so the change in velocity with time becomes

$$\frac{d\mathbf{v}}{dt} = \frac{1}{m}(\text{pressure gradient} + \text{gravity} + \text{drag}),\tag{7.1}$$

where $\mathbf{v}$ is the vector of parcel velocity and m is the mass of the parcel of air.

Note, we neglect Coriolis force because it is small for the canopy thermal-slope flows.

A word about derivatives

In Eq. (7.1), we introduce $\frac{d\mathbf{v}}{dt}$, which is the derivative of velocity with respect to time. It is important to note that $\mathbf{v}$ is a three component vector: $\mathbf{v} = u\mathbf{i} + v\mathbf{j} + w\mathbf{k}$ with the components u, v, w in x, y, z directions. This means that to fully express Eq. (7.1) we would need three equations each taking a rather long form.

In other equations you will see the notation $\frac{\partial}{\partial x}$ or similar. These are partial derivatives used to represent the parts of the total process that are relevant.

For example, in Eq. (7.11) you see $\frac{\partial \bar{u}}{\partial t}$, which represents the change in the mean u-component over time.

To further simplify the governing Eq. (7.1), we assume that there is no flow ($v = 0$), and no variation ($\partial/\partial y = 0$) in the y direction as in Figure 7.8, resulting in a two-dimensional model. The coordinate system in Figure 7.8 following the parcel is called a Lagrangian coordinate system, while the coordinate system fixed at a point in the space is called an Eulerian coordinate system. We can use an Eulerian coordinate system to define F_{dx} and F_{dz} as the x-component and z-component of drag force $\mathbf{F}_d$.

Next, we will consider the *Boussinesq approximation*, which assumes that variations of air density are small,

$$\rho = \bar{\rho} + \rho',\tag{7.2}$$

here $\bar{\rho}$ is mean air density and ρ' is a fluctuation (see Chapter 2 for a discussion of representing variables in this way). We also can split the pressure P into mean $\bar{P}$ and fluctuation p' parts,

$$P = \bar{P} + p'.\tag{7.3}$$

The mean pressure $\overline{P}$ and density $\overline{\rho}$ satisfy the hydrostatic equation

$$\frac{d\overline{P}}{dz} = -\overline{\rho}g = -\frac{g}{\overline{\alpha}}.$$ (7.4)

and we can then define our moving parcel in terms of gravity and α, the specific volume of air. The Boussinesq approximation (Boussinesq, 1877) considers that density variations are only important in the term multiplied by gravity.

To be convenient for description of slope flow, we rotate the coordinate system to align the z direction normal to the slope (Figure 7.9). In the new rotated coordinate system, we define $\widetilde{u}$ as the velocity component in the $\widetilde{x}$ coordinate direction and $\widetilde{w}$ as the velocity component in the $\widetilde{z}$ coordinate direction. Combining all the above, the full governing equations become

$$\begin{pmatrix} \dfrac{d\widetilde{u}}{dt} \\ \dfrac{d\widetilde{w}}{dt} \end{pmatrix} = \begin{pmatrix} \cos\phi & -\sin\phi \\ \sin\phi & \cos\phi \end{pmatrix} \begin{pmatrix} F_{dx} \\ \dfrac{\alpha'}{\overline{\alpha}}g + F_{dz} \end{pmatrix}.$$ (7.5)

Our two-dimensional canopy thermal slope model is taking shape now. We have velocity governing equations in good condition, but what do we do about temperature? McNider (1982) incorporated ambient stratification into the layer parcel model. He assumed that the parcel potential temperature varies due to the local warming rate L_c for the parcel. The ambient stratification is sketched in Figure 7.10 and described by

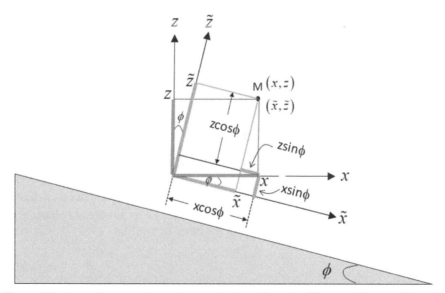

Figure 7.9 A coordinate transformation rotating the z direction. We start with z parallel to the gravity **g**, and finish with z normal to the slope.
Credit: Chuixiang Yi.

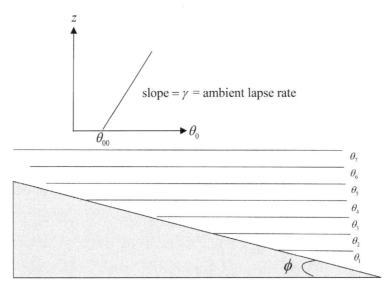

Figure 7.10 The uniform distribution of ambient potential temperature.
Credit: Chuixiang Yi.

$$\theta_0(z) = \theta_{00} + \gamma z, \tag{7.6}$$

where γ is the ambient lapse rate and θ_{00} is the potential temperature at $z = 0$. McNider (1982) assumed that variations in the parcel potential temperature, θ, along the slope are all due to the constant ambient lapse rate. For a uniform slope, that means $\partial/\partial \tilde{x} = 0$. Thus, the continuity equation for an incompressible flow simplifies the governing-energy Eq. (7.5) to,

$$\frac{\partial \tilde{u}}{\partial t} = \tilde{F}_{dx} - g\frac{\alpha'}{\overline{\alpha}} \sin\phi, \tag{7.7}$$

$$\frac{\partial \theta}{\partial t} = \tilde{u}\gamma \sin\phi + L_c. \tag{7.8}$$

Next, we need to translate $\alpha'/\overline{\alpha}$, (changes in volume) into temperature variations. Replacing each variable in the ideal gas equation with their mean and fluctuation components, and assuming that in most boundary layer situations, the pressure term, $p'/\overline{P}$, is an order of magnitude less than that of the other terms (Stull, 1988) we come to

$$\frac{\alpha'}{\overline{\alpha}} \approx \frac{T'}{\overline{T}} \approx \frac{\theta'}{\theta_0}, \tag{7.9}$$

Then Eq. (7.7) becomes

$$\frac{\partial \widetilde{u}}{\partial t} = \widetilde{F}_{dx} - g\frac{(\theta - \theta_0)}{\theta_0}\sin\phi, \tag{7.10}$$

and we now have defined motion in terms of drag, potential temperature, and slope angle.

We can add vegetation in as well. Yi (2009) assumed that Fleagle-McNider's drainage parcel concept is valid within a canopy that has a uniform leaf area density, a, and drag coefficient, c_D, based on the data measured at Niwot Ridge AmeriFlux site (Yi et al., 2005). Although the data indicate wind speed and potential temperature were relatively uniform in a drainage layer within canopy (Figure 8 in Yi et al., 2005), for generality, we use parcel-layer average of velocity $\overline{u}$ and potential temperature $\overline{\theta}$ instead of $\widetilde{u}$ and θ in the governing equations:

$$\frac{\partial \overline{u}}{\partial t} = -g\frac{(\overline{\theta} - \theta_0)}{\theta_0}\sin\phi - c_D a\overline{u}|\overline{u}|, \tag{7.11}$$

$$\frac{\partial \overline{\theta}}{\partial t} = \overline{u}\gamma\sin\phi + \overline{L}_c, \tag{7.12}$$

$\overline{L}_c$ is the average heat flux from the canopy layer. The second term in Eq. (7.11) represents a canopy drag (Yi, 2008; De Roo and Banerjeea, 2018),

$$\overline{F}_{dx} = -c_D a\overline{u}|\overline{u}|, \tag{7.13}$$

the minus symbol indicates that the drag direction is always opposite to the wind direction. Although we attempt to include vegetation effect on drainage flows, we want to clarify that this model is also valid for a bare slope. For a bare slope, it represents a ground surface drag ($\widetilde{F}_{dx} = -k''\overline{u}|\overline{u}|$). In both the models of Fleagle (1950) and McNider (1982), they assumed that the surface friction is proportional to wind speed, i.e., $\overline{F}_{dx} = -k\overline{u}$. In the present model, we prefer to use the velocity-squared law to represent land surface drag. Taylor (1916) was the first person to use $\widetilde{F}_{dx} = -k''\overline{u}|\overline{u}|$ to describe Earth's surface friction. Since Taylor's investigation, $\widetilde{F}_{dx} = -k''\overline{u}|\overline{u}|$ has been tested and widely used (e.g., Deacon, 1949; Sutcliffe, 1936; Sutton, 1953). Therefore, Eq. (7.13) will be qualitatively valid in both cases with and without canopy cover on the slope. Without canopy, the $c_D a$ can be attributed to k'' that represents the property of rough surface to cause the surface drag for the fluid passing over it (Sutton, 1953).

In the canopy thermal-slope flow model Eqs. (7.11) and (7.12), canopy drag and buoyancy are two major forces to determine the basic behavior of the parcel's downslope motion. This is illustrated in Figure 7.11. Under cold inflow condition (Figure 7.11A), the direction of the buoyancy force $-g\sin\phi(\overline{\theta} - \theta_0)/\theta_0$ is

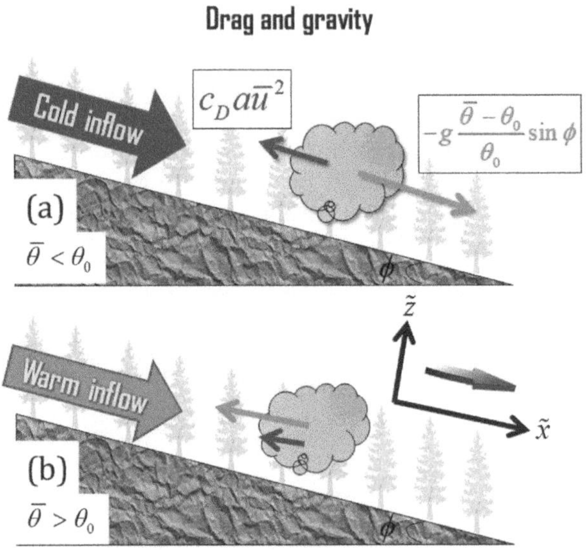

Drag and gravity

Figure 7.11 The illustration of force directions of gravity (buoyancy) and drag exerted on the parcel of air by plants: (A) cold inflows, and (B) warm inflows. The *3D arrow* in the coordinate indicates parcel movement direction (downslope). The *blue arrow* indicates drag force with the magnitude of $c_D a \bar{u}^2$ and direction that is always opposite to the movement direction. The *red arrow* represents gravity (buoyancy) caused by the difference between parcel temperature $\bar{\theta}$ and ambient temperature θ_0. The direction of buoyancy forcing is downslope under cold canopy flow conditions and upslope under warm canopy flow conditions. Credit: Chuixiang Yi modified Figure 3 of his paper (Yi, 2009). Licensed.

downslope because the parcel temperature $\bar{\theta}$ is lower than environmental temperature θ_0, and its magnitude is constant and much larger than canopy drag $-c_D a \bar{u}^2$ initially because the inflow speed $\bar{u}$ is slow, hence the parcel speed is accelerated. As $\bar{u}$ increases, adiabatic compression warming is enhanced through the term $\bar{u}\gamma \sin \phi$ in Eq. (7.12). Thus, as the parcel accelerates, the canopy drag force becomes larger and buoyancy becomes smaller. Eventually, the drag is increased to equal the buoyancy force in magnitude and opposite in direction, and then canopy flows move downslope at a constant speed $\bar{u}_s$, i.e., steady state.

Under warm inflow condition (Figure 7.11B), the buoyancy force is persistently upslope. Given downslope inflows in Figure 7.11B, both drag and buoyancy forces are initially upslope and result in deceleration of parcel speed. Eventually, canopy flows will stop moving downslope at some point and begin moving upslope due to the net force. This net force acts on canopy flows in the upslope direction, which is why the downslope inflow is able to result in upslope flows over the lower part of the slope under the warm canopy-flow condition. The cooperation and competition between canopy drag and buoyancy forces might cause oscillation region of canopy thermal-slope flows. McNider (1982) obtained the exact oscillation solution from the governing equation with a simple friction force ($\overline{F}_{dx} = -k'\overline{u}$). However, the oscillation regime of canopy flows has not been confirmed by rigorous observations. We believe that the oscillation regime of canopy thermal-slope flows is more likely to occur during the transition time between day and night. The importance of the transition time between day and night was emphasized for the CO_2 transport over complex terrain by Sun et al. (2007).

7.4.1 Maximum speed of katabatic flows

In this section, we examine the steady-state behavior of katabatic (downslope) flows that are governed by the balance between buoyancy and drag forces. At the beginning, drag force is smaller than the buoyancy force because wind speed is low, accelerating the parcels. A steady state is eventually reached when these two forces are exactly balanced by strengthening of the drag force due to the acceleration and weakening of the downward buoyancy force due to warming from an adiabatic compression.

Mathematically, the steady states of katabatic flows can be determined by setting $\partial \overline{u}_s / \partial t = 0$ and $\partial \overline{\theta}_s / \partial t = 0$, yielding

$$0 = -\Delta\theta_s \frac{g \sin \phi}{\theta_0} - c_D a \overline{u}_s^2, \tag{7.14}$$

$$0 = \overline{u}_s \gamma \sin \phi - R_c \left(\frac{\overline{\theta}_s - \overline{\theta}_c}{\overline{\theta}_0 - \theta_c} \right), \tag{7.15}$$

where $\Delta\theta_s = \overline{\theta}_s - \theta_0$ is the temperature difference between the parcel and the surroundings and is called potential temperature deficit. The subscript, s, refers to steady states. Eqs. (7.14) and (7.15) are algebraic equations that are easily solved. The value of steady-state wind speed of katabatic flows is determined by

$$\overline{u}_s = \left(-1 + \sqrt{1 + \eta \sin^{-3}\alpha} \right) \eta_0 \sin^2\alpha, \tag{7.16}$$

where

$$\eta = \frac{4 c_D a R_c^2 \theta_0}{\gamma^2 g (\theta_0 - \theta_c)}, \tag{7.17}$$

and

$$\eta_0 = \frac{\gamma g (\theta_0 - \theta_c)}{2 c_D a R_c \theta_0}. \tag{7.18}$$

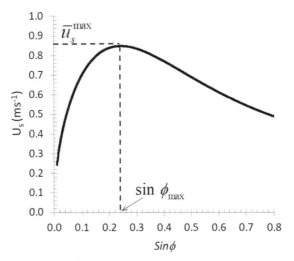

Figure 7.12 Wind speed of katabatic flows versus the *sin* of slope angle ϕ. The wind speed are the values of steady-state of katabatic flows predicted by Eq. (7.16) with the values of model parameters: $g=9.8\,\mathrm{m\,s^{-2}}$, $\gamma=2\times10^{-3}\,\mathrm{K\,m^{-1}}$, $R_c=1.122\times10^{-3}\,\mathrm{K\,s^{-1}}$, $(\theta_0-\theta_c)/\theta_0=0.1$, $c_D=0.15$, and $a=0.5\,\mathrm{m^{-1}}$. Details of derivation can be found in Chen and Yi (2012). Credit: Chuixiang Yi.

Katabatic flows attain a maximum speed at a certain value of $\sin\phi$ as the other parameters are fixed, as shown in Figure 7.12. The maximum katabatic wind speed is located at a specific slope ϕ_m, which is opposite to the view that a deeper slope would produce stronger katabatic flows. Let us examine the phenomenon in detail to get to the truth of the matter. The relationship between katabatic flow speed and slope angle is

Thus, the optimal conditions of maximum katabatic flows are

$$l_c(V_T)^{-2}\sin^3\phi_m = b, \tag{7.19}$$

where $l_c=(c_d a)^{-1}$ is a so-called canopy length scale (Belcher et al., 2003), $V_T=R_c/\gamma$ the thermal factor that is defined by the ratio of parcel cooling rate R_c to ambient lapse rate γ, and $b=(2g)^{-1}[\theta_0/(\theta_0-\theta_c)]$ taken as a constant with fixation of $\theta_0/(\theta_0-\theta_c)$ at 10 following Davis and McNider (1997). Using l_c and V_T, the maximum katabatic flows ($\bar{u}_{sm}$) can be expressed as a function of thermal factors and the canopy length scale,

$$\bar{u}_s^{max} = g'(l_c V_T)^{\frac{1}{3}}, \tag{7.20}$$

where $g' = \sqrt[3]{0.25g(\theta_0-\theta_c)/\theta_0}$.

What this tells us is the position ($\sin\phi_m$) of maximum katabatic flows can be shifted from a small slope angle to a large slope angle by increasing the thermal factor (V_T) or

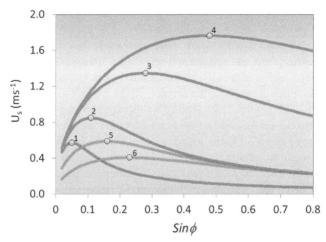

Figure 7.13 Wind speed of katabatic flows versus the *sin* of slope angle ϕ under different conditions of thermal control V_T and canopy control l_c. The wind speeds are the values of steady-state katabatic flows predicted by Eq. (7.16). The values of model parameters: (1) $g = 9.8 \, \text{m s}^{-2}$ and $(\theta_0 - \theta_c)/\theta_0 = 0.1$ are the same for all six curves; (2) $l_c = 13.333$ is the same for the curves 1, 2, 3, and 4, but V_T takes values of 0.0575, 0.187, 0.561, and 1.683 for them respectively; and (3) $V_T = 0.187$ is the same for curves 5 and 6, but l_c is 4.444 and 1.4815, respectively. Details of the derivation can be found in Chen and Yi (2012).
Credit: Chuixiang Yi modified Figure 1 from Chen and Yi (2012). Licensed.

decreasing the canopy length scale (l_c). For cases with a similar canopy-length scale (l_c = constant), a weak cooling rate or a strong stratification favors the occurrence of maximum katabatic flows on a gentle slope (see examples 1 and 2 of $\bar{u}_s^{\max}$ in Figure 7.13), while a stronger cooling rate or a weaker stratification favors the occurrence of maximum katabatic flows on a steep slope (see examples 3 and 4 of $\bar{u}_s^{\max}$ in Figure 7.13). For cases with a similar thermal condition (V_T = constant), although decreasing the canopy length scale causes the position of the maximum katabatic flows to shift toward a larger slope angle (see examples 5 and 6 of $\bar{u}_s^{\max}$ in Figure 7.13), canopy control (l_c) on maximum katabatic flows is weaker than thermal control (V_T), in comparison with examples 2, 5, and 6 with 1–4 of maximum katabatic flows shown in Figure 7.13. Canopy effects on katabatic flows become diminished with increasing slope angles and can be ignored when the slope angle is greater than 30 degrees (Figure 7.13). Regardless of whether it is controlled by thermal canopy factors, katabatic wind speed is more sensitive to slope angle on a smaller slope angle than on a large slope angle (Figure 7.13).

Katabatic flows are not governed by slope angle (gravity) alone but are controlled synergistically with slope cooling, ambient stratification, and vegetation structure. The existence of maximum katabatic flow can be understood mainly as a result of competition between gravity and buoyancy. An increased slope causes the increasing of the gravitational acceleration on the drainage parcel. Meanwhile, the increased velocity of the drainage parcel will be warmed by adiabatic compression heating. Consequently, the potential temperature deficit ($\Delta\theta = \bar{\theta} - \theta_0$) is reduced and hence the buoyancy term becomes smaller. Thus, there is a point in the mid-range of slope where maximum katabatic flow can approach when all the other parameters, such as, lapse rate, cooling rate, leaf area density, and drag coefficient (or surface drag coefficient, k'', for the case without canopy), are fixed.

The condition of maximum katabatic flow production can be used to resolve the paradox that maximum katabatic flows can occur on either steep or gentle slopes. As illustrated in Figure 7.13, the location of maximum katabatic flows shifts from a small slope angle to a large slope angle with increasing thermal factor $V_T = R_c/\gamma$. An increase in the thermal factor results from either increasing of the cooling rate, R_c, or decreasing of the lapse rate, γ. However, the thermal factor is more sensitive to variation of lapse rate than those of cooling rate because nighttime cooling rate is proportional to $(273 + T)^4$ from the Stefan-Boltzmann law (T is temperature in Celsius). To a good approximation by fixing the cooling rate, the values of the thermal factor are inversely proportional to lapse rate, which are lower with strong stratification and higher with weak stratification. Thus, Eq. (7.19) predicts that under strong stable stratification maximum katabatic flows occur on gentle slopes, while under weak stratification maximum katabatic flows occur on steep slopes. These theoretical predictions provide insights into why some studies reported the occurrence of strong katabatic flows on gentle slopes, while others indicated the occurrence of maximum katabatic flows on steep slopes. It can be verified that all the ambient stratifications were stronger in the reported cases of occurrence of maximum katabatic flows on gentle slopes (e.g., Zhong and Whiteman, 2008; Axelsen and van Dop, 2009), and were weak in the reported cases of occurrence of maximum katabatic flows on steep slopes (e.g., Davies et al., 1995; Horst and Doran, 1986; Princevac et al., 2008). The optimal condition indicates that gentle slopes are optimal for katabatic flow development under highly stably stratified air and steep slopes are optimal under weak or near neutral stratification (Chen and Yi, 2012). Canopy effect is significant in the control of katabatic flows only on gentle slopes. The instability analysis of the steady states of katabatic flows can be found in De Roo and Banerjea (2018) and Yi (2009).

7.4.2 Super-stable layer

During a calm and clear nighttime, katabatic flows within canopy persistently exist (Figure 7.14). Due to the nature of the katabatic flows, they pick up high CO_2 concentration air respired from the soils and drain it downslope, hence the term "drainage flow." Eddy covariance tower systems discussed in Chapter 4 cannot properly account for the high CO_2 air carried away by drainage flows (Feigenwinter et al., 2010; Yi et al., 2008). The significant uncertainties in eddy-covariance measurements caused by drainage flows have been a long-standing problem amongst eddy-flux tower networks (Baldocchi, 2003; Finnigan, 2008). In this section, we explore underlying reasons for the puzzle by a stability analysis of the air within canopy in terms of using a simple Richardson number.

Figure 7.14 A schematic of a super-stable layer within canopy. This super-stable layer is defined by Richardson number Ri (Yi et al., 2005).
Credit: Chuixiang Yi and Jielun Sun.

The gradient Richardson number (Salcido et al., 2020) is defined as:

$$Ri = \frac{\frac{g}{\theta}\frac{\partial \bar{\theta}}{\partial z}}{\left(\frac{\partial \bar{u}}{\partial z}\right)^2}. \tag{7.21}$$

This number is used to study the conditions for transitions between laminar flow and turbulent flow:

Laminar flow becomes turbulent when $Ri < R_c$.
Turbulent flow becomes laminar when $Ri > R_T$.

Theoretical and laboratory research suggest that the value of R_c is between 0.21 and 0.25 and $R_T = 1.0$.

Before we discuss the Richardson number in detail, let us examine the typical profiles of wind speed and temperature within canopy. Many observational studies have shown that wind profile within forest canopy is "S"-shaped (i.e., Shaw, 1977; Denmead and Bradley, 1985; Yi, 2008). The S-shaped profile (Figure 7.14) refers to a secondary wind maximum that is often observed within the trunk space of forests and a secondary minimum wind speed in the region of greatest foliage density. The second wind maximum is not only related to the vegetation structure (Massman, 1997; Massman and Weil, 1999; Poggi et al., 2004) but also to terrain slope and nocturnal radiative cooling of the slope surface (Fleagle, 1950; Manins and Sawford, 1979; Doran and Horst, 1981; Mahrt, 1982). The variation of temperature is less in the lower part of canopy but decreases quickly in the upper part of canopy (Figure 7.14). At the level of greatest foliage density, wind speed is minimum $\partial \overline{u}/\partial z = 0$, and temperature gradient is significant $\partial \overline{\theta}/\partial z \neq 0$, thus the Richardson number tends toward infinity. This means that the layer is very stable and is called the super-stable layer (Yi et al., 2005; Xu et al., 2015).

The super-stable layer acts as a "lid" that minimizes vertical exchange between lower and upper canopy layers, essentially uncoupling the nighttime near-surface layer from the layers near the top of the canopy (Figure 7.14). The super-stable layer is characterized by: (1) greatest leaf area density, (2) maximum canopy drag resistance, (3) minimum wind speed, (4) minimum vertical exchanges, and (5) maximum horizontal CO_2 gradients. The observational data indicate the maximum horizontal CO_2 gradients are highest during the nighttime in the summer (Figure 7.15). The super-stable layer divides the canopy layer into two layers: (1) below the super-stable layer that is dominated by horizontal exchanges of CO_2, water vapor and heat, and (2) above the super-stable layer is dominated by vertical exchanges with the atmosphere. The sensors on the tower above the canopy are insensitive to what occurs

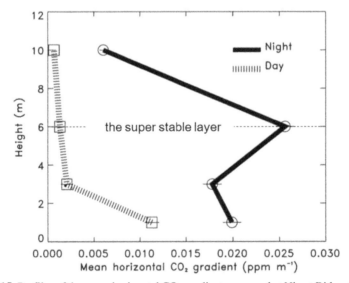

Figure 7.15 Profiles of the mean horizontal CO_2 gradient measured at Niwot Ridge AmeriFlux site in Rocky Mountain in summer. The number of 30-min averaging periods used in each profile is 1,709 for the night, and 1,001 for the day. The error bars indicate ± 1 S.D. of the mean (Yi et al., 2008).

Credit: Chuixiang Yi modified Figure 2 from Yi et al. (2008). Licensed.

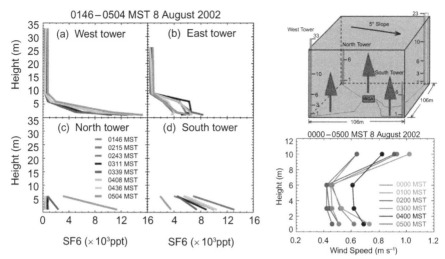

Figure 7.16 Half hour average of SF_6 concentrations at four towers during the period 0146-0504 MST on August 8, 2002. SF_6 was released continuously during the observation period from a 200-m line source that was located 0.2 m above the ground surface. The line source was located 50 m upslope from the USGS tower, 125 m upslope from the North Tower and South Tower, and 200 m upslope from the CU tower. During the observation period we observed the typical "S"-shaped wind profile. Details can be found in Yi et al. (2008). Credit: Chuixiang Yi modified Figures 9 and 10 from Yi et al. (2005). Licensed.

below the super-stable layer because the air above and below the super-stable layer are almost isolated by the super-stable layer (Figure 7.14).

The existence of the super-stable layer was observed by SF_6 experiments. As shown in Figure 7.16, a sulfur hexafluoride (SF_6) tracer was released continuously at a steady rate from a point source located at 0.2 m above the forest floor approximately 50 m upslope from the USGS tower in 2001. In 2002, SF_6 was released from a 200 m line source deployed cross-slope at 0.2 m height and 50 m upslope from the USGS tower in Niwot Ridge AmeriFlux site. A fast response, continuous SF_6 analyzer was operated with the existing CO_2 ambient profiler system to measure SF_6 concentrations at 14 locations on towers (CU, USGS, ST, NT) in 2001 and 2002. The profiler sequentially samples for 30 s at each position and the measured concentrations are combined into 30-min averages. During a 3.5-h observation period on the night of August 8, 2002 (Figure 7.16), when atmospheric stability was relatively high, drainage flows were strong, and the SF6 release line and tower-mounted sensors were well calibrated and working at optimal performance, we made observations of vertical dispersion at all four observation towers (Figure 7.16). There was clear, sharp stratification among even the lowest observation layers on the USGS tower only 50 m downslope from the release line, indicating the presence of a thin downslope flow with little upward mixing. By the time the tracer-laden air reached the lower towers, there was evidence of some vertical mixing to the 6-m layer, but essentially none in the layers above 8 m, even at the lowermost tower 200 m downslope from the release line. We interpret this as evidence of a thin "slab" of drainage flow, carrying tracer downslope, with little vertical dispersion above the canopy layer with maximal leaf area density. The existence of a super-stable layer has also been confirmed by different datasets (e.g., stable carbon isotope data; Schaeffer et al., 2008).

7.5   Mountain wind patterns

In Section 7.2 we said that when prevailing winds meet a mountain range, the air is usually forced to rise over the terrain. This is usually true, but not always. The path an incoming air parcel will take upon encountering a mountain depends on the wind speed, the wind direction, and the potential temperature of the air. In the earth's atmosphere, temperature is usually the most dominant factor in determining whether the air parcel is more or less dense than the surrounding atmosphere, since warm air is often significantly less dense than cold air. As a result, an air parcel's condition is measured by referring to its **thermal stability**. This is simply a comparison of the strength of the parcel's lapse rate: how quickly does the temperature increase (or decrease) as altitude increases? An air parcel with a potential temperature that increases with height – a positive vertical potential temperature gradient – is called "stable." An air parcel with a negative vertical potential temperature gradient is called "unstable," and air with potential temperature that is constant with height is "neutral."

If the wind speed and the vertical potential temperature gradient of an air parcel are known, and the mountain geography is known, the path the air will take upon reaching the mountain can be predicted. The simplest cases are where the air parcel is either strongly stable or strongly unstable. In a stable atmosphere, where the vertical potential temperature gradient is large and positive, the air tends to flow around the obstruction horizontally, focusing through a pass or along a valley. Weakly stable air will be deflected vertically in a wave pattern called **lee waves**, rising and falling at a predictable frequency after passing the mountaintop (see Figure 7.17).

Lee waves form in stable conditions, when an air parcel meets an obstruction and is deflected upwards into warmer air. The cold air rebounds back downwards where it meets colder air, causing it to deflect back upwards since the buoyancy of the parcel overcomes both the gravitational pull and the momentum of the first deflection. This repeating cycle of deflection causes a wave pattern leeward of the obstruction: lee waves. The Brunt-Väisälä equation describes the frequency of the lee waves:

$$f = \frac{1}{2\pi} \sqrt{\frac{g}{\theta} \frac{d\theta}{dz}}, \tag{7.22}$$

with acceleration due to gravity g at $9.8\,\mathrm{m\,s^{-2}}$. Since potential temperature cannot be negative, the square root is a real number when the potential temperature gradient is positive, i.e., in a stable atmosphere. If the potential temperature gradient is negative

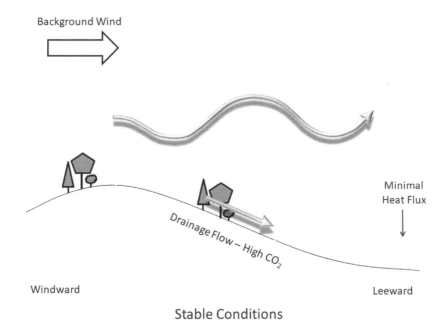

Figure 7.17 Lee waves as air moves past a slope under stable conditions.
Credit: Eric Kutter.

(unstable atmosphere), the equation returns a complex number for the frequency, suggesting a circle or a spiral pattern instead of a wave.

In unstable conditions, the air parcel deflects upwards on the windward side of the mountain and causes a spinning recirculation pattern leeward of the mountain. Recirculation will be discussed in greater detail in Section 7.5.

Imagine an air parcel reaching the foot of a mountain. If the atmosphere is unstable, the air parcel is *less dense* than the air above it, priming the air for an upwards journey. As that air parcel is deflected up by the slope, the air above it is even denser in comparison to the air parcel, further increasing the parcel's buoyancy as it continues up, without appreciably changing the gravitational force pulling the air parcel down. If that air parcel is in a stable atmosphere when it reaches the foot of the mountain, it is *more dense* than the air above it, making it much more likely to deflect sideways around the mountain instead of up into progressively less and less dense air, simply because the buoyancy force pushing the air parcel upwards is significantly less than the gravitational force pulling the air parcel downwards.

In neutral conditions (often assumed to be a potential temperature gradient of less than a magnitude of 1 K per 100 m vertically), the path the air parcel takes from the foot of a mountain is a complex, heterogeneous combination of the stable and unstable behaviors that can be difficult to predict. When the potential temperature of the air is constant with height, a difference in buoyancy is not the controlling factor that it was in stable or unstable atmospheres. Instead, the momentum (i.e., wind speed) of the incoming air parcel and the exact geometry of the slope and other topographic features determine how closely the air parcel adheres to the stable versus unstable behavior.

The importance of wind speed will be no surprise to those students who are familiar with the Pasquill Stability Classes (see Chapter 10). Qualitatively, a light breeze in a very stable atmosphere encountering a steep slope is likely to produce air flow patterns that horizontally deflect around the obstacle, while a strong wind, unstable atmosphere, hitting a gentle slope will result in upwards deflection of the air parcel, with flow separation and recirculation leeward of the hill. Quantitatively, a meteorological version of the Froude number is used to predict wind flow patterns considering thermal stability, obstacle size, and wind speed. This modified Froude number can be thought of as the ratio of momentum forces divided by the buoyant forces acting on an air parcel. When wind impacts an obstacle of height h, we can determine the modified Froude number as:

$$Fr = \frac{\overline{u}}{fh},$$ (7.23)

where f is the Brunt-Väisälä frequency (Salcido et al., 2020) from Eq. (7.22), and $\overline{u}$ is the average wind speed horizontally along the incoming wind flow path. When the Froude number is much less than 1.0, indicating a combination of light winds, stable air, and a tall obstacle, the air often deflects horizontally around the object, rather than rising over it (see Figure 7.18). In cases where the Froude number is much larger

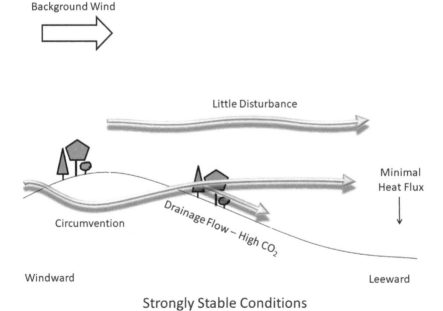

Figure 7.18 Wind circumventing a slope during strongly stable conditions, with a Froude number **much** less than 1.0.
Credit: Eric Kutter.

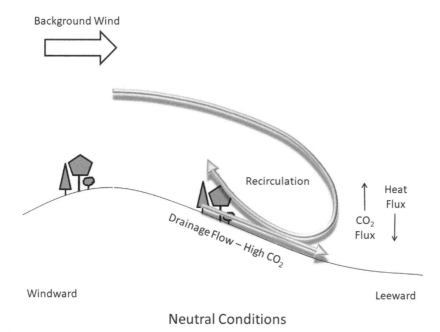

Figure 7.19 Flow for Froude number much larger than 1.0, with partial recirculation mixed with drainage flow, causing a separation between below canopy and above canopy flow. Credit: Eric Kutter.

than 1.0, such as during strong winds and near-neutral air impinging on a short hill, the air parcel will deflect upwards over the obstacle. Leeward of the hill, flow separation and partial recirculation can occur with large Froude numbers (see Figure 7.19). For a critical Froude number near 1.0, the atmospheric conditions and the terrestrial geometry reach a resonance that causes lee waves to form leeward of the obstacle. Figure 7.17 shows a sketch of wind flow patterns for a Froude number less than 1.0.

The modified Froude number can also be used as a rough approximation of the height of the air column that is disturbed by lee waves. Since our Froude number is a ratio of momentum forces and buoyancy forces, the ratio is also approximately equal to the amount of air that flows over the top of the slope (momentum-controlled) divided by the amount of air that flows around the hill instead (buoyancy-controlled):

$$Fr = \frac{z_{\text{lee}}}{h}, \tag{7.24}$$

where z_{lee} is the initial amplitude of the lee wave, and h is the obstruction height.

For unstable atmospheres the Brunt-Vaisala frequency turns complex, and Eq. (7.22) inappropriate. The pressure drop leeward of the slope causes full recirculation during unstable conditions, with increased vertical mixing that can last for many hours.

7.6   Recirculation

A recirculation zone is a microclimate-scale spinning vortex aligned vertically, rather than the horizontal vortices regularly seen on synoptic scales. While recirculation regularly occurs leeward of an abrupt change in topography, the phenomenon can also be caused by gentle slopes under certain atmospheric conditions. Recirculation is associated with vertical shear and increased turbulence, sometimes causing ground-level wind flow that is opposite in direction to the synoptic flow. This air circulation influences vertical mixing and the fluxes of energy and trace gases, thereby impacting the regional climate and the propagation of air pollution.

When wind passes an obstruction, fluid dynamics predicts a pressure drop on the far surface of the obstacle, in atmospheric terms leeward of the incoming wind. Air will tend to flow from higher pressure regions toward lower pressure regions. Because the air parcel was in motion it has some momentum, resulting in a curved deflection downwards around the pressure drop. It is difficult to take a photograph of this phenomenon in action in the atmosphere. However, accepting differences in boundary conditions and the direction of gravity, Figure 7.20 shows the principle of recirculation in water.

In a stable or near-neutral atmosphere that curved deflection could rebound upwards and repeat as lee waves. However, in an unstable atmosphere denser cold air is located above less dense warm air. Since the buoyancy is not sufficient, the air parcel does not rebound upwards. Instead, it continues to curve around the pressure drop until reaching the slope surface and turning upslope, opposite to the synoptic wind direction. If the atmosphere is strongly unstable, the recirculation vortex will overwhelm the downslope katabatic flow as well (see Figure 7.21). This condition

Figure 7.20 Photograph of river sediment eddies showing recirculation in Eddington River, Lake Te Anau, NZ.
Photo credit: Jordy Hendrikx (National Institute of Water and Atmospheric Research).

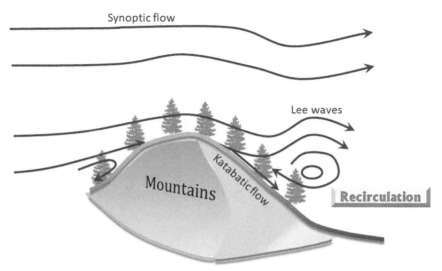

Figure 7.21 Schematic of recirculation in near-neutral conditions, with a Froude number larger than 1.0, but not approaching infinity.
Credit: Chuixiang Yi.

is relatively common during daytime on sloped surfaces and has been observed to last for many hours.

A neutral atmosphere causes a complex hybrid of stable and unstable behaviors. The recirculation is not strong enough in neutral conditions to overwhelm the katabatic downslope flow. The result is a recirculation vortex that turns upslope at a certain transition height above the slope surface, dependent on many factors, notably including wind speed, topography, and the density of the cold drainage flow. Beneath the transition height, the katabatic flow continues down the slope. Note that this situation causes a disconnect between the atmosphere above the transition height and the atmosphere below, separated by a **super-stable layer** of air with zero horizontal or vertical motion.

Recirculation is studied because of its influence on the exchange of energy and matter between the atmosphere and the terrestrial biosphere. A vertical vortex spinning around in the lee of a mountain causes the atmosphere to be well mixed vertically in terms of trace gases, such as carbon dioxide and water vapor. A recirculation vortex transfers energy upwards, as the warm near-surface air is exchanged with the denser, colder air aloft (Figure 7.22).

Quantitatively, atmospheric fluxes can be measured using the eddy covariance method described in Chapter 4. The technique begins from the mass conservation equation, imagining a three-dimensional chunk of the atmosphere and following the motion of carbon dioxide. The difficulty of measuring all those parameters at once necessitates some assumptions to simplify the technique into a practical tool for real-world scenarios. Firstly, on relatively short time periods (an hour or less), the carbon dioxide concentration does not change rapidly in the atmosphere, so the first term is assumed to be zero. Over flat terrain with no vegetation or structures, the carbon

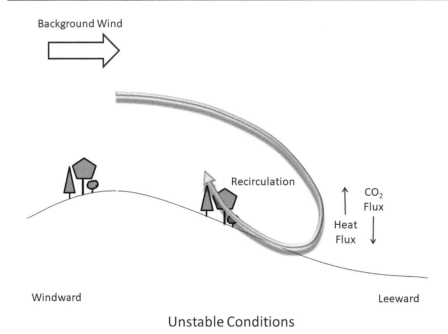

Figure 7.22 Schematic recirculation leeward of a hill in unstable atmospheric conditions. Credit: Eric Kutter.

dioxide would be uniformly distributed horizontally and there would be no average vertical wind speed, so several of the advection terms are also zero. There would also be no horizontal flux divergence in the case of uniform, slopeless terrain. The simplified mass balance tells us that the vertical flux divergence is equal to the source/sink term. Making it very convenient for researchers to measure key parameters directly, including the net ecosystem exchange of carbon dioxide.

However, using the popular eddy covariance method to study sloped regions causes mountainous problems because of the assumptions made to simplify the technique. On a slope, the horizontal and vertical advection terms are not zero, and the horizontal flux divergence terms may also be non-zero.

Taking advection as an example, the horizontal advective flux of carbon dioxide is:

$$F^c_{adv,h} = \frac{1}{V_m} \int_0^h \bar{u} \frac{\partial \bar{c}}{\partial x} dz, \tag{7.25}$$

where V_m is the molar volume of dry air and h is the reference height. We can write a similar equation for vertical advective energy flux:

$$F^\theta_{adv,v} = \rho C_p \int_0^h \bar{w} \frac{\partial \bar{\theta}}{\partial z} dz. \tag{7.26}$$

In order to calculate the flux, we need to know the horizontal wind speed and the carbon dioxide gradient at multiple heights and multiple locations along the flow path. That gradient is difficult to determine if the measurement path is through recirculating air. The gradient must be measured across a relatively small dx, so the researcher can be sure that the wind measured at one location on the slope is the same parcel that then passed the second location on the slope. The added complexity required to use eddy covariance over sloped terrain partially undermines its practicality.

Qualitatively, the critical point is that advective flux is the movement of energy or mass due to the mean air flow. Simply put, if the temperature of the air parcel changes along the flowpath, then the wind is advecting energy from one location to another. In order for a full understanding of an environment to be reached, that advective flux must be accounted for. The same is true of advective flux of a scalar: if the carbon dioxide concentration (or moisture content!) of the air parcel changes along the flowpath either horizontally or vertically, then the wind is causing an advective flux.

With that in mind, you can see conceptually why the equations for advective flux have that form. They include integrations throughout the air column, since we are studying an air parcel, not a single molecule. Within the integral the equations have a wind speed (vertical or horizontal, as appropriate), multiplied by the change in the key parameter over the flowpath being studied. Vertical wind speed times the vertical potential temperature gradient is a measure of how much energy the wind has carried upwards or downwards using our air parcel. The final factor is the conversion term before the integral. For vertical energy flux, the units we receive through the integration would be $K\,m\,s^{-1}$ – a clumsy unit system for critical thinking. Instead, since sunlight is often measured in terms of $W\,m^{-2}$, we use the air density and the air heat capacity to convert those units into $W\,m^{-2}$. Given Eq. (7.25) for horizontal advective flux of carbon dioxide, can you explain why the term V_m^{-1} is included? Can you write the equation for horizontal advective energy flux?

Gu et al. (1999) modified the procedure set forth by Budyko (1958) to determine the degree of error in energy flux measurements in terms of an energy flux balance ratio (EBR):

$$EBR = \frac{F_H + \lambda E}{R_n - G},$$
(7.27)

where R_n is the net radiation, G is ground heat flux, F_H is sensible heat flux, and λE is latent heat flux, all units in $W\,m^{-2}$. Looking at the equation, you can see that the denominator is direct sources and sinks of energy into and out of our air parcel, while the numerator is what effects these energy sources have on the environment in the air parcel. In real-world scenarios, measurements yield a balanced EBR very close to 1.0 only for flat, homogeneous terrain.

Measurements from any environment with a slope, with structures, or mixed vegetation will return an unbalanced EBR. In the hills and mountains of this chapter, Eq. (7.27) is not complete because the starting equations are not complete (assumptions were made to simplify them). In complex terrain, the exchange of energy and trace gases between the atmosphere and the terrestrial biosphere is not dependent solely on turbulent flux. Including the net horizontal and vertical advective fluxes in the numerator is critical to fully characterizing a site.

Net ecosystem exchange (NEE) of carbon dioxide is mentioned in this textbook often since it is a critical piece of information regarding the interaction between the terrestrial biosphere and the atmosphere. Scientists use the EBR as a test to determine if the *in-situ* observations at a project site are high-quality, defensible measurements. It is currently not practical to measure carbon dioxide in an air parcel to the extent that we could test out a CO_2 Balance Ratio (CBR) to partner with our EBR in determining data quality. However, if we use the equations herein to calculate the fluxes at our project site, and the energy fluxes are not balanced, the implication is that the carbon dioxide fluxes are also unbalanced.

7.7 Summary

In this chapter we presented the major challenges and approaches to understanding complex terrain impacts on the boundary layer. Starting from base principles we showed how our thinking over flat terrain can be modified to address the vertical temperature changes associated with slopes. For further information on mountain meteorology, we recommend Whiteman (2000).

References

Ahrens, C.D., 2012. Essentials of Meteorology: An Invitation to the Atmosphere, sixth ed. Brooks/Cole Cengage Learning, Belment, USA.

Aubinet, M., Heinesch, B., Yernaux, M., 2003. Horizontal and vertical CO_2 advection in a sloping forest. Bound.-Layer Meteorol. 108, 397–417.

Axelsen, S., van Dop, H., 2009. Large-eddy simulation of katabatic winds. Part 2: Sensitivity study and comparison with analytical models. Acta Geophys. 57, 837–856.

Baldocchi, D., 2003. Assessing the eddy covariance technique for evaluating carbon dioxide exchange rates of ecosystems: past, present and future. Glob. Change Biol. 9, 1–14.

Belcher, S.E., Jerram, N, Hunt, J.C.R., 2003. Adjustment of a turbulent boundary layer to a canopy of roughness elements. J. Fluid Mech. 488, 369–398.

Boussinesq, J., 1877. Theorie de l'ecoulement tourbillant. Mem. Pres. Acad. Sci. Paris 23, 56–58.

Bromwich, D.H., Cassano, J.J., Klein, T., Heinemann, G., Hines, K.M., Steffen, K., Box, J.E., 2001. Mesoscale modeling of katabatic winds over greenland with the polar MM5*. Mon. Weather Rev. 129, 2290–2309.

Budyko, M.I., 1958. The Heat Balance of the Earth's Surface. Translated by N. A. Stepanova, U.S. Dept. of Commerce. 259 pp.

Burns, S.P., Sun, J., Lenschow, D.H., Oncley, S.P., Stephens, B.B., Yi, C., Anderson, D.E., Hu, J., Monson, R.K., 2011. Atmospheric stability effects on wind fields and scalar mixing within and just above a subalpine forest in sloping terrain. Bound.-Layer Meteorol. 138, 231–262. https://doi.org/10.1007/s10546-010-9560-6.

Chen, H., Yi, C., 2012. Optimal control of katabatic flows within canopies. Q. J. R. Meteorol. Soc. 138, 1676–1680. https://doi.org/8.1002/qj.1904.

Davies, T.D., Palutikof, J.P., Guo, X., Berkofsky, L., Halliday, J.A., 1995. Development and testing of a two-dimensional downslope wind model. Bound.-Layer Meteorol. 73, 279–297.

Davis, A.M.J., McNider, R.T., 1997. The development of Antarctic katabatic winds and implications for the coastal ocean. J. Atmos. Sci. 54, 1248–1261.

De Roo, F., Banerjeea, T., 2018. Can a simple dynamical system describe the interplay between drag and buoyancy in terrain-induced canopy flows? J. Atmos. Sci. https://doi.org/10.1175/JAS-D-17-0161.1.

Deacon, E.L., 1949. Vertical diffusion in the lowest layers of the atmosphere. Q. J. R. Meteorol. Soc. 75, 89–103.

Denmead, O.T., Bradley, E.F., 1985. Flux-gradient relationships in a forest canopy. In: Hutchison, B.A., Hicks, B.B. (Eds.), The Forest-Atmosphere Interaction. D. Reidel Publ. Co, pp. 421–442.

Doran, J.C., Horst, T.W., 1981. Velocity and temperature oscillations in drainage winds. J. Appl. Meteorol. 20, 360–364.

Duynkerke, P.G., 1999. Turbulence, radiation and fog in Dutch stable boundary layers. Bound.-Layer Meteorol. 90, 447–477.

Feigenwinter, C., Bernhofer, C., Eichelmann, U., Heinesch, B., Hertel, M., Janous, D., Kolle, O., Lagergren, F., Lindroth, A., Minerbi, S., Moderow, U., Molder, M., Montagnani, L., Queck, R., Rebmann, C., Vestin, P., Yernaux, M., Zeri, M., Ziegler, W., Aubinet, M., 2008. Comparison of horizontal and vertical advective CO_2 fluxes at three forest sites. Agric. For. Meteorol. 148 (1), 12–24.

Feigenwinter, C., Montagnani, L., Aubinet, M., 2010. Plot-scale vertical and horizontal transport of CO_2 modified by a persistent slope wind system in and above an alpine forest. Agric. For. Meteorol. 150, 665–673.

Finnigan, J.J., 2008. An introduction to flux measurements in difficult conditions. Ecol. Appl. 18 (6), 1340–1350.

Fleagle, R.G., 1950. A theory of air drainage. J. Meteorol. 7, 227–232.

Froelich, N.J., Schmid, H.P., 2007. Flow divergence and density flows above and below a deciduous forest. Part II. Below-canopy thermotopographic flows. Agric. For. Meteorol. 138, 29–43.

Goulden, M.L., Munger, J.W., Fan, S.M., Daube, B.C., Wofsy, S.C., 1996. Measurements of carbon sequestration by long-term eddy covariance: methods and a critical evaluation of accuracy. Glob. Change Biol. 2, 169–182.

Gu, L., Fuentes, J.D., Shugart, H.H., Staebler, R.M., Black, T.A., 1999. Responses of net ecosystem exchanges of carbon dioxide to changes in cloudiness: results from two North American deciduous forests. J. Geophys. Res. 104, 31421–31434.

Haiden, T., Whiteman, C.D., 2005. Katabatic flow mechanisms on a low-angle slope. J. Appl. Meteorol. 44, 113–126.

Helbig, M., et al., 2021. Integrating continuous atmospheric boundary layer and tower-based flux measurements to advance understanding of land-atmosphere interactions. Agric. For. Meteorol. 307. https://doi.org/10.1016/j.agrformet.2021.108509.

Hodur, R.M., 1997. The Naval Research Laboratory's Coupled Ocean/Atmosphere Mesoscale Prediction System (COAMPS). Mon. Weather Rev. 125, 1414–1430.

Horst, T.W., Doran, J.C., 1986. Nocturnal drainage flow on simple slopes. Bound.-Layer Meteorol. 34, 263–286.

Mahrt, L., 1982. Momentum balance of gravity flows. J. Atmos. Sci. 39, 2701–2711.

Manins, P.C., Sawford, B.L., 1979. A model of katabatic winds. J. Atmos. Sci. 105, 1011–1025.

Massman, W.J., 1997. An analytical one-dimensional model of momentum transfer by vegetation of arbitrary structure. Bound.-Layer Meteorol. 83, 407–421.

Massman, W.J., Lee, X., 2002. Eddy covariance flux corrections and uncertainties in long-term studies of carbon and energy exchanges. Agric. For. Meteorol. 113, 121–144.

Massman, W.J., Weil, J.C., 1999. An analytical one-dimensional second-order closure model of turbulence statistics and the Lagrangian time scale within and above plant canopies of arbitrary structure. Bound.-Layer Meteorol. 91, 81–107.

Massom, R.A., Eicken, H., Hass, C., Jeffries, M.O., Drinkwater, M.R., Sturm, M., Worby, A.P., Wu, X.R., Lytle, V.I., Ushio, S., Morris, K., Reid, P.A., Warren, S.G., Allison, I., 2001. Snow on Antarctic sea ice. Rev. Geophys. 39, 413–445.

McNider, R.T., 1982. A note on velocity fluctuations in drainage flows. J. Atmos. Sci. 39, 1658–1660.

Montagnani, L., Manca, G., Canepa, E., Georgieva, E., Acosta, M., Feigenwinter, C., Janous, D., Kerschbaumer, G., Lindroth, A., Minach, L., Minerbi, S., Molder, M., Pavelka, M., Seufert, G., Zeri, M., Ziegler, W., 2009. A new mass conservation approach to the study of CO_2 advection in an alpine forest. J. Geophys. Res. 114, D07306. https://doi.org/10.1029/2008JD010650.

Poggi, D., Katul, G.G., Albertson, J.D., 2004. A note on the contribution of dispersive fluxes to momentum transfer within canopies. Bound.-Layer Meteorol. 111, 615–621.

Princevac, M., Hunt, J.C.R., Fernando, H.J.S., 2008. Quasi-steady katabatic winds on long slopes and in wide valleys: hydraulic theory and observations. J. Atmos. Sci. 65, 627–643.

Salcido, A., Celada-Murillo, A.-T., Carreón-Sierra, S., Castro, T., Peralta, O., Salcido-González, R.-S., Hernández-Flores, N., Tamayo-Flores, G.-A., Martínez-Flores, M.-A., 2020. Estimations of the Mexicali Valley (Mexico) mixing height. Atmosphere 11, 505.

Schaeffer, S.M., Anderson, D.E., Burns, S.P., Monson, R.K., Sun, J., Bowling, D.R., 2008. Canopy structure and atmospheric flows in relation to the d13C of respired CO_2 in a subalpine coniferous forest. Agric. For. Meteorol. 148, 592–605.

Shaw, R.H., 1977. Secondary wind speed maxima inside plant canopies. J. Appl. Meteorol. 16, 514–521.

Staebler, R.M., Fitzjarrald, D.R., 2004. Observing subcanopy CO_2 advection. Agric. For. Meteorol. 122, 139–156.

Steiner, M., Bousquet, O., Houze, R.A., Smull, B.F., Mancini, M., 2003. Airflow within major Alpine river valleys under heavy rainfall. Q. J. R. Meteorol. Soc. 129, 411–431.

Stull, R.B., 1988. An Introduction to Boundary Layer Meteorology. Kluwer Academic Publishers. 670 pp.

Sun, J., Burns, S.P., Delany, A.C., Oncley, S.P., Turnipseed, A.A., Stephens, B.B., Lenschow, D.H., LeMone, M.A., Monson, R.K., Anderson, D.E., 2007. CO_2 transport over complex terrain. Agric. For. Meteorol. 145, 1–21.

Sutcliffe, R.C., 1936. Surface resistance in atmospheric flow. Q. J. R. Meteorol. Soc. 62, 3–12.

Sutton, O.G., 1953. Micrometeorology: A Study of Physical Processes in the Lowest Layers of the Earth's Atmosphere. McGraw-Hill Book Company. 333 pp.

Taylor, G.I., 1916. Skin friction of the wind on the earth's surface. Proc. Roy. Soc. 92, 196–199.

Whiteman, C.D., 2000. Mountain Meteorology: Fundamentals and Applications. Oxford University Press.

Xu, X., Yi, C., Kutter, E., 2015. Stably stratified canopy flow in complex terrain. Atmos. Chem. Phys. 15, 7457–7470. https://doi.org/10.5194/acp-15- 7457-2015.

Yi, C., 2008. Momentum transfer within canopies. J. Appl. Meteorol. Climatol. 47, 262–275. doi:8.1175/2007JAMC1667.1.

Yi, C., 2009. Instability analysis of terrain-induced canopy flows. J. Atmos. Sci. 66, 2134–2142. https://doi.org/10.1175/2009JAS3005.1.

Yi, C., Davis, K.J., Bakwin, P.S., Berger, B.W., Marr, L., 2000. The influence of advection on measurements of the net ecosystem-atmosphere exchange of CO_2 from a very tall tower. J. Geophys. Res. 105, 9991–9999.

Yi, C., Monson, R.K., Zhai, Z., Anderson, D.E., Lamb, B., Allwine, G., Turnipseed, A.A., Burns, S.P., 2005. Modeling and measuring the nocturnal drainage flow in a high-elevation, subalpine forest with complex terrain. J. Geogr. Res. 110, D22303. https://doi.org/10.1029/2005JD006282.

Yi, C., Anderson, D.E., Turnipseed, A.A., Burns, S.P., Sparks, J.P., Stannard, D.I., Monson, R. K., 2008. The contribution of advective fluxes to net ecosystem exchange in a high-elevation, subalpine forest. Ecol. Appl. 18, 1379–1390.

Yi, C., Pendall, E., Ciais, P., 2015. Focus on extreme events and the carbon cycle. Environ. Res. Lett. 10, 070201. doi:8.1088/1748-9326/10/7/070201.

Zhong, S.Y., Whiteman, C.D., 2008. Downslope flows on a low-angle slope and their interactions with valley inversions. Part II: Numerical modeling. J. Appl. Meteorol. Climatol. 47, 2039–2057. https://doi.org/10.1175/2007JAMC1670.1.

Further reading

Bohren, C.F., Albrecht, B.A., 1998. Atmospheric Thermodynamics. Oxford University Press, New York.

Foken, T., 2008. The energy balance closure problem: an overview. Ecol. Appl. 18, 1351–1367.

Massom, R., Coauthors., 2001. Snow on Antarctic sea ice. Rev. Geophys. 39, 413–445.

If a tree falls: The role of vegetative environments in boundary layer fluxes

Gil Bohrer and Theresia Yazbeck

Department of Civil, Environmental and Geodetic Engineering, The Ohio State University, Columbus, OH, United States

8.1 Introduction

Trees cover around 30% of the earth's surface (Carlowicz, 2012), thus they are crucial elements of the lower part of the planetary boundary layer interacting with the atmospheric surface layer. Tree canopies act as a porous medium where wind flow can penetrate. Within the canopy leaf, surfaces interact with the atmosphere by adding friction (extracting momentum), intercepting solar radiation, and facilitating the exchange of heat, water vapor, CO_2, and other chemicals and particles between the ecosystem and atmosphere. This chapter discusses canopies contribution to the atmospheric boundary layer dynamics through roughness, energy balance, and water budget.

8.2 Canopy aerodynamics

Unlike the ground surface, vegetation canopies are porous, meaning that wind is not restricted to only move above but also through tree canopies. Aerodynamic effects of canopies, as porous media, can be split into three scales: (1) Leaf scale – where an aerodynamic resistance formed by the leaf surface influences the exchange of momentum and gases from the leaf to the air through a leaf-surface roughness layer of few millimeters thick, (2) Canopy scale – where reduced connectivity for air flow from the open atmosphere above the forest leads to development of distinct canopy-air conditions inside the forest that affect the flux exchanges between vegetation and the air, and (3) Large scale – where the canopy as a whole generates drag within the planetary boundary layer, thus altering the wind profile throughout the boundary layer and the net exchange of gases and energy between the vegetation-covered land surface and the open atmosphere. We will start by discussing the large scale.

Conceptual Boundary Layer Meteorology. https://doi.org/10.1016/B978-0-12-817092-2.00007-2

8.2.1 Surface roughness parameters

Tree canopies generate drag and are therefore sinks for momentum. The rate at which momentum of the free flow above the canopy is lost toward the surface is determining the rate at which the wind slows as it approaches the surface and effectively determines the vertical profile of wind speed. Based on Monin–Obukhov similarity theory (MOST see Chapter 3), the vertical profile of wind speed follows a logarithmic curve. There are two key characteristic properties associated with the surface that describe how the surface, including the tree canopies, modifies the vertical profile of wind speed:

(1) Surface displacement height – It describes the height above ground below which the near-surface air, and in the case of forests, the air inside the canopy, becomes detached from the free flow above. In this context, a "detached" air volume means that small perturbations to wind speed above do not penetrate below that depth. An intuitive example is provided by the rustling of leaves created by wind gusts above the canopy that shake leaves at the top parts of the canopy, while down by the forest floor, you can often hear the leaves rustle but cannot feel the wind gusts that drive them. This is because you are walking below the aerodynamic displacement height and surrounded by "detached" air volume.

(2) Surface roughness length – It describes the virtual height above the ground surface (or the surface displacement height) at which the flow will asymptote to zero. While the air does not actually come to a stop, the roughness length parameter specifies the degree of curviness of the shape of the logarithmic profile of wind speed above the roughness length. Both displacement height and roughness length, collectively known as the surface roughness parameters, vary with the type and characteristics of the vegetation. Every ecosystem will have a large-scale roughness length and displacement height associated with it. A corn field would have a smaller roughness length than an oak forest but larger than a grass field (Figure 8.1).

These two surface-roughness parameters allow us to reduce the complex description of a fully detailed forest canopy to a simplistic set of two numbers. Much work was done to relate canopy characteristics to their outcome on the roughness parameters.

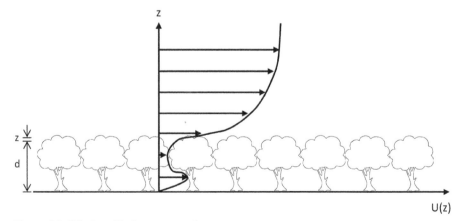

Figure 8.1 Wind profile in tree canopies.
Credit: Authors.

The most common approximation relates the average canopy-top height, h, to the roughness length, z_0, and displacement height, d, as:

$$d = \frac{2}{3}h$$

$$z_0 = 0.123h$$

Since the roughness parameters are important for modeling the surface fluxes and heat and momentum exchange, many studies were conducted to understand the small-scale interactions between z_0 and d and canopy structures through meteorological measurements, remote sensing, and numerical experiments. For example, it was shown that higher leaf area index can further increase z_0 and d (Lindroth, 1993). Seasonal changes (going from full leaf cover to leaf off) can also modify the roughness parameters, often leading to a decrease of z_0 but an increase of d in the winter (Maurer et al., 2013; Nakai et al., 2008). Other canopy structural characteristics, such as tree spacing, vertical leaf density profile, variation of tree top heights, and gap fraction can also change the roughness parameters (Maurer et al., 2015).

8.2.2 Friction velocity

The friction velocity, symbolized with u^*, is a measure of the strength of turbulence generated by drag from the surface. Wind slows down as it is dragged by the surface, and in cases where the surface is covered by vegetation, the air layer just above the canopy is therefore faster. Shear develops between two air layers of different speeds. This shear creates turbulent mixing. u^* is a measure for the turbulence intensity generated by this shear. It is defined as the vertical flux of horizontal momentum, and described by the following equation:

$$u^* = \sqrt[4]{\overline{u'w'}^2 + \overline{v'w'}^2}$$

where u, v, and w are the horizontal, transversal, and vertical wind velocity components (m/s), respectively (Chapter 2). An overbar marks a large-scale or long-term average (typically 30 min) and a prime marks an instantaneous turbulent deviation from that long-term average. u^* is scaled to have the units of wind speed (most commonly, m/s). It is approximately constant within the vertical range of the free mixing boundary layer, starting roughly 2–4 times the height of the canopy above the ground and extending until the capping inversion, an area of rapid increase in the potential temperature determined by the rate of surface heat flux. Turbulent mixing, estimated through u^*, drives the exchanges of momentum between the atmosphere and the surface, as well as other surface fluxes such as sensible heat flux, latent heat flux (the flux of water vapor), and all other gas fluxes.

To understand the role of u^* in driving fluxes, we need to consider the nature of atmospheric surface fluxes. By definition, a flux is the rate of flow of something per unit of reference area. It is estimated as a concentration difference of something (chemical mass, heat, etc.) times the velocity of the flow that moves that concentration (the advective velocity), divided by the area over which the flow occurs. For example, consider the flux of a pollutant in a river. That flux will be estimated by multiplying the concentration difference of the pollutant upstream (high concentration) and downstream (low concentration) by the river flow velocity (in this example, the river flow provides the advective velocity) divided by the cross-section area of the river. However, for atmospheric surface fluxes moving upward to the atmosphere, the mean vertical wind velocity, w, is zero, which means that the mean vertical wind velocity is not the advective velocity that drives surface fluxes. Instead, surface fluxes are moved by the short-term turbulent fluctuations of the wind, quantified through u^*. u^* is rarely measured, but, as we explain above, strongly needed to predict fluxes from canopies. MOST formulates the relationship between u^*, the horizontal wind speed (which is measured more commonly in weather stations, and is easier to estimate in models), and the surface roughness parameters. Therefore, through altering the effective surface roughness, vegetation canopies modify all surface fluxes. Furthermore, predicting the effects of canopy structure on the surface roughness parameters is critical to our ability to model surface fluxes. For a more in-depth view of fluxes and MOST see Chapters 2 and 3, respectively.

8.2.3   The big-leaf representation of the canopy

Consistent with the large-scale approach to parameterize surface roughness, a basic conceptualization of the canopy-air interaction in large-scale models is by considering the canopy as a virtual homogeneous "big leaf" without vertical structure. That virtual big leaf exchanges heat and gas fluxes with the air above. The large-scale roughness length and displacement height of the forest are assigned as the aerodynamic properties of the big leaf. They govern the exchange of momentum between the wind above and the surface and determine the resulting u^*. Further fluxes (water vapor, CO_2, heat) are driven by the turbulent mixing, by the energy budget of the surface, and by other parameters that represent the biological properties of the big leaf. The big leaf represents the effective average of each property of the canopy averaged over the large-scale volume (area and depth) of the forest. The big-leaf representation allows large-scale models to represent the effective properties of the canopy even though they cannot resolve the small-scale details of air flows inside the canopy. The obvious benefit of this representation is its simplicity. *The cost is losing the vertical details of canopy structure and neglecting their potential contribution to fluxes.* For example, the surface albedo (i.e., the fraction of the incoming shortwave radiation that is reflected by the canopy) is defined for a big-leaf surface as the ratio between the incoming radiation above the canopy to the reflected radiation that exits the canopy. It does not consider the consequences of the fact that some of the radiation is reflected from leaves at the top of the canopy, while other portions of the radiation are reflected from leaves deeper inside the canopy, and even from the ground surface under the canopy. It also ignores the gradients of light, temperature, humidity, and CO_2 concentrations along the canopy height. These may interact non-linearly, leading to different overall flux rate when summing the local flux contributions in each location in the canopy in comparison to assuming an average flux rate times the total leaf area.

8.2.4 Canopy-air approaches

Improvements of the simple big leaf include a two-leaf approach where the upper virtual leaf layer represents flux exchanges by leaves that are fully exposed to light, and a second lower virtual layer represents shade leaves that are only exposed to diffuse, indirect solar radiation. Big-leaf approaches do not directly account for the fact that most flux exchange occurs between the vegetation and the air within the canopy space, and not directly with the open air above the canopy. Therefore, realism to canopy representation can be developed by adding one virtual air layer within the canopy space that accounts for the exchanges between leaf and canopy air and, in a separate step, between canopy-air space and the free atmosphere above the canopy. The next level of improved realism in canopy representation utilizes this virtual canopy-air concept. In models using the canopy-air concept, surface fluxes do not move directly from the canopy to the atmosphere. Instead, fluxes are exchanged between the canopy and the virtual canopy air, and then between the canopy air and the free air above the forest (i.e., the atmospheric roughness layer). The virtual space of the canopy air is characterized by different conditions than the average for the atmospheric grid cell above the canopy. Typically, the canopy air will be warmer and more humid than the air above the canopy, and the turbulent mixing inside the canopy air is weaker. Further improvements include a small number of multiple leaf layers located at different heights throughout the canopy air. These layers can partially shade the leaf layers below them, representing different tree age-size groups (see review by Fisher et al., 2018). While the canopy-air approach introduces more realism by resolving different conditions above and within the canopy, it is still limited, since it allows only a single set of atmospheric conditions governing the canopy air interaction.

8.2.5 Canopy-air vertical profile

A more explicit approach that resolves turbulence and light attenuation in discrete layers throughout the canopy was developed and could replace the canopy-air approach. Wind speed and turbulent mixing in each height above ground along the canopy can be resolved as a function of the wind speed above the canopy, the leaf surface roughness, and the leaf area density profile (distribution of leaf density over different heights throughout the canopy) (Massman, 1997; Massman and Weil, 1999). This approach assumes that the vertical distributions of vegetation properties, including leaf density profile, are constant over the horizontal extent of a limited patch that is resolved by a single large-scale horizontal grid cell in the model, or in models that allow it, a single sub-grid tile of similar vegetation plant-functional type. This assumption allows large-scale models to treat the canopy at each location and for each vegetation type as a set of vertical profiles of atmospheric conditions and interacting vegetation properties. Such approach was incorporated in models that resolve dispersion within a canopy (Nathan and Katul, 2005). It is possible to further use these profiles of wind, turbulence, radiation, and vegetation properties to determine the corresponding vertical profile of canopy-air temperature and humidity (Harman and Finnigan, 2008). Bonan et al. (2018) developed a new approach for large-scale modeling of canopy fluxes that utilizes the resolved vertical profiles of canopy-air properties and moves away from the big-leaf approach.

8.2.6   Canopy resistance

Consider the path of a flux of any material or energy that is moving from the vegetation to the atmosphere. That property must first move from within the plant to the plant surface. That movement is limited by the conductance of the plant to the specific movement (conductance is the inverse of resistance). For example, the thermal conductance of the leaves controls the movement of heat from within the leaf to the leaf surface. Similarly, stomatal conductance controls the movement of water vapor and CO_2 between the leaf surface and the parenchyma tissue in the leaf. Stomata are openings in the leaf that allow gas exchange through diffusion. They open to allow photosynthesis when light and water are available, and close when photosynthesis is not favorable. Next, the flux must move through a thin air layer that surrounds the leaf surface. The aerodynamic resistance of the leaf surface layer is determined by the roughness of the leaf skin and by the shape and smoothness of the leaves. After crossing the leaf-skin aerodynamic roughness layer, the flux reaches the canopy-air space. It must then move upward through the canopy-air space and mix with the free air above the canopy – the atmospheric roughness layer.

At each step, movement is driven by the concentration difference of the property that is moving (e.g., temperature difference, humidity difference) while the structure and properties of the canopy interact with the aerodynamic flow of the air to create some resistance to the flux. Figure 8.2 illustrates the resistances governing

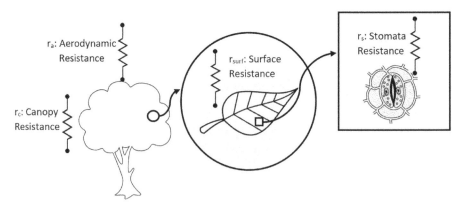

Figure 8.2 Sequential resistances sketch governing canopy-air exchange: aerodynamic resistance, canopy resistance at tree level (stem and crown), leaf level where surface resistance governs, and stomatal resistance, which is at the level of the leaf internal physiology. Credit: Authors.

canopy-air exchange through the four different scales: within plant, leaf skin, canopy air, and above-canopy atmospheric roughness layer. Different modeling approaches at different levels of canopy structure detail differ in the number of steps they resolve for the flux movement, and consequently, the number of independent resistance terms and concentration gradients they resolve. For example, big-leaf models only account for the gradients of heat, water vapor, and CO_2 between the open air and the leaf surface and use one effective "bulk canopy" resistance and a stomatal resistance to restrict fluxes from within the plant to the air. Alternatively, canopy-air models will utilize stomatal resistance and leaf aerodynamic conductance with the concentration gradient between plant and canopy air to resolve the flux to the canopy air, and then the gradient between canopy air and open atmosphere and the canopy roughness to resolve the flux from the canopy air to the open atmosphere.

8.3   Canopy energy budget

Vegetation plays a significant role in the energy budget of the earth. Being dark green, forest canopies typically have a darker color than bare soil or grasslands, that is, they reflect less and absorb more incoming solar radiation and consequentially their albedo is lower than that of bare soil or grasslands. Forest albedo typically ranges from 0.11 for tall forests to 0.18 depending on forest type and density. The conservation of energy determines that energy that hits the surface is not lost but can change form. The net amount of radiation that is absorbed by the surface (considering both short-waves, i.e., visible light, and longwaves, i.e., infrared radiation) is then redistributed by the surface into different energy fluxes. The Earth's surface, including forests, emits sensible heat flux (i.e., flux of energy in the form of warm air), latent heat flux (i.e., the energy required to evaporate water from the wet surface into a flux of water vapor to the air), and heat flux into the deeper layers of the soil. The surface energy budget can be formulated as:

$$(1 - \alpha)S \downarrow + L \downarrow - L \uparrow = H + \lambda E + G$$

where α is albedo, $S \downarrow$ is downwelling shortwave radiation, and $L \downarrow$ and $L \uparrow$ are the downwelling and upwelling longwave radiation, respectively. The left-hand side of this equation provides the net surface energy. The right-hand side provides the surface fluxes: H is sensible heat flux, λE is the latent heat flux, which is the flux of water vapor by evapotranspiration, E, multiplied by the latent heat of vaporization, λ. G is the soil heat flux. The ratio between sensible and latent heat flux is called the Bowen ratio. The Bowen ratio is dynamic and changes during the time of day, seasons, and based on the surface and vegetation properties. For example, seasonal changes of the forest have significant effects on vegetation contribution to the energy budget. In the winter at deciduous forests, leafless trees lead to a brighter surface with higher albedo and less radiation absorption, however, most of the energy that is emitted is in the form of sensible heat flux and evapotranspiration is very low, leading to a high Bowen ratio. When leaves start growing in the spring, Bowen ratio starts increasing until it reaches its maximum at full leaf state. Any property of the vegetation that affects its ability to evaporate water will affect Bowen ratio. For example, while they absorb more energy, forest canopies are typically not hotter than bare soil areas under similar conditions. The reason for this is that forest canopies have a large leaf area that can evaporate water and therefore more of the surface energy will be converted to latent heat flux, compared to bare soil that will emit less water vapor and convert more of the surface energy to sensible heat flux, which heats the air. Similarly, the stomata sensitivity to soil moisture will affect Bowen ratio. When the soil is dry, trees may not find enough water to evaporate and will close stomata to prevent dehydration. This increase in stomata resistance to water-vapor flux will shift the Bowen ratio toward more sensible heat flux (Figure 8.3). The absolute and relative strengths of latent and sensible heat fluxes to the atmosphere determines the larger-scale structure of the atmosphere. Sensible heat flux drives the daily growth of the atmospheric boundary layer, and air temperature gradients drive global circulation patterns. Latent heat flux provides the water vapor for cloud formation. Therefore, by changing the surface properties that govern momentum and energy fluxes, forest canopies have large effects on local weather and global atmospheric circulation and climate (e.g., Werth and Avissar, 2002, 2004, 2005a, b).

An interesting example of the effects of forest canopies on surface fluxes are forests in arid environments that can create a canopy conveyor effect. The darker forests (compared to bare desert soil or sparse shrubland) absorb more incoming radiation.

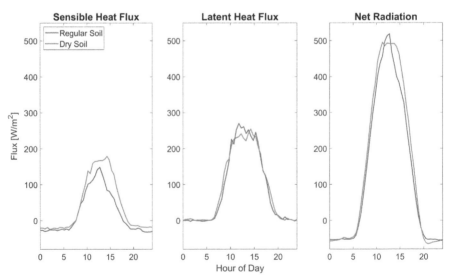

Figure 8.3 Summer 2019 daily average sensible and latent heat fluxes and net radiation for dry and regular soil conditions in a University of Michigan Biological Station (UMBS) forest (data from https://ameriflux.lbl.gov/, Site-ID US-UMB). During days of dry soil, the trees limit stomata conductance to prevent dehydration. As a result, sensible heat increases and latent heat decreases, especially when radiation is maximal around midday.
Credit: Chapter Authors.

The dry arid soil cannot support transpiration during most days and the trees have relatively low latent heat flux. Hypothetically, this should lead to very hot surface and canopy air. However, because the forest is sparse but still has higher roughness than the surrounding bare desert, turbulence mixing is greater over the forest, leading to more effective sensible heat flux that mixes the excess surface heat upwards, effectively cooling the canopy air (Banerjee et al., 2017; Eder et al., 2015).

8.3.1 Forest large eddy simulations

At the higher end of the resolution scale, LES provide explicit high-resolution calculation of airflow inside and above the canopy. Global and regional atmospheric models are too coarse to resolve turbulence, as turbulence is characterized by eddies (circular movements of air) that are typically smaller than a few hundred meters. LES work at a high resolution (meters to tens of meters) and therefore can resolve some eddies, while some kinetic energy is still unresolved in eddies that are smaller than the grid spacing of the model. While large-scale models parameterize for all turbulence kinetic energy, LES include parameterization schemes to account for the energy of the small (sub-grid scale) eddies but resolve the larger ones (Maronga, 2020). The high resolution of LES provides an advantage for forest simulations and they are capable of explicitly resolving the flow patterns inside the forest canopy. To account for the effects of the canopy, LES treat the canopy as a porous medium where each voxel (a resolved grid box, the 3D version of a pixel) contains a prescribed area of leaves with a prescribed skin roughness. While these simulations still do not resolve individual leaves, they do represent the canopy as a realistically heterogeneous space of air and vegetation and do not need to assume virtual canopy air or horizontal homogeneity (Yazbeck et al., 2021). The disadvantage of LES is that they are computationally intensive and cannot be used over large areas and long times, and they need detailed high-resolution description of the canopy structure, which is rarely available.

8.3.2 Canopy gas fluxes

Forests act as a major source (and sinks) for different gases, mainly H_2O, CO_2, and VOCs. Plant and soil bacteria respiration lead to O_2 uptake and CO_2 emission. However, when sunlight is available, photosynthesis takes place, and plants uptake CO_2, fixing it into carbohydrates used for plant growth, and then emitting O_2 and water vapor. This daily cycle makes the plants active drivers of CO_2 diurnal fluctuations in the atmosphere. In addition to the daily cycle, plants' emissions are subjected to seasonal cycles as well; during winter seasons, water fluxes are reduced to low values due to the low leaf area and closed stomata. Particularly, CO_2 uptake is stopped and rather than having a daily cycle of uptake (negative flux) and emission (positive flux), as during the summer, wintertime CO_2 flux becomes monotonously positive. Figures 8.4 and 8.5 illustrate the diurnal and seasonal cycles of CO_2 concentration and flux and water vapor flux from field measurements of the University of Michigan Biological Station (UMBS) forest (Gough et al., 2016).

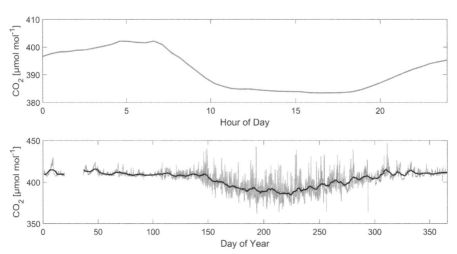

Figure 8.4 Mean daily profile of CO_2 concentration during summer 2016 (*upper panel*) and full-year, half-hourly time series of 2016 CO_2 concentration (*lower panel*) at the University of Michigan flux tower (US-UMB). CO_2 uptake by vegetation reduces CO_2 concentration during the day, and during the summer. The *thick black line* shows the weekly moving average of CO_2 concentration throughout the year.
Credit: Chapter Authors.

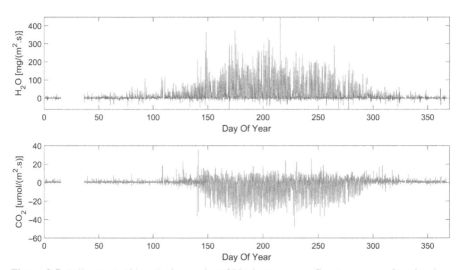

Figure 8.5 Full-year, half-hourly time series of 2016 water vapor flux (*upper panel*) and carbon flux (*lower panel*) at the University of Michigan flux tower (US-UMB). In leaf-on season, carbon and water fluxes are active. In leaf-off season, insignificant fluxes take place.
Credit: Chapter Authors.

In addition to its contribution to short-term daily and seasonal cycles of CO_2 and water vapor, vegetation plays an important role in the long term through the carbon cycle. Trees are estimated to store around 45% of all land carbon (Liu et al., 2018), in addition to being sinks and sources of carbon through CO_2 and methane fluxes. That being said, interventions such as deforestation do not only eliminate a major sink of CO_2 but also release a large amount of stored carbon.

Vegetation is a major source of VOCs such as terpene (the typical pine forest smell) and isoprene. It is estimated that around two-thirds of the VOCs in the air are from trees and plants (Naranjo, 2011). Trees emit VOCs through the stomata of their leaf skins and bark to attract pollinators and repel insects and animals. Once in the air, VOCs react with other airborne chemicals such as ozone, where some of these reaction products contribute to air pollution. It should be noted that tree burning emits significant amounts of black carbon and VOCs in a short period of time, which makes burning a major contributor to high levels of VOCs in regions of potential forest burning events.

8.3.3 Conclusions

Trees are intricately involved in surface fluxes to the atmosphere. They change the surface energy that is available for surface energy fluxes, they control the Bowen ratio, modify the surface resistance and wind patterns that drive fluxes, and emit and uptake water, CO_2, and other chemicals. Different models, limited to different degrees by scale, resolution, and computational time, make different assumptions about canopy representation, from a simple big leaf to explicit high resolution of flows in the canopy. A deeper understanding of the role of vegetation in the energy and flux budgets is crucial for improving weather and climate modeling, energy balance, and forest management. Thus, this field is attracting research interest to improve the present knowledge and develop more accurate models of the vegetation response to environmental stressors driving surface fluxes. Field measurements are performed all over the globe to improve our knowledge in that field. One of the most common techniques for measurement is eddy covariance (EC), which measures heat, vapor, and gas fluxes. Results of EC measurements are provided to the public through data networks such as AmeriFlux and FLUXNET. More global measurements are done through remote sensing (e.g., MODIS satellites), though usually operating at coarser spatiotemporal scales than EC.

References

Banerjee, T., De Roo, F., Mauder, M., 2017. Explaining the convector effect in canopy turbulence by means of large-eddy simulation. Hydrol. Earth Syst. Sci. 21 (6), 2987–3000.

Bonan, G.B., et al., 2018. Modeling canopy-induced turbulence in the Earth system: a unified parameterization of turbulent exchange within plant canopies and the roughness sublayer (CLM-ml v0). Geosci. Model Dev. 11 (4), 1467–1496.

Carlowicz, M., 2012. Seeing Forests for the Trees and the Carbon: Mapping the World's Forests in Three Dimensions. Available at: https://earthobservatory.nasa.gov/features/Fore stCarbon. Accessed October 9, 2019.

Eder, F., De Roo, F., Rotenberg, E., Yakir, D., Schmid, H.P., Mauder, M., 2015. Secondary circulations at a solitary forest surrounded by semi-arid shrubland and their impact on eddy-covariance measurements. Agric. For. Meteorol. 211–212, 115–127. Available at https://doi.org/10.1016/j.agrformet.2015.06.001.

Fisher, R.A., et al., 2018. Vegetation demographics in earth system models: a review of progress and priorities. Glob. Chang. Biol. 24 (1), 35–54.

Gough, C., Bohrer, G., Curtis, P., 2016. AmeriFlux US-UMB Univ. of Mich. Biological Station. Available at http://ameriflux.lbl.gov/sites/siteinfo/US-UMB.

Harman, I.N., Finnigan, J.J., 2008. Scalar concentration profiles in the canopy and roughness sublayer. Bound. Lay. Meteorol. 129 (3), 323–351.

Lindroth, A., 1993. Aerodynamic and canopy resistance of short-rotation forest in relation to leaf area index and climate. Bound. Lay. Meteorol. 66 (3), 265–279.

Liu, X., et al., 2018. Tree species richness increases ecosystem carbon storage in subtropical forests. Proc. Biol. Sci. 285 (1885).

Maronga, B., et al., 2020. Overview of the PALM model 6.0. Geosci. Model Dev. 13, 1335–1372. https://doi.org/10.5194/gmd-13-1335-2020.

Massman, W.J., 1997. An analytical one-dimensional model of momentum transfer by vegetation of arbitrary structure. Bound. Lay. Meteorol. 83 (3), 407–421.

Massman, W.J., Weil, J.C., 1999. An analytical one-dimensional second-order closure model of turbulence statistics and the Lagrangian time scale within and above plant canopies of arbitrary structure. Bound. Lay. Meteorol. 91 (1), 81–107.

Maurer, K.D., Hardiman, B.S., Vogel, C.S., Bohrer, G., 2013. Canopy-structure effects on surface roughness parameters: observations in a Great Lakes mixed-deciduous forest. Agric. For. Meteorol. 177, 24–34. Available at https://doi.org/10.1016/j.agrformet.2013.04.002.

Maurer, K.D., Bohrer, G., Kenny, W.T., Ivanov, V.Y., 2015. Large eddy simulations of surface roughness parameter sensitivity to canopy-structure characteristics. Biogeosci. Discuss. 11 (11), 16349–16389.

Nakai, T., et al., 2008. Parameterisation of aerodynamic roughness over boreal, cool- and warm-temperate forests. Agric. For. Meteorol. 148 (12), 1916–1925.

Naranjo, L., 2011. Volatile Trees. EARTHDATA. Available at https://earthdata.nasa.gov/learn/sensing-our-planet/volatile-trees.

Nathan, R., Katul, G.G., 2005. Foliage shedding in deciduous forests lifts up long-distance seed dispersal by wind. Proc. Natl. Acad. Sci. U. S. A. 102, 8251–8256.

Werth, D., Avissar, R., 2002. The local and global effects of Amazon deforestation. Geophys. Res. Lett. 107.

Werth, D., Avissar, R., 2004. The regional evapotranspiration of the Amazon. J. Hydrometeorol. 5 (1), 100–109.

Werth, D., Avissar, R., 2005a. The local and global effects of African deforestation. Geophys. Res. Lett. 32 (12), 1–4.

Werth, D., Avissar, R., 2005b. The local and global effects of Southeast Asian deforestation. Geophys. Res. Lett. 32 (20), 1–4.

Yazbeck, T., Bohrer, G., Vines, C., Frederik, D.R., Matthias, M., Bhavik, B., 2021. Effects of spatial heterogeneity of leaf density and crown spacing of canopy patches on dry deposition rates. Agric. For. Meteorol. 306. https://doi.org/10.1016/j.agrformet.2021.108440, 108440.

But we build buildings: Urban boundary layer

Kodi L. Berry
NOAA National Severe Storms Laboratory, Norman, OK, United States

As humans, we build structures in which to work and live. The size, shape, location, and even material of those structures alter the boundary layer. Disruption of atmospheric flow happens with any building; even a single farmhouse will provide a barrier to the wind. For the discussion here, we define the difference between rural and urban quite simply: Rural areas have more natural surface cover than urban areas. Urbanization radically changes the surface and atmospheric properties of a region. The overall magnitude of the impact of urbanization on the boundary layer depends on building height and density.

In Chapter 8, the concepts of displacement height and surface roughness were discussed; they still apply in urban environments, but a more regular spacing and often higher heights impact the boundary layer structure differently. These concepts are presented in Sections 9.1 and 9.2.

The shift from natural to humanmade surfaces alters the surface energy balance, and it is this change that makes urban boundary layers truly unique. The specifics of this are discussed in Section 9.3. This chapter concludes with a section on urban heat islands, a well-known climate impact of urbanization.

9.1 The structure of the urban boundary layer

As the landscape transitions from rural to urban, the urban boundary layer develops. The urban boundary layer is defined as the part of the planetary boundary layer that is impacted by the presence of the urban surface below (Arnfield, 2003; Figure 9.1). The urban boundary layer extends from roof level to a height where the impacts of the urban surface are no longer apparent (Oke, 1988). If the size of the urban area is sufficiently large, the urban boundary layer may include the entire depth of the planetary boundary layer (Schmid et al., 1991).

The urban boundary layer includes the roughness layer, surface layer, and mixed layer. The urban canopy layer is the layer from the ground to roof level where airflow and energy exchanges are dominated by small-scale characteristics and processes (Oke, 1988). The flow within the roughness layer is complex and strongly depends on the individual trees and buildings. The nature of the urban surface creates a roughness layer with a depth several times the average building height (Roth, 2000). For more discussion on the various sublayers in the boundary layer, see Chapter 1.

Conceptual Boundary Layer Meteorology. https://doi.org/10.1016/B978-0-12-817092-2.00004-7

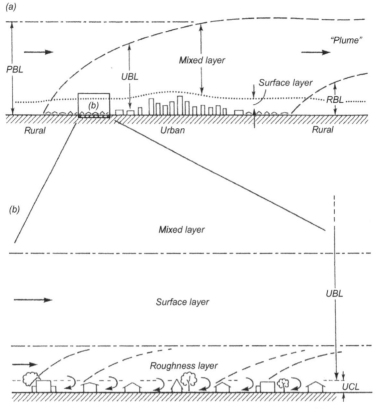

Figure 9.1 Illustration of boundary layer structures over a city (Oke, 1988).

9.2   Aerodynamic structure of the urban boundary layer

Buildings obstruct the wind and induce significant changes in airflow within the urban boundary layer. If an isolated building is oriented normal to the wind, air is deflected over the top, down the front, or around the sides of the building and produces lee eddies (Figure 9.2). The bulk of the flow is displaced over the building, which results in increased wind speeds in the displacement zone, labeled B in Figure 9.2A. After passing over the building, the wind speed decreases and cannot immediately fill the increased volume downwind of the building. The flow separates and becomes more turbulent in the wake zone (D in Figure 9.2A). The cavity zone (C in Figure 9.2A), immediately behind the building, is characterized by low pressure that produces lee eddies or vortices. The isolated building impacts the flow at least three times the building height vertically, three times the building height upstream of the building, and five to ten times the building height in the horizontal distance downstream of the building (Oke, 1987).

The flow pattern of an urban area can be thought of as the combined effects of many individual buildings, which depends significantly on building density. If the distance

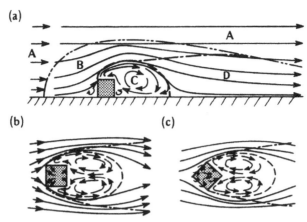

Figure 9.2 Flow patterns around a sharp-edged building. Vertical cross-section of (A) streamlines *(solid lines)* and flow zones *(dashed lines and letters)*. Plan view of streamlines around a building (B) normal and (C) diagonal to the flow (Oke, 1987).

between buildings is relatively large, the flow patterns are similar to those of isolated buildings (Figure 9.3A). When more closely spaced, the lee vortex of each building interacts with those downwind and results in a complicated airflow pattern (Figure 9.3B). When buildings are even closer, the airflow skims over the top of the buildings and creates a vortex in the street canyon (Figure 9.3C). However, if

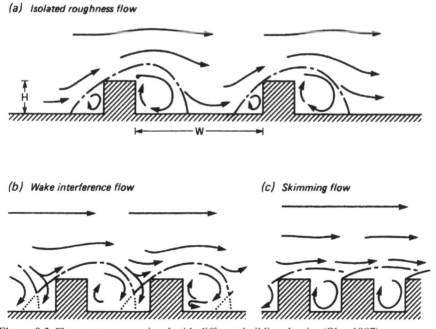

Figure 9.3 Flow patterns associated with different building density (Oke, 1987).

the airflow is parallel to the street, winds are channeled through the street canyon at speeds faster than those over a flat surface area.

The lee vortices that develop in urban areas are commonly characterized by increased vertical velocities and reverse flow at the surface (Kastner-Klein et al., 2004). However, when pitched roofs are used in wind tunnel studies, the canyon vortex does not develop (Rafailidis, 1997; Kastner-Klein et al., 2004). As a result, wind speed values within the canyon are dramatically reduced.

Overall, wind speeds in the urban canopy are less than those in rural environments at the same height (Changnon et al., 1971; Landsberg, 1981; Oke, 1987). However, studies have shown that when wind speed values drop below a threshold of 2.5–$5.0 \, \mathrm{m \, s^{-1}}$, referred to as the critical wind speed, winds are stronger in the city than in rural areas (Chandler, 1965; Bornstein and Johnson, 1977; Wong and Dirks, 1978; Shreffler, 1979a; Landsberg, 1981; Hildebrand and Ackerman, 1984). Below the critical wind speed, increased vertical mixing in the urban atmosphere and pressure gradients induced by the urban heat island produce stronger winds within the city, whereas above the critical wind speed, frictional forces reduce wind speed values within the city (Chandler, 1965; Wong and Dirks, 1978).

While wind speed is reduced within the urban canopy, wind profile data reveal that maximum wind speeds occur at roof top level. As such, strong momentum transfer occurs at the rooftop level (Graham, 1968; Castro et al., 2006) and the resultant shear layer produced may shelter the canopy flow from that above (Coceal et al., 2006). At the same time, eddies larger than the thickness of the shear layer aid in the transport of heat, moisture, and pollutants from within the urban canopy layer to the roughness layer (Castro et al., 2006). From rooftop level to two to five times the building height, the flow is complex and depends strongly on the individual roughness elements, while the mean wind profile well above the roof top level is described by the logarithmic wind law.

The increased surface roughness in cities results in the increased importance of the mechanical production of turbulence as opposed to buoyant production (Shea and Auer, 1978; Hildebrand and Ackerman, 1984; Roth and Oke, 1995). The increased drag and turbulence due to the roughness elements produces a deep layer of frictional influence whereby wind speed values are reduced compared with those at the same height in the surrounding rural areas. As wind speed values decrease in urban areas, the airflow converges over the city and results in rising motion and divergence aloft (Hildebrand and Ackerman, 1984). Wind tunnel studies have demonstrated that this effect is intensified in the presence of slanted roofs, as opposed to flat roofs. In case of slanted roofs, the retardation of the horizontal winds at roof level is accompanied by a simultaneous transfer of momentum to the vertical flow (Rafailidis, 1997). The changes in speed across urban areas can also induce changes in wind direction due to changes in the strength of the Coriolis force. In the Northern Hemisphere, as air enters the urban area it slows and backs. Upon exiting the urban area, the winds veer and increase in speed.

To understand the exchanges of momentum, heat, moisture, and pollutants between the urban surface and the atmosphere, the turbulent structure of the urban boundary layer must be quantified (Roth, 2000). Roth and Oke (1995) and Roth (2000) identified four processes primarily responsible for modifying the turbulence structure of the urban boundary layer:

- An intense shear layer near the top of the canopy, where mean kinetic energy of the flow is converted into turbulent kinetic energy and results in high turbulence intensity.
- Wake diffusion (Thom et al., 1975), or the efficient mixing of momentum, heat, and moisture generated by turbulent wakes behind individual roughness elements. The size of these turbulent wakes is related to the dimensions of the roughness elements.
- Form drag due to bluff-bodies, or pressure differences across individual roughness elements, which impacts the transport of momentum to the surface and has no analog in the transport of heat or mass (Thom, 1972).
- Sources and sinks of momentum, heat, and water vapor organized in three-dimensional arrays and not necessarily collocated, which results in the development of a complex system of discrete and localized heat and mass plumes.

Comprehensive comparisons between different studies on urban boundary layer turbulence are complicated by several factors: (a) the variety of statistics and normalizations utilized, (b) the strong dependence of turbulence on urban morphology and fetch conditions, (c) the focus on the height variation of turbulence based on a single profile, (d) the lack of data on air flows around buildings, and (e) the lack of knowledge of the upstream conditions (Roth, 2000; Eliasson et al., 2006).

9.3 Surface energy balance

Urbanization drastically modifies the surface energy balance of a city compared to that of rural areas (Yap and Oke, 1974; White et al., 1978; Kalanda et al., 1980), which in turn impacts stability conditions within the boundary layer, thermodynamic properties, and mixing layer height.

The energy balance of a building-air volume can be expressed as

$$Q^* + Q_F = Q_H + Q_E + \Delta Q_S + \Delta Q_A$$

where Q^* is net all-wave radiation flux; Q_F is the anthropogenic heat flux or heat flux due to combustion; Q_H is the sensible heat flux; Q_E is the latent heat flux; ΔQ_S are the heat storage changes in the ground, buildings, and air within the volume; and ΔQ_A is the advection of sensible and latent heat through the sides of the building-air volume (Oke, 1987). Each component of the urban surface energy balance is discussed in the following sections.

9.3.1 Shortwave radiation

Net all-wave radiation (Q^*) is the most important energy exchange because it represents the limit to the available energy source or sink (Oke, 1987). The Q^* can be expressed as

$$Q^* = K \downarrow + K \uparrow + L \downarrow + L \uparrow$$

where K↓ is downwelling shortwave radiation, K↑ is upwelling shortwave radiation, L↓ is downwelling longwave radiation, and L↑ is upwelling longwave radiation. As defined, Q^* is positive when the surface gains energy and negative when the surface loses energy.

In 1970, the consensus was that the average city received 15–20% less downwelling shortwave radiation than unpolluted rural areas (Landsberg, 1970; Oke, 1988). This estimate remains consistent for industrial cities where coal-burning or industrial processing is prevalent (Oke, 1988). However, legislation such as the Clean Air Act resulted in decreased pollution levels over many US cities (Peterson and Stoffel, 1980). As a result, more recent studies estimated that urban sites received 2–10% less downwelling shortwave radiation than rural sites (Bergstrom and Peterson, 1977; White et al., 1978; Oke and McCaughey, 1983; Oke, 1988; Christen and Vogt, 2004). When analyzed by season, radiation depletion by aerosols was greater during the winter than summer due, in part, to a greater solar path length in winter (Changnon et al., 1971; Peterson and Stoffel, 1980).

Urban areas are characterized by lower values of surface albedo, which decreases upwelling shortwave radiation. Several aspects of the urban canopy contribute to a decreased albedo for urban areas. For example, as building height and density increases, urban street canyons capture more solar radiation due to multiple reflections from the canyon walls (Aida, 1982; Aida and Gotoh, 1982; Brest, 1987; Oke, 1988; Arnfield, 1988; Christen and Vogt, 2004). Materials such as asphalt for roads and tar on roofs have lower values of albedo than rural surfaces (Oke, 1988). Albedo is also impacted by the presence or absence of vegetation. Suburban land use has an albedo similar to that of rural land use due to the amount of plant cover (White et al., 1978; Brest, 1987). The presence of snow during winter months can result in large differences between urban and rural values of albedo. The urban albedo is significantly lower in the presence of snow than rural areas due to several factors: snow removal, soiling of snow by cars and pollution, snow-free vertical walls, and faster snow melt within the city (Brest, 1987; Oke, 1987, 1988; Christen and Vogt, 2004).

Suggested values of albedo for urban and suburban land use are 0.14 and 0.15, respectively (Oke, 1988). However, observed values of mean albedo for urban centers are as low as 0.08–0.10 compared to values of approximately 0.20 for rural areas (Offerle et al., 2003; Christen and Vogt, 2004; Lemonsu et al., 2004). Albedo of the urban canopy does display a diurnal variation whereby values reach a minimum at solar soon and increase with increasing solar zenith angle. As a result, a diurnal variation of albedo of 3–4% (Aida, 1982) is possible.

Because aerosols decrease downwelling shortwave radiation and albedo differences decrease upwelling shortwave radiation, the urban-rural differences in net shortwave radiation are not large. However, the balance or imbalance of these effects varies with urban geometry and construction material. Thus, some urban-rural differences in net shortwave radiation are significant and dependent upon location (Christen and Vogt, 2004).

9.3.2 Longwave radiation

Urbanization alters the infrared radiative properties of the surface and the atmosphere (Oke, 1988). For example, observations demonstrate that urban pollution enhances downwelling longwave radiation at the surface, with urban-rural differences peaking between 2% and 25% on clear nights (Oke and Fuggle, 1972; Aida and Yaji, 1979; Estournel et al., 1983; Nunez et al., 2000). At the same time, Christen and Vogt (2004) found that urban values of downwelling longwave radiation were lower than rural values during the day due to drier air within the urban boundary layer. Thus, because downwelling longwave radiation is affected by urban boundary layer temperature, humidity, and aerosol composition and concentration, daytime trends in the urban-rural differences of downwelling longwave radiation can vary.

The infrared radiative properties of the land surface are altered by large differences between urban-rural surface characteristics. As such, the surface temperature of urban areas tends to be greater than rural zones (Christen and Vogt, 2004). For example, satellite and *in-situ* measurements have shown that urban surface temperatures are 1–5°C higher than those over croplands (White et al., 1978; Jin et al., 2005). In addition, peak surface heating within urban areas occurs approximately 1 h after solar noon, with residential land use having lower surface temperatures than industrial or commercial land use (White et al., 1978). As a result of the increased thermal capacity of urban areas, more upwelling longwave radiation is emitted from urban surfaces than rural. Urban-rural differences in downwelling longwave radiation are strongest in the evening and decrease throughout the night due to radiation trapping within urban street canyons.

9.3.3 Net radiation

Overall, the attenuation of downwelling shortwave radiation by aerosols in urban areas tends to be offset by the enhancement of downwelling longwave radiation. Similarly, the reduction in upwelling shortwave radiation due to decreased albedo is generally offset by the increase in upwelling longwave radiation. The net outcome results in only slight differences in net radiation received over urban, suburban, and rural surfaces (Peterson and Stoffel, 1980; Oke and Fuggle, 1972; White et al., 1978; Kalanda et al., 1980; Oke and McCaughey, 1983; Cleugh and Oke, 1986; Oke, 1988; Christen and Vogt, 2004).

9.3.4 Anthropogenic heat flux

Anthropogenic heat flux is defined as the release of heat due to the combustion of fuels and electric heating (Oke, 1988). The average anthropogenic heat flux depends on the average energy use by individuals as well as the population density of a city. As a result, the largest values of anthropogenic heat fluxes are often found in the urban core of cities with cold climates (Klysik, 1996). Due to the dependence on human activity, anthropogenic heat flux often exhibits diurnal, weekly, and annual patterns, with peak periods in the morning and evening of weekdays and during winter.

Because anthropogenic heat flux is difficult to measure, it is often omitted from the observed surface energy balance with the assumption that it is embedded within the other energy fluxes (Oke and Cleugh, 1987; Grimmond and Oke, 2002). However, in energy balance studies, it is important to assess the spatial and temporal variability of anthropogenic heat flux to determine the significance of its role in boundary layer processes. For example, heat added to the atmosphere from traffic and subway vents primarily impacts the urban canopy layer, whereas heat injected from smokestacks and chimneys impacts the urban boundary layer (Oke, 1988). Thus, anthropogenic heat flux must be simulated using models or inventory methods based on traffic and energy consumption (Grimmond, 1988; Schmid et al., 1991; Sailor and Lu, 2004) or measurements of the components of the surface energy balance (Pigeon et al., 2003; Offerle et al., 2005).

9.3.5 Storage heat flux

Storage heat flux is defined as the total heat uptake or release from the urban system and includes sensible and latent heat changes in the air, buildings, vegetation, and ground extending from above roof level to a depth in the ground where net heat exchange is negligible (Oke and Cleugh, 1987). Because the thermal conductivity of most building materials is higher and the heat capacity is lower than those of rural soils (Landsberg, 1981; Oke and Cleugh, 1987; Oke, 1988), storage heat flux of urban areas is large in comparison with the ground heat flux at the surface measured in rural areas. As such, storage heat flux is considered to be a key factor in the formation of the urban heat island.

In contrast to rural surfaces, measurement of storage heat flux in the urban environment is complicated by the large number of surface types, orientations, and interactions within the urban canopy (Oke and McCaughey, 1983; Oke, 1988; Christen and Vogt, 2004). Urban-rural comparisons have revealed that storage heat flux values at urban sites were significantly higher than at rural sites (Christen and Vogt, 2004; Offerle et al., 2006a). Because, the urban canopy represents a much larger three-dimensional surface than rural areas, the urban environment can store more heat than the rural surface. For example, as the day progresses, building surfaces that were previously shaded become illuminated as the sun angle decreases. Thus, the vertical structures become an efficient heat sink when horizontal surfaces are already in equilibrium (Christen and Vogt, 2004). As a result, the daytime storage heat flux into buildings is counterbalanced by nocturnal release of storage heat flux in the form of upwelling longwave radiation, sensible heat flux, and latent heat flux. This nocturnal release reaches a maximum 1–2 h after sunset and decreases throughout the night.

9.3.6 Turbulent heat fluxes

The general focus of the urban surface energy balance is on the magnitude, sign, temporal variability, and spatial variability of turbulent latent and sensible heat fluxes (Kalanda et al., 1980). Often it is assumed that latent heat flux is significantly lower in urban areas than in rural areas due to the abundance of impervious construction materials and sparse vegetation (Peterson, 1969; Christen and Vogt, 2004). However, surface energy balance studies have illustrated that urban and suburban areas can have significant latent heat flux (Oke, 1979; Suckling, 1980; Kalanda et al., 1980; Oke and McCaughey, 1983; Cleugh and Oke, 1986; Oke, 1988; Grimmond and Oke, 1995; Spronken-Smith et al., 2000; Moriwaki and Kanda, 2004; Grimmond et al., 2004; Offerle et al., 2006a).

Due to reduced water availability, sensible heat flux is the most significant heat flux in the urban environment (Nunez and Oke, 1977; Cleugh and Oke, 1986; Grimmond and Oke, 1995; Christen and Vogt, 2004; Pearlmutter et al., 2005). Urban-rural comparisons revealed that urban daytime values of sensible heat flux are typically twice as large as rural values (Hildebrand and Ackerman, 1984; Christen and Vogt, 2004). Relative to net radiation, sensible heat flux accounts for approximately 50–95% of available energy in the urban core, 35–50% in suburban neighborhoods, and 30% or less in rural areas during the summer (Grimmond et al., 2004; Christen and Vogt, 2004; Offerle et al., 2006a). However, during the winter these values can approach 130–140% due to the contribution by anthropogenic heat flux (Offerle et al., 2006b).

Overall, the diurnal cycle of sensible heat flux resembles net radiation early in the day but typically peaks 1–2 h later than net radiation; this trend varies with each city (Grimmond and Oke, 1995). Further, sensible heat flux retains a tail of relatively high values into the late afternoon and remains positive after net radiation becomes negative in the evening due to the heat storage, which is similar to desert-like environments (Yap and Oke, 1974; Kalanda et al., 1980; Oke, 1988; Grimmond and Oke, 1995; Grimmond et al., 2004; Offerle et al., 2006b). Unlike rural and suburban surfaces, often sensible heat flux remains positive, or directed away from the surface, throughout the night. As a result, the urban boundary layer is often near neutral or unstable (Yap and Oke, 1974; Kalanda et al., 1980; Oke, 1988; Christen and Vogt, 2004; Grimmond et al., 2004). The positive sensible heat flux values are considered a significant contributor to the development of the urban heat island (Yap and Oke, 1974).

The lack of vegetation in urban areas can result in seasonal changes in sensible heat flux that are minimal when compared to rural seasonal trends (Offerle et al., 2006b). However, due to the impact of anthropogenic heat flux, observations of sensible heat flux in winter can be greater than other periods of the year (Offerle et al., 2006b).

Recent work has revealed that urban-rural trends in latent heat flux are more complicated than once thought. Typically, the latent heat flux values measured in urban areas are relatively small due to small fractions of vegetation cover, limited open water, and enhanced run-off reducing the availability of surface water (Christen and Vogt, 2004; Grimmond et al., 2004). However, latent heat flux is not negligible. Sources of latent heat flux in urban and suburban areas include dewfall, irrigation (lawns, gardens, golf courses, and cemeteries), open water (lakes, swimming pools, and ponds), street cleaning, and leakage from water mains and sewers tapped by deep-rooting trees (Kalanda et al., 1980; Oke and McCaughey, 1983; Grimmond et al., 2004). In addition, Moriwaki and Kanda (2004) found that urban construction materials, such as concrete and asphalt, can absorb water and serve as a significant source of latent heat flux.

During daytime, latent heat flux is approximately 20% of net radiation in the urban core, 30% in suburban neighborhoods, and 60% in rural areas (Christen and Vogt, 2004). However, practices such as irrigation in suburban neighborhoods, have a significant impact on latent heat flux. Kalanda et al. (1980) found that even under mild drought conditions, much of the available energy in the suburban area was partitioned into latent heat flux due to lawn irrigation. Furthermore, a downward flux or horizontal advection of sensible heat can enhance evaporation and result in an oasis effect whereby the hourly and daily values of latent heat flux exceed net radiation (Oke, 1979; Suckling, 1980; Oke and McCaughey, 1983; Cleugh and Oke, 1986). The oasis effect occurs when warm dry air, likely heated by pavement and buildings, is advected across an irrigated lawn or garden. A large moisture gradient develops and creates a high potential for evaporation. In addition to suburban neighborhoods, the oasis effect can be observed in urban parks. Spronken-Smith et al. (2000) observed that a wet park evaporated three times more water than the surrounding residential neighborhood, and 1.3–1.4 times more than an irrigated rural sod farm.

Sparse or patchy urban vegetation can impact latent heat flux because it transpires at a higher rate than a completely vegetated surface. For example, the radiation and turbulence around an isolated tree increases latent heat flux at the leaf surface over most of the tree versus a tree where radiation and turbulence affect evaporation mainly at the upper part of the tree within a homogeneous forest (i.e., increased surface area available for latent heat flux; Moriwaki and Kanda, 2004).

Latent heat flux is also an important energy sink during the morning hours as water from irrigation and dewfall are evaporated. Once the water sources are exhausted, sensible heat flux becomes dominant and remains so throughout the day (Oke and Cleugh, 1987). However, a secondary peak in latent heat flux may occur in the evening due to lawn irrigation (Grimmond and Oke, 1995). Further, the amount of vegetation cover exerts some control over the diurnal cycle of latent heat flux. At heavily built sites, latent and sensible heat fluxes increase simultaneously, whereas at vegetated sites, sensible heat flux lags latent heat flux (Offerle et al., 2006a). During the nighttime, latent heat flux is often away from the surface (Kalanda et al., 1980; Christen and Vogt, 2004; Offerle et al., 2006a). Despite the day-to-day variability of fluxes, many cities display similarities in the timing of peaks and changes in sign of the fluxes (Grimmond and Oke, 1995).

9.3.7 Advection of turbulent fluxes

Direct measurement of the advection of latent and sensible heat through the sides of the building-air volume is very difficult over a complex surface such as the urban canopy. However, past studies at the street canyon-scale demonstrated that advection of turbulent fluxes depends on the wind speed and amount of energy available to the canyon system (Nunez and Oke, 1977). Because of the difficulty associated with obtaining measurements, most measurement programs assume advective influences are negligible (Oke, 1988). However, horizontal variations in latent and sensible heat fluxes in the presence of a mean flow induce advection (Schmid et al., 1991). As such, advection of turbulent fluxes can be estimated from the spatial analysis of the variability of latent and sensible heat fluxes.

Ching et al. (1983) determined that the magnitude and impact of thermal advection is important and can dominate the variability in latent and sensible heat fluxes observed in an urban area. Because land use exerts significant control on latent and sensible heat fluxes, large differences in the heat fluxes were found over different land use types for different seasons (Ching et al., 1983).

9.4   Urban heat island

The most notable impact of urbanization is the urban heat island effect where air temperature in the urban canopy is warmer compared to the rural surroundings. The magnitude of the urban heat island varies in time, in space, and according to the large-scale weather conditions (Kim and Baik, 2005; Fast et al., 2005). The urban heat island intensity, the difference between urban air temperatures and that of the surrounding rural environment, is largest during clear, calm conditions (Ackerman, 1985; Oke, 1987; Kim and Baik, 2005). The urban-rural differences are smallest during cloudy, windy conditions (Vukovich et al., 1976; Fast et al., 2005).

Figure 9.4 shows a vertical cross-section of a typical urban heat island. The rural-suburban boundary is characterized by a steep increase in air temperature. A weaker temperature increase occurs toward the urban core. Often, the urban temperature maximum is co-located with or slightly downwind of the urban core.

The urban heat island intensity is defined according to the availability of air temperature observations. For example, when a limited number of observation sites are available, the urban heat island intensity is often calculated as the difference between the background rural and maximum urban temperatures (Oke, 1973). However, when numerous observation sites are available, urban heat island intensity can be calculated as the difference between mean urban and mean rural temperatures. Although the urban heat island is defined according to air temperature, studies have investigated the urban heat island based on remotely sensed land surface temperatures (Figure 9.5; Roth et al., 1989; Epperson et al., 1995; Jin et al., 2005).

The urban heat island intensity varies across several time scales. After sunset, the urban heat island intensity grows rapidly and reaches a maximum approximately 3–5 h later (Ackerman, 1985; Vukovich et al., 1979; Wanner and Hertig, 1984; Kim and Baik, 2005; Fast et al., 2005). Once at peak intensity, it decreases slowly throughout the night, but may still be detectable at sunrise in large cities (Landsberg, 1981). As a result, minimum temperature values in urban areas may be warmer than rural minimum temperatures (Cayan and Douglas, 1984; Comrie, 2000; Baker et al., 2002). After sunrise, the urban heat island intensity rapidly decreases due to the slower warm-up of the urban area relative to rural areas as a result of the high heat capacity (the amount of heat necessary to change the temperature) of urban materials (e.g.,

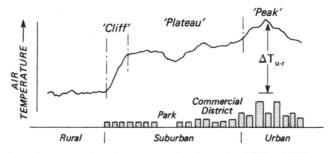

Figure 9.4 Vertical cross-section of a typical urban heat island (Oke, 1987).

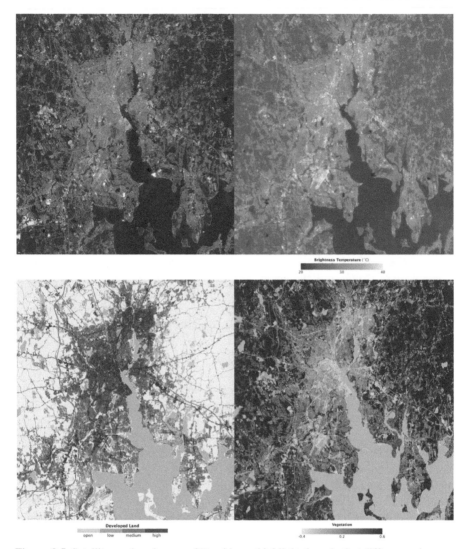

Figure 9.5 Satellite-produced maps of Providence highlight the role that differences in development patterns and vegetation cover can have on the magnitude of a city's urban heat island.
Credit: NASA/Earth Observatory.

concrete, asphalt). The urban heat island intensity may be negative at midday when the rural surroundings are warmer than the urban core (Unwin, 1980; Ackerman, 1985; Wanner and Hertig, 1984; Peterson, 2003; Kim and Baik, 2005). A negative urban heat island intensity is also called an urban cool island.

On a weekly time scale, studies have detected a stronger urban heat island on weekdays than weekends due to heavy traffic and/or high-commercial activities (Kim and Baik, 2005). The urban heat island intensity also varies with season, with the largest values occurring during the summer (Wanner and Hertig, 1984; Ackerman, 1985). A second maximum in urban heat island intensity can be found during the winter as a result of the increased impacts of anthropogenic heating.

The magnitude of the urban heat island intensity is primarily related to urban geometry, construction material, and the amount of anthropogenic heat released (Oke, 1987). Urban geometry, in the form of building height and density (i.e., the distance between buildings), plays a significant role in the trapping of radiation and reducing wind speeds within the urban canopy layer. Building and traffic heat serve as anthropogenic heat sources. Pollution traps heat radiated from the surface. The construction materials of the urban canopy increase the heat storage and decrease evapotranspiration. The changes in the surface energy balance combine to impact the development of the urban heat island, with some factors being more important according to the time of day. The overall size of the urban area has a minimal impact on the magnitude of the urban heat island intensity (Atkinson, 2003).

The time and space variations of urban-rural humidity differences are smaller and more complex than those of temperature. For mid-latitude cities, the urban canopy air is typically drier during the day and more humid during the night (Hage, 1975; Changnon, 1981; Hildebrand and Ackerman, 1984; Oke, 1987; Ackerman, 1987). During the summer, daytime humidity values are higher in rural areas due to evapotranspiration from the surface, with maximum humidity values occurring at sunset (Hilberg, 1978). The impermeable surfaces within the city convert precipitation into run-off rapidly via sewer systems and reduce surface evaporation. Urban areas have limited vegetation and larger roughness elements (i.e., trees and buildings), which limit transpiration and enhance vertical mixing, respectively (Sisterson and Dirks, 1978). While combustion, open water (pools, ponds, canals), and irrigation provide moisture to the atmosphere, the impacts are not enough to counteract the reduced evapotranspiration due to vegetated surfaces being replaced with impervious surfaces (Brazel and Balling, 1986). As a result of expanding impervious surfaces and rapid run-off, the urban heat island can be accompanied by a daytime dry "island," even in arid climates (Hilberg, 1978; Brazel and Balling, 1986).

Rural humidity decreases during the night due to dew formation and reaches a minimum at sunrise (Hilberg, 1978). In the urban canopy at night, anthropogenic moisture, reduced dewfall, and stagnant airflow combine to maintain a more humid atmosphere and produce a moisture "island" similar to that of temperature (Oke, 1987), with maximum humidity values occurring after midnight (Hilberg, 1978). During the winter, daytime humidity values can be higher in the city than in rural areas, especially when the ground is covered with snow or ice, vegetation is dormant, and anthropogenic fluxes of moisture are significant (Hage, 1975; Oke, 1987). If a daytime urban heat island is present, a portion of the urban-rural humidity difference is attributed to the increased air temperature (Landsberg and Maisel, 1972).

Figure 9.6 illustrates the temperature profile of the urban boundary layer during the day and night. During the day (Figure 9.6A), the influence of the urban canopy can extend up to 1.5 km (0.9 miles) above the surface. At night, the urban boundary layer contracts to 0.1–0.5 km (0.1–0.3 miles). Limited information is available regarding the vertical profile of humidity in the urban boundary layer.

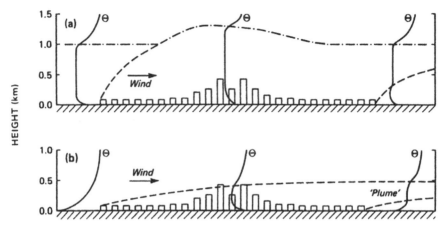

Figure 9.6 Profiles of potential temperature illustrating the thermal structure of the urban boundary layer in a large city during the (A) day and (B) at night (Oke, 1987).

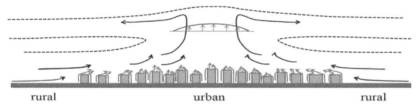

rural   urban   rural

Figure 9.7 The multi-scale nature of convection over urban areas.
Credit: Wang, X., Li, Y., 2016. Predicting urban heat island circulation using CFD. Build. Environ. 99, 82–97.

In combination with the urban heat island, an urban-rural circulation pattern develops in which low-level air converges into the urban area from all directions (Figure 9.7; Changnon et al., 1971; Shreffler, 1979a,b; Landsberg, 1981; Oke, 1987; Eliasson and Holmer, 1990). It was originally thought that the strength of the urban heat island circulation was directly related to the strength of the urban heat island (Chandler, 1965). However, Shreffler (1979a,b) found that a stronger circulation occurs with weak urban heat islands during the daytime than with strong urban heat islands during the nighttime. Studies have also shown that atmospheric stability plays a key role in determining the strength of the urban heat island circulation (Vukovich and Dunn, 1978; Draxler, 1986). During the daytime, heating in the boundary layer is distributed through a much deeper layer than the nocturnal boundary layer, and as a result, lower pressure and faster horizontal and vertical winds at the surface contribute to a more intense urban heat island circulation (Vukovich and Dunn, 1978). The urban heat island circulation is strongest in the early afternoon and weakens near sunset as the urban heat island strengthens (Shreffler, 1979b; Landsberg, 1981). The rising motion due to the urban heat island and frictional effects can cause the urban boundary layer to "dome" over the city by approximately 250 m (820 ft) in the daytime (Oke, 1987) while downwind of the city the airflow sinks over the rural land surface.

References

Ackerman, B., 1985. Temporal march of the Chicago heat island. J. Clim. Appl. Meteorol. 24, 547–554.

Ackerman, B., 1987. Climatology of Chicago area urban-rural differences in humidity. J. Clim. Appl. Meteorol. 26, 427–430.

Aida, M., 1982. Urban albedo as a function of the urban structure – a model experiment. Bound.-Layer Meteorol. 23, 405–413.

Aida, M., Gotoh, K., 1982. Urban albedo as a function of the urban structure – a two-dimensional numerical simulation. Bound.-Layer Meteorol. 23, 415–424.

Aida, M., Yaji, M., 1979. Observations of downward atmospheric radiation in the Tokyo area. Bound.-Layer Meteorol. 16, 453–465.

Arnfield, A.J., 1988. Validation of an estimation model for urban surface albedo. Phys. Geogr. 9, 361–372.

Arnfield, A.J., 2003. Two decades of urban climate research: a review of turbulence, exchange of energy and water, and the urban heat island. Int. J. Climatol. 23, 1–26.

Atkinson, B.W., 2003. Numerical modeling of urban heat-island intensity. Bound.-Layer Meteorol. 109, 285–310.

Baker, L.A., Brazel, A.J., Selover, N., Martin, C., McIntyre, N., Steiner, R.S., Nelson, A., Musacchio, L., 2002. Urbanization and warming of Phoenix (Arizona, USA): impacts, feedbacks and mitigation. Urban Ecosyst. 6, 183–203.

Bergstrom, R.W., Peterson, J.T., 1977. Comparison of predicted and observed solar radiation in an urban area. J. Appl. Meteorol. 16, 1107–1115.

Bornstein, R.D., Johnson, D.S., 1977. Urban-rural wind velocity differences. Atmos. Environ. 11, 597–604.

Brazel, S.W., Balling Jr., R.C., 1986. Temporal analysis of long-term atmospheric moisture levels in Phoenix, Arizona. J. Clim. Appl. Meteorol. 25, 112–117.

Brest, C.L., 1987. Seasonal albedo of an urban/rural landscape from satellite observations. J. Clim. Appl. Meteorol. 26, 1169–1187.

Castro, I.P., Cheng, H., Reynolds, R., 2006. Turbulence over urban-type roughness: deductions from wind-tunnel measurements. Bound.-Layer Meteorol. 118, 109–131.

Cayan, D.R., Douglas, A.V., 1984. Urban influences on surface temperatures in the southwestern United States during recent decades. J. Clim. Appl. Meteorol. 23, 1520–1530.

Chandler, T.J., 1965. The Climate of London. Hutchinson, London. 292 pp.

Changnon Jr., S.A. (Ed.), 1981. METROMEX: A Review and Summary. Meteorological Monographs, American Meteorological Society. No. 40, 181 pp https://link.springer.com/book/10.1007/978-1-935704-29-4#bibliographic-information.

Changnon, S.A., Huff, F.A., Semonin, R.G., 1971. METROMEX: an investigation of inadvertent weather modification. Bull. Am. Meteorol. Soc. 52, 958–967.

Ching, J.K.S., Clarke, J.F., Godowitch, J.M., 1983. Modulation of heat flux by different scales of advection in an urban environment. Bound.-Layer Meteorol. 25, 171–191.

Christen, A., Vogt, R., 2004. Energy and radiation balance of a central European city. Int. J. Climatol. 24, 1395–1421.

Cleugh, H.A., Oke, T.R., 1986. Suburban-rural energy balance comparisons in summer for Vancouver, B.C. Bound.-Layer Meteorol. 36, 351–369.

Coceal, O., Thomas, T.G., Castro, I.P., Belcher, S.E., 2006. Mean flow and turbulence statistics over groups of urban-like cubicle obstacles. Bound.-Layer Meteorol. 121, 491–519.

Comrie, A.C., 2000. Mapping a wind-modified urban heat island in Tucson, Arizona (with comments on integrating research and undergraduate learning). Bull. Am. Meteorol. Soc. 81, 2417–2431.

Draxler, R.R., 1986. Simulated and observed influence of the nocturnal urban heat island on the local wind field. J. Clim. Appl. Meteorol. 25, 1125–1133.

Eliasson, I., Holmer, B., 1990. Urban heat island circulation in Göteborg, Sweden. Theor. Appl. Climatol. 42, 187–196.

Eliasson, I., Offerle, B., Grimmond, C.S.B., Lindqvist, S., 2006. Wind fields and turbulence statistics in an urban street canyon. Atmos. Environ. 40, 1–16.

Epperson, D.L., Davis, J.M., Bloomfield, P., Karl, T.R., McNab, A.L., Gallo, K.P., 1995. Estimating the urban bias of surface shelter temperatures using upper-air and satellite data. Part II: estimation of the urban bias. J. Appl. Meteorol. 34, 358–370.

Estournel, C., Vehil, R., Guedalia, D., Fontan, J., Druilhet, A., 1983. Observations and modeling of downward radiative fluxes (solar and infrared) in urban/rural areas. J. Clim. Appl. Meteorol. 22, 134–142.

Fast, J.D., Torcolini, J.C., Redman, R., 2005. Pseudovertical temperature profiles and the urban heat island measured by a temperature datalogger network in Phoenix, Arizona. J. Appl. Meteorol. 44, 3–13.

Graham, I.R., 1968. An analysis of turbulence statistics at Fort Wayne, Indiana. J. Appl. Meteorol. 7, 90–93.

Grimmond, C.S.B., 1988. An Evaporation-Interception Model for Urban Areas. Ph.D. Dissertation, The University of British Columbia, Vancouver, BC. 206 pp.

Grimmond, C.S.B., Oke, T.R., 1995. Comparison of heat fluxes from summertime observations in the suburbs of four North American cities. J. Appl. Meteorol. 34, 873–889.

Grimmond, C.S.B., Oke, T.R., 2002. Turbulent heat fluxes in urban areas: observations and a local-scale urban meteorological parameterization scheme (LUMPS). J. Appl. Meteorol. 41, 792–810.

Grimmond, C.S.B., Salmond, J.A., Oke, T.R., Offerle, B., Lemonsu, A., 2004. Flux and turbulence measurements at a densely built-up site in Marseille: heat, mass (water and carbon dioxide), and momentum. J. Geophys. Res. 109, D24101. https://doi.org/10.1029/2004JD004936.

Hage, K.D., 1975. Urban-rural humidity differences. J. Appl. Meteorol. 14, 1277–1283.

Hilberg, S.D., 1978. Diurnal temperature and moisture cycles. In: Summary of METROMEX, Vol. 2: Causes of Precipitation Anomalies, pp. 25–42. Ill. State Water Survey Bull. 63.

Hildebrand, P.H., Ackerman, B., 1984. Urban effects on the convective boundary layer. J. Atmos. Sci. 41, 76–91.

Jin, M., Dickinson, R.E., Zhang, D.-L., 2005. The footprint of urban areas on global climate as characterized by MODIS. J. Clim. 18, 1551–1565.

Kalanda, B.D., Oke, T.R., Spittlehouse, D.L., 1980. Suburban energy balance estimates for Vancouver, B.C., using the Bowen Ratio-Energy Balance approach. J. Appl. Meteorol. 19, 791–802.

Kastner-Klein, P., Berkowicz, R., Britter, R., 2004. The influence of street architecture on flow and dispersion in street canyons. Meteorog. Atmos. Phys. 87, 121–131.

Kim, Y.-H., Baik, J.-J., 2005. Spatial and temporal structure of the urban heat island in Seoul. J. Appl. Meteorol. 44, 591–605.

Klysik, K., 1996. Spatial and seasonal distribution of anthropogenic heat emissions in Lódz, Poland. Atmos. Environ. 30, 3397–3404.

Landsberg, H.E., 1970. Micrometeorological temperature differentiation through urbanization. In: Urban Climate. World Meteorological Organization (WMO), pp. 129–136. WMO Tech. Note, No. 108.

Landsberg, H.E., 1981. The Urban Climate. Academic Press. 275 pp.

Landsberg, H.E., Maisel, T.N., 1972. Micrometeorological observations in an area of urban growth. Bound.-Layer Meteorol. 2, 365–370.

Lemonsu, A., Grimmond, C.S.B., Masson, V., 2004. Modeling the surface energy balance of the core of an old Mediterranean city: Marseille. J. Appl. Meteorol. 43, 312–327.

Moriwaki, R., Kanda, M., 2004. Seasonal and diurnal fluxes of radiation, heat, water vapor, and carbon dioxide over a suburban area. J. Appl. Meteorol. 43, 1700–1710.

Nunez, M., Oke, T.R., 1977. The energy balance of an urban canyon. J. Appl. Meteorol. 16, 11–19.

Nunez, M., Eliasson, I., Lindgren, J., 2000. Spatial variations of incoming longwave radiation in Göteborg, Sweden. Theor. Appl. Climatol. 67, 181–192.

Offerle, B., Grimmond, C.S.B., Oke, T.R., 2003. Parameterization of net all-wave radiation for urban areas. J. Appl. Meteorol. 42, 1157–1173.

Offerle, B., Grimmond, C.S.B., Fortuniak, K., 2005. Heat storage and anthropogenic heat flux in relation to the energy balance of a central European city centre. Int. J. Climatol. 25, 1405–1419.

Offerle, B., Grimmond, C.S.B., Fortuniak, K., Pawlak, W., 2006a. Intraurban differences of surface energy fluxes in a central European city. J. Appl. Meteor. Climatol. 45, 125–136.

Offerle, B., Grimmond, C.S.B., Fortuniak, K., Klysik, K., Oke, T.R., 2006b. Temporal variations in heat fluxes over a central European city centre. Theor. Appl. Climatol. 84, 103–115.

Oke, T.R., 1973. City size and the urban heat island. Atmos. Environ. 7, 769–779.

Oke, T.R., 1979. Advectively-assisted evapotranspiration from irrigated urban vegetation. Bound.-Layer Meteorol. 17, 167–173.

Oke, T.R., 1987. Boundary Layer Climates, second ed. Routledge. 435 pp.

Oke, T.R., 1988. The urban energy balance. Prog. Phys. Geogr. 12, 471–508.

Oke, T.R., Cleugh, H.A., 1987. Urban heat storage derived as energy balance residuals. Bound.-Layer Meteorol. 39, 233–245.

Oke, T.R., Fuggle, R.F., 1972. Comparison of urban/rural counter and net radiation at night. Bound.-Layer Meteorol. 2, 290–308.

Oke, T.R., McCaughey, J.H., 1983. Suburban-rural energy balance comparisons for Vancouver, B.C.: an extreme case? Bound.-Layer Meteorol. 26, 337–354.

Pearlmutter, D., Berliner, P., Shaviv, E., 2005. Evaluation of urban surface energy fluxes using an open-air scale model. J. Appl. Meteorol. 44, 532–545.

Peterson, J.T., 1969. The Climate of Cities: A Survey of Recent Literature. Publication No. AP-59, National Air Pollution Control Administration, U.S. Government Printing Office, Washington, DC. 48 pp.

Peterson, T.C., 2003. Assessment of urban versus rural in situ surface temperatures in the contiguous United States: no difference found. J. Clim. 16, 2941–2959.

Peterson, J.T., Stoffel, T.L., 1980. Analysis of urban-rural solar radiation data from St. Louis, Missouri. J. Appl. Meteorol. 19, 275–283.

Pigeon, G., Lemonsu, A., Masson, V., Durand, P., 2003. Sea-town interactions over Marseille – Part II: Consequences on atmospheric structure near the surface. In: Preprints, 5th Int. Conf. on Urban Climate, Lodz, Poland. Int. Assoc. Urban Climate, 3.4.

Rafailidis, S., 1997. Influence of building areal density and roof shape on the wind characteristics above a town. Bound.-Layer Meteorol. 85, 255–271.

Roth, M., 2000. Review of atmospheric turbulence over cities. Q. J. R. Meteorol. Soc. 126, 941–990.

Roth, M., Oke, T.R., 1995. Relative efficiencies of turbulent transfer of heat, mass, and momentum over a patchy urban surface. J. Atmos. Sci. 52, 1863–1874.

Roth, M., Oke, T.R., Emery, W., 1989. Satellite-derived urban heat islands from three coastal cities and the utility of such data in urban climatology. Int. J. Remote Sens. 10, 1699–1720.

Sailor, D.J., Lu, L., 2004. A top-down methodology for developing diurnal and seasonal anthropogenic heating profiles for urban areas. Atmos. Environ. 38, 2737–2748.

Schmid, H.P., Cleugh, H.A., Grimmond, C.S.B., Oke, T.R., 1991. Spatial variability of energy fluxes in suburban terrain. Bound.-Layer Meteorol. 54, 249–276.

Shea, D.M., Auer, A.H., 1978. Thermodynamic properties and aerosol patterns in the plume downwind of St. Louis. J. Appl. Meteorol. 17, 689–698.

Shreffler, J.H., 1979a. Urban-rural differences in tower-measured winds, St. Louis. J. Appl. Meteorol. 18, 829–835.

Shreffler, J.H., 1979b. Heat island convergence in St. Louis during calm periods. J. Appl. Meteorol. 18, 1512–1520.

Sisterson, D.L., Dirks, B.A., 1978. Structure of the daytime urban moisture field. Atmos. Environ. 12, 1943–1949.

Spronken-Smith, R.A., Oke, T.R., Lowry, W.P., 2000. Advection and the surface energy balance across an irrigated urban park. Int. J. Climatol. 20, 1033–1047.

Suckling, P.W., 1980. The energy balance microclimate of a suburban lawn. J. Appl. Meteorol. 19, 606–608.

Thom, A.S., 1972. Momentum, mass and heat exchange of vegetation. Q. J. R. Meteorol. Soc. 98, 124–134.

Thom, A.S., Stewart, J.B., Oliver, H.R., Gash, J.H.C., 1975. Comparison of aerodynamic and energy budget estimates of fluxes over a pine forest. Q. J. R. Meteorol. Soc. 101, 93–105.

Unwin, D.J., 1980. The synoptic climatology of Birmingham's heat island, 1965–1974. Weather 35, 43–50.

Vukovich, F.M., Dunn, J.W., 1978. A theoretical study of the St. Louis heat island: some parameter variations. J. Appl. Meteorol. 17, 1585–1594.

Vukovich, F.M., Dunn III, J.W., Crissman, B.W., 1976. A theoretical study of the St. Louis heat island: the wind and temperature distribution. J. Appl. Meteorol. 15, 417–440.

Vukovich, F.M., King, W.J., Dunn III, J.W., Worth, J.J.B., 1979. Observations and simulations of the diurnal variation of the urban heat island circulation and associated variations of the ozone distribution: a case study. J. Appl. Meteorol. 18, 836–854.

Wanner, H., Hertig, J.-A., 1984. Studies of urban climates and air pollution in Switzerland. J. Clim. Appl. Meteorol. 23, 1614–1625.

White, J.M., Eaton, F.D., Auer Jr., A.H., 1978. The net radiation budget of the St. Louis metropolitan area. J. Appl. Meteorol. 17, 593–599.

Wong, K.K., Dirks, R.A., 1978. Mesoscale perturbations on airflow in the urban mixing layer. J. Appl. Meteorol. 17, 677–688.

Yap, D., Oke, T.R., 1974. Sensible heat fluxes over an urban area—Vancouver, B.C. J. Appl. Meteorol. 13, 880–890.

Further reading

Brown, M.J., 2000. Urban parameterizations for mesoscale meteorological models. In: Boybeyi, Z. (Ed.), Mesoscale Atmospheric Dispersion. WIT Press, pp. 193–255.

Wang, X., Li, Y., 2016. Predicting urban heat island circulation using CFD. Build. Environ. 99, 82–97. https://doi.org/10.1016/j.buildenv.2016.01.020.

Coming and going: Transport and tracking

Brian Viner
Atmospheric Technologies Group, Savannah River National Laboratory, Aiken, SC, United States

10.1 Introduction

The atmosphere is filled with many small objects that are moved around by the wind. These objects can come from natural sources such as ash from fires or volcanoes or anthropogenic sources such as pollutants emitted from cars or factories. Sometimes, it can be difficult to picture the need for atmospheric dispersion modeling. While some plumes such as smoke from a fire are easily visible (Figure 10.1), many particles are too small or too few in concentration to be seen. Sand from the Saharan desert often migrates through the atmosphere to be deposited in the southeastern United States, but the only visible evidence of this in the atmosphere is often a haze as well as more vivid sunrises across the Atlantic Ocean as viewed from Florida (Figure 10.2). Other times, we're interested in the movement of chemical gas species through the atmosphere, which are invisible but can have harmful consequences in large quantities.

Atmospheric transport and dispersion describe the process by which particles or gases in the atmosphere move. While gases refer to the transport of chemicals or materials in their vapor form, particles refer to objects in the solid or liquid phase. Another defining characteristic is that particles typically have sufficient mass that gravity has a measurable effect on their movement. While the force of gravity pushing downward may not be sufficient to bring an object to the surface, such as the case of liquid water droplets in clouds where the gravitational settling is insufficient to overcome the turbulent vertical motions of the atmosphere, it has a measurable influence on the movement of the particle. Conversely, gases are light enough that the motion of the atmosphere is the dominant influence on their movement.

The wind is the primary meteorological driver in determining where things go in the atmosphere, but other factors such as temperature and humidity also impact the fate of airborne gases and particles through processes of buoyant vertical motions or controlling the rate at which chemical or physical transformations take place within the atmosphere. Additionally, particles can undergo changes such as clumping together or settling to the surface based on their size, while gases can interact with other chemical species of the atmosphere to change their composition and behavior in the environment.

Beyond meteorology and the environmental sciences, fate and transport modeling is important to several professions and fields, many of which involve

Figure 10.1 Smoke plumes from a prescribed burn. These plumes initially rise quickly vertically due to the convective motions generated by the heat of the fire before cooling and moving with the atmospheric wind.
Credit: US Forest Service – Savannah River.

human health concerns. Forest managers often use prescribed burning to maintain forest health but must know where smoke plumes from the fires will travel to. Allergists and those who suffer seasonal allergies are interested in understanding the triggers that lead to releases of pollens or spores into the atmosphere and how quickly they will be removed from the atmosphere or deposited to the surface. Industry may be interested in performing assessments of effluents from factories to ensure they will be diluted in the atmosphere to limit impacts to nearby populations. In all these cases, experts rely on atmospheric transport modeling to assess where airborne particles will travel to, what their concentration will be downwind, and whether they pose impacts to humans or environments.

Atmospheric transport modeling can also be performed in both a predictive manner as well as a forensic manner. In many cases, we're interested in predicting where an airborne plume of material will move from a known source type and location. Many times, this is done to ensure that releases of material will meet government-mandated regulatory requirements or not exceed industry standards for releases of a specific chemical or aerosol. Other times, however, modeling is performed when a harmful aerosol has been detected in the atmosphere and we need to identify its source. In these cases, models are run "backwards in time" to identify potential regions where the detected material came from.

As in many endeavors, we can better describe the process we are interested in if we have more information. In atmospheric dispersion modeling, this usually requires taking additional measurements, which may include additional sampling locations or taking more samples at a single location to provide a time series of concentration measurements. These may be combined with additional meteorological measurements, which can provide a clearer understanding of the wind or other atmospheric variables to describe the fate and transport of airborne particles or gases we are interested in.

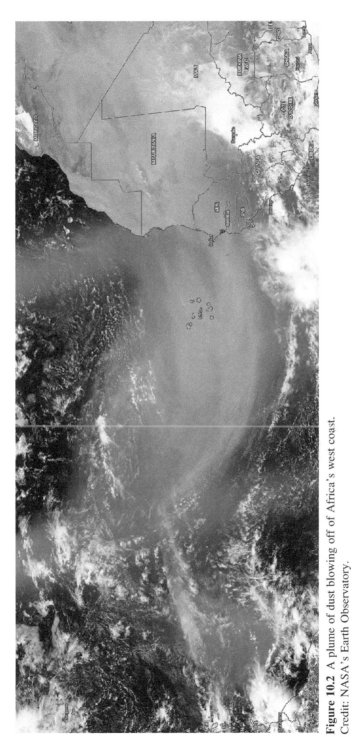

Figure 10.2 A plume of dust blowing off of Africa's west coast.
Credit: NASA's Earth Observatory.

This chapter will focus on the key components needed to perform atmospheric transport modeling. Understanding the meteorological state of the atmosphere is a key component that describes the fate and transport of material in the atmosphere. Also important, however, is making sure we understand what must be considered about the source of the release and how the underlying environment will act to potentially remove material from the atmosphere. The chapter will first provide a brief history of atmospheric dispersion studies that have led to the framework and models that are widely used today. This will be followed by a description of the primary elements necessary to model atmospheric transport: understanding the source, understanding how atmospheric structure affects dispersion, and the models we use to simulate and study airborne plumes. Finally, the chapter will discuss the processes that lead to the removal of airborne gases and particles from the atmosphere.

10.2 History of atmospheric transport studies

Studies of how particles and gases move in the atmosphere have been undertaken for around 100 years. Initially, experiments to understand atmospheric dispersion were focused simply on understanding the nature of turbulence in the atmosphere. The first studies of atmospheric transport modeling were conducted with the goal of understanding how the turbulent portion of the wind affected the growth and spread of the plume. This is a key concept still studied today because even the most complex models cannot resolve all the turbulent motions of the atmosphere to their smallest scales and fluctuations.

Early studies by Taylor (1921), Roberts (1923), and Richardson (1926) focused on simple methods of exploring plume behavior by releasing various tracer particles to the atmosphere and observing their behavior as they moved through the air. These experiments, though simple, formed the basis for atmospheric dispersion modeling that we still use today, such as Richardson's power law for estimating how rapidly a plume spreads horizontally as a function of the size of the plume.

From the 1940s through the 1960s, many atmospheric transport studies were focused on defense and atomic weapons testing. Many of these tests were conducted over small scales, with measurements taken at distances less than a kilometer from their source, but attention was given to understanding how dispersion was affected by different environments. Comparisons of day vs night, low wind conditions vs high wind conditions, and understanding how local environments such as coastal regions affected airborne transport were all topics that began to be considered during this time.

Another key development during that period was the development of atmospheric stability classes by Pasquill (1961). These stability classes were designed to describe the magnitude of turbulence in the atmosphere, which in turn drives the rate of plume diffusion or how much the plume grows. These stability classes, which will be described in more detail in a later section, are still widely used today to predict the growth of airborne plumes as they move through the atmosphere.

In the 1970s, the focus shifted again following passage of the Clean Air Act in 1968. During the 1970s and 1980s, several atmospheric transport studies were concerned with industrial or other pollution sources. Emissions in the United States began to be monitored and regulated by the Environmental Protection Agency. This led to the development of one of the most widely used atmospheric dispersion models, the Industrial Source Complex (ISC3) model and its successor, the American Meteorological Society/Environmental Protection Agency Regulatory Model (AERMOD). AERMOD is a widely distributed Gaussian transport model used to ensure that industrial sources of regulated gases and particulate emissions are within prescribed standards. A variety of supporting software packages have been developed around the AERMOD framework to support more complex atmospheric transport processes including the local-scale turbulence that can be caused by air moving around buildings and as well as accounting for the influence of local topography on atmospheric conditions.

In the last decade, atmospheric dispersion models and studies have focused on a range of scales ranging from local to global considerations. This has been facilitated by the expanding growth and availability of computing power, which has opened up new areas of research by allowing for global-scale problems or highly complex local-scale issues to be studied in relatively short periods. Improvements in meteorological instrumentation since 1990, such as the development and wider availability of sonic anemometers that can measure the turbulent fluctuations in the wind at frequencies up to 20 times per second, have also allowed studies to improve our characterization of small scales of turbulence and how complex building or topography geometry influences the movement and diffusion of airborne plumes.

An area that has received increased attention since 2000 is understanding particle transport in urban canyons. There are several concerns around cities ranging from the movement of pollution particles to the dispersion of hazardous material from terrorist attacks that can now be studied given the advances in measurements, modeling, and computing power.

There has also been considerable support across the international modeling community to improve atmospheric dispersion modeling. Large experiments such as the European Tracer Experiment in 1994 have brought together atmospheric transport scientists from around the globe to test different modeling configurations against a single field experiment and draw conclusions regarding how well these models can be expected to perform in a variety of scenarios. Another large experiment was the Atmospheric Studies in Complex Terrain campaign in 1991 whose goal was to examine nocturnal drainage flows in complex terrain to understand the transition between large-scale flow and local-scale, topographically driven flow.

10.3   Understanding the source

The location where an effluent enters the atmosphere is referred to as the source in atmospheric transport modeling. Sources can be industrial, such as from factories or power plants, or they can be environmental, such as the release of pollen from trees or methane from cows. The complexity of sources can range from single locations (a factory with a single smokestack), multiple locations (an industrial complex with many smokestacks), or even mobile locations (emissions from cars as they drive through cities). Regardless of the source, understanding the behavior of the source is a key component to predicting downwind concentrations and effects. If the source term is not characterized properly, then the predictions of the effluents and their movement and fate in the atmosphere cannot be accurate either. The description of the source term contains three main components: whether the source should be modeled as a point source or an area source, the height of the release, and whether the source strength or rate of effluent is changing over time.

Most effluent sources can be considered as a single location source, which we refer to as a point source. In models, a point source represents the effluent release location as a specific location in space regardless of the exact geometry of the source (is it a small or large smokestack, what shape is it, etc.). Point sources are typically used when the source of a release is sufficiently small relative to the distance scale that we're modeling over that it can be considered a single point. Since most applications of dispersion modeling are concerned with predictions of dispersion over a kilometer to tens of kilometers, sources such as those coming from a building or factory can often be considered a single point release. The other consideration in determining whether a source can be considered as a point source is whether the computational grid is sufficiently larger than the source area.

The computational grid describes the spacing of our grid cells in our numerical model (see Chapter 5 for more information on modeling grids). It is impractical to try and predict a concentration at every point in space and time, so instead we try to predict it at a discrete number of points and time intervals. The grid is usually arranged as a rectangular grid of evenly spaced predictive points (Figure 10.3). However, there have been developments to improve models by implementing a grid that can change size and shape to improve the resolution of the modeled plume where the gradients of the plume concentration are largest, such as near the concentration maxima, or around the edges of the plume where concentrations are very small. Many meteorological computational grids are on the order of tens of meters to kilometers; even very fine-scale modeling typically has grid spacing of at least a few meters. If the source is smaller than the modeling grid (such as a factory smokestack that is 1 m in diameter being modeled with a grid that is 20-m wide), then the source can be considered a point source because it matters little exactly where in the grid the modeled effluent is placed since it will be subject to essentially the same meteorological conditions within the grid.

Area releases are required when the source region is much larger, such as spanning hundreds of meters or being distributed over multiple model grid cells. Smoke plumes from wildfires are a classic example of when an area source is needed. Even though we may be interested in modeling the airborne transport of the smoke plume over hundreds of miles, the extent of the file may cover tens or hundreds of square miles itself. A smaller example of when an area source may be appropriate may be a chemical spill. The spill may only encompass a hundred square meters, but the primary concern in this case may be health of the local population within a kilometer of the spill or less. Any time that the size of the spill is comparable within an order of magnitude of

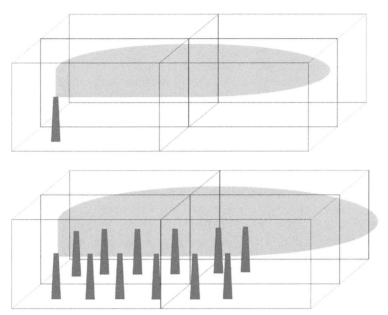

Figure 10.3 Examples of computational grids that contain sources that would be treated as a point source (top) and as an area source (bottom).

the distance that we're interested in or when the size of the release would be expected to cover multiple grid cells within a meteorological or dispersion model, an area source should be considered. In these cases, we can choose to model a series of point sources, or we can initialize a representative concentration of pollutants throughout the region of a model. The drawback of an area source is that it generally assumes that the release is consistent over the entire region described by the area source. If the release rates are changing, we have the option of prescribing an average release rate over the area or choosing to model instead as a grid of point sources, though the latter option can become costly in terms of computational power and time.

After determining the source location and whether it should be treated as a point or area source, we must consider what the height of the source should be. The appropriate release height represents the elevation within the atmosphere where the plume's vertical motion is only due to the atmosphere's turbulent vertical motions and not due to any inertial motion of the plume itself. In most cases, the height of the release may be easy to select as it would be known and can be measured (such as the height of a smokestack or of an average car's exhaust) as long as the release is not being given an initial inertia and is not substantially different in temperature than the surrounding air. In these cases, we simply need to identify the height of the emission release point in the model. However, cases where the plume has some overlying inertial or buoyant contribution to its height require us to determine an effective release height that is different than the actual release height.

The first case is when the release is being forcibly emitted, such as being blown through a factory smokestack. The effect of forcible emission produces an inertia on the plume as it enters the atmosphere, which will act to push it upward in the atmosphere. The result is a plume that reaches a height higher than the emission release point before the atmosphere becomes the dominant force on the plume's motion. Typically, the effluent in these scenarios has a known vertical momentum that will artificially move it upward in the atmosphere before its movement is dominated by the ambient atmospheric conditions.

The other case is when the plume initially has a temperature that is substantially different than the ambient air temperature (Figure 10.4). In these cases, the plume will initially have a buoyancy that will cause it to rise or fall until its temperature comes into equilibrium with the air around it. More difficult to determine than the case with a known initial momentum because buoyancy is dependent on the vertical temperature structure of the atmosphere and source temperatures can vary extensively. For example, smoke plumes from fires are significantly hotter than the surrounding atmosphere and may rise thousands of feet in the air before coming to equilibrium. Even though the smoke source is at the surface, it wouldn't make sense to model a smoke plume as a surface source since the plume rises so high. In these cases, we need to determine the effective source height of the plume and use that as the input to our model.

10.4 The role of atmospheric stability

Atmospheric stability is a description of how turbulent the atmosphere is (see Chapter 6 for more on this). Turbulence tends to have two primary sources in the atmosphere: mechanically generated turbulence and buoyancy-generated turbulence.

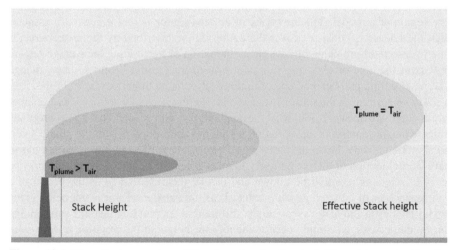

Figure 10.4 When the temperature of the plume leaving the stack (T_{plume}) is greater than the temperature of the ambient air (T_{air}), the plume will have a natural buoyancy and rise. When we model dispersion in the case of a naturally buoyant plume, we need to account for the change.

Mechanically generated turbulence is driven by changes in wind speed or wind direction with height or caused by friction of the air as it moves along the ground. The difference between faster and slower moving air causes air to overturn, much like the breaking of waves in the ocean. We frequently see the mechanical generation of turbulence near the Earth's surface where friction with the ground causes wind speeds to drop very quickly relative to winds higher in the atmosphere.

Buoyancy-generated turbulence is driven by the density differences between warm and cold air in the atmosphere. Where warmer air is underneath colder air, the warmer air will rise and the colder air will sink due to their density differences, leading to mixing of the atmosphere, which we measure as turbulence. In conditions where the colder air is underneath warmer air, buoyancy-generated turbulence is limited. Stable atmospheres tend to occur at night when the atmospheric temperature profile is characterized by cooling air near the surface, while warmer residual daytime air lies higher in the atmosphere. Unstable atmospheres tend to occur during the daytime when the ground and the air near the ground is warming through solar radiation, making it warmer relative to the air higher in the atmosphere. For more details on turbulence and how it is quantified see Chapter 2.

Airborne plumes are often constrained to the planetary boundary layer (PBL) because they are subject to the mixing of the atmosphere. During the daytime, the development of the PBL can range from a few hundred meters to about 2.5 km depending on the season and cloud conditions. During winter months and when skies are cloudier, less solar radiation reaches the earth's surface, leading to lower rates of warming for the atmosphere and the PBL is shallower. Conversely, during summer months and sunny conditions, greater warming occurs and turbulent mixing driven by thermal buoyancy reaches a greater depth of the atmosphere. Above the PBL, atmospheric conditions are determined by larger-scale regional- and global-scale circulations. As a result, the transition between the PBL and the atmosphere above it tends to act as a cap, preventing the transport of particles from the PBL to the air above it in great quantity. For most applications of atmospheric transport modeling, the plumes have sources at or near the ground. This generally means that the plumes will be confined to the PBL.

At night, the cooling of the earth's surface due to lack of incoming solar radiation leads in turn to a cooling of the lowest levels of the atmosphere, forming a nocturnal stable layer near the surface. The PBL that was created during the daytime will continue to persist above the stable layer and is referred to as a residual mixed layer. Because the stable layer at the surface is cooler than the air in the residual mixed layer, the top of the residual layer acts as a barrier, preventing significant mixing between the stable layer and the residual mixed layer. A plume that resides in the residual PBL will have difficulty mixing into the nocturnal stable layer and may be prevented from reaching the surface. Likewise, a plume that is released into the nocturnal stable layer will be constrained there, possibly leading to elevated concentrations. In modeling applications that are concerned with identifying worst-case conditions, typically defined as the highest potential concentration, releases into very stable layers are usually used. These scenarios are used to help drive emergency-response modeling and regulatory modeling, which share the common goal of determining the potential worst-case impacts that a release may have on the environment or population.

The mean wind speed of the atmosphere can be readily determined and tends to not vary significantly over a scale of kilometers, but the stability of the atmosphere is determined by the magnitude of turbulence within the PBL, and this is more difficult to determine and can exhibit greater variation in time and space relative to the mean wind speed. It's important to understand the stability of the atmosphere, however, because this determines how the plume is dispersed (spreads out in the vertical and horizontal directions); effectively this determines the volume of air that the plume is spread out over. Unstable conditions will produce greater turbulence leading to increased dispersion and lower concentrations within the plume by spreading it out over a greater area; stable conditions will produce the opposite, leading to higher concentrations by confining the plume to a smaller volume of air.

Measuring or modeling the rapidly fluctuating wind conditions of the atmosphere on a scale small enough to capture the very fine scales of motion that impact atmospheric transport is difficult and generally impractical to do in most scenarios. The turbulent motions that cause a plume to disperse and grow as it moves downwind can occur on scales ranging from centimeters to the size of the boundary layer on the order of a kilometer (see Chapter 2 for a deeper discussion of turbulence). While there is now meteorological instrumentation that can provide direct measurements of turbulence, this hasn't always been the case. Historically, scientists have relied on estimates of atmospheric stability to drive plume models and determine the horizontal and vertical spreads of the plume. These estimates are based on empirical evidence gathered from many field studies and incorporate several potential predictors including measurements of the average and standard deviations of horizontal and vertical wind speed, temperature changes with height, cloudiness estimates, and time of day.

These estimates have been traditionally linked to the Pasquill-Gifford (P-G) stability classes, which were first developed by Pasquill (1961) and later modified by Gifford (1976). These classes are used to prescribe the turbulent characteristics of the atmosphere on a plume, in the form of plume diffusion parameters (Figure 10.5), when measurements or modeling predictions of turbulence are not

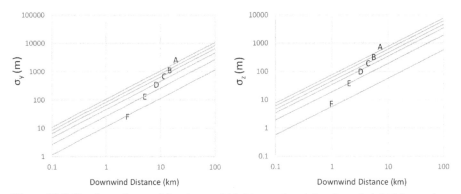

Figure 10.5 The dependence of the plume width (sigma-y) and the plume height (sigma-z) on the atmospheric stability and distance downwind. These factors are used in Gaussian dispersion models to estimate puff characteristics during transport using the equations in Pasquill (1974).

available. In the P-G system, stability is broken into seven classes labeled A through G. The A, B, and C classes refer to unstable, daytime atmospheres with A being the most unstable. The E, F, and G classes refer to stable, nighttime atmosphere where little mixing is taking place; the G class refers to an extremely stable atmosphere with little mixing and is often eliminated, combining these cases with the F class. The D class is a neutral atmosphere that is often found near sunrise and sunset. The neutral and stable classes are often dominated by mechanically generated turbulence while the unstable cases are often dominated by buoyantly driven turbulence. The appropriate stability class to use for a scenario can be empirically determined by using other meteorological measurements. Common systems for determining P-G class will include relating stability to other quantities such as the standard deviation of the vertical or horizontal wind direction, cloud cover, temperature, or wind speed. A detailed account of approaches for determining P-G stability is given by the US Environmental Protection Agency (2000) for atmospheric dispersion applications, which bases the estimates on a range of potential meteorological measurements, including cloud cover, measured turbulence, or temperature.

While the preceding description is appropriate for most atmospheric dispersion cases, there are some unique situations that need to be addressed in dispersion modeling. One such example is modeling the plume from a volcanic eruption. In this case, the force of the eruption can be so large that the material being injected into the atmosphere can reach well beyond PBL and even into the stratosphere before the vertical motion of the eruption is overtaken by the motion of the atmosphere. Another example is in thunderstorms where strong updrafts or downdrafts can stretch vertically across many kilometers. In these cases, applying a stability class or assumptions about the PBL's limits on atmospheric dispersion are not applicable, which can eliminate some atmospheric dispersion modeling techniques such as the Gaussian dispersion models described in a later section. However, these situations do not occur frequently and are not applicable in most applications of dispersion modeling.

10.5 Modeling transport

10.5.1 Meteorological modeling

The atmosphere is a complex structure, particularly near the surface. To quantify the movement and concentration of airborne constituents, we can use meteorological instruments and atmospheric air samplers (Figure 10.6) to tell us about the relevant meteorological quantities or to measure the concentration of particles or gases in the atmosphere. However, meteorological variables and airborne concentrations vary over small scales relative to the movement of plumes, which can travel tens to hundreds of kilometers. It is impractical to believe that we can provide sufficient measurements to provide fine-scale detail of the complex structure of the plume and the processes that dictate airborne transport and dispersion processes.

To understand these processes, we often rely on using numerical models to simulate fate and transport in the atmosphere. For nearly a century, the idea of

Figure 10.6 Example of an air sampler. The sampler is generally designed with an air pump that draws in air and a filter that captures the gases or particulates of interest, which can later be identified and analyzed.

understanding how airborne particles and gases move in the environment has led to developments that we put into practice in today's world. Through numerous field experiments and detailed analysis of atmospheric chemical and physical processes in controlled environments, relationships and behaviors have been identified that can be included in our predictions of how wind-borne particles or gases behave in the atmosphere.

The primary factor that influences the transport and dispersion of gases and particles is wind. The mean wind speed and direction determines the predominant direction and speed of airborne constituents as a whole, but the small, turbulent motions of the wind are equally important in understanding how quickly the plume spreads vertically and horizontally relative to the mean wind direction. Turbulence refers to the rapid changes in wind speed or direction that occur around the mean wind speed and direction. The wind is rapidly fluctuating, so it is useful to separate the wind speed at any given time up into a mean wind component (denoted by the overbar) and a turbulent wind component (denoted by the prime symbol):

$$U = \bar{u} + u'$$

The mean wind component can be computed as the average wind speed over a time period and can be easily measured and applied to atmospheric dispersion problems.

The turbulent portion changes rapidly and cannot be readily applied to dispersion problems in a direct manner due to its rapidly changing nature. For that reason, atmospheric transport models generally predict the direction a plume travels using the mean wind and estimate the impacts of the turbulent motions by applying a variety of methods to predict the diffusion of the plume (i.e., how the plume grows and spreads in the lateral and vertical directions).

The first factor in modeling atmospheric transport is to have a good representation of the movement and characteristics of the atmosphere that will influence the transport and downwind concentrations of an effluent. Meteorological models come in a variety of packages and applications, and choosing the correct model depends on the application and goals of the modeling. Atmospheric transport modelers are always interested in more detail, and that often means obtaining the highest resolution, or smallest modeling grid, that is reasonable for the application.

The grid cells of a model are defined by the three-dimensional space they represent. The model parameters are predicted for the center of the grid cell and are considered representative for the entire grid cell. The smaller our model grid cell is, the more exact the representation is for a particular point in space. Conversely, the smaller the grid cell is, the more of the grid cells we need to cover a model region. When performing modeling, we need to consider what is reasonable for our application, which involves considerations of how much detail is required to obtain a good prediction while balancing how long it will take to make the prediction. In purely research applications, time may not be a strong constraint. In operational work or in the event of a facility emergency where there is an active release of a potentially harmful effluent, every second counts and it may be worth sacrificing a little detail in the specific details of the plume to get a timely prediction. While the prediction may not be exactly accurate, the models used for emergency response are often designed to produce a conservative result; that is, the models are designed to produce an overprediction of the potential consequences to ensure that human health is protected even if it means ordering protective actions to a larger region than may be necessary.

The scale of meteorological modeling can range from global scales all the way down to micro-scales. Examples of applications that require a global scale of modeling include volcanic eruptions or the release of nuclear material from the Fukushima-Daiichi power plant in Japan (Figure 10.7). In these cases, enough material can be released to the atmosphere that there will be measurable concentrations and potential impacts to human health around the globe. The modeling for these problems requires a global circulation model that predicts meteorological conditions around the globe, but generally at a low resolution with grid cells on the order of tens of kilometers. This sacrifices specific details of the plume, but much like climate modeling, which is more interested in changes to the general behavior of the weather as it evolves over long periods of time, global transport modeling is more interested in the general behavior of the plume rather than the specific concentration at specific points in space and time.

At the other end of the modeling spectrum, we may be interested in very high-resolution modeling of the plume where we can look for specific behaviors of a plume over very small time and distance scales. In these cases, we often require very fine-scale weather predictions and will make use of fine-scale meteorological modeling

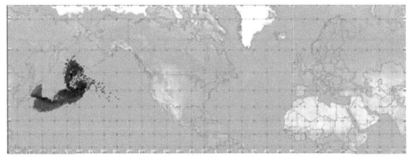

March 13, 2011 00:00 UTC

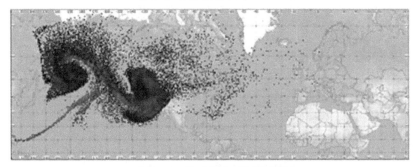

March 18, 2011 06:00 UTC

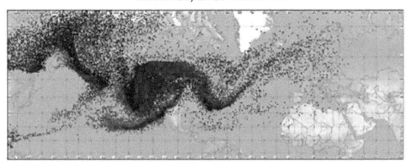

March 20, 2011 12:00 UTC

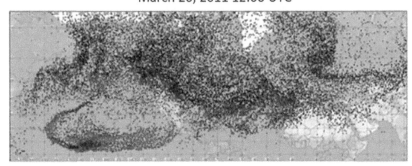

March 26, 2011 18:00 UTC

Figure 10.7 Example of global dispersion modeling following the accident at the Fukushima-Daiichi power plant in Japan.
Taken from Povinec et al. (2013).

techniques such as large-eddy simulation (see Chapter 5) and computational fluid dynamic models, which make predictions of wind and other relevant variables as small as every meter or less.

Regardless of the scale in modeling, the techniques we use to predict the fate and transport of airborne particles or effluents remains largely the same.

10.5.2 Gaussian models

The simplest form of dispersion model is the Gaussian model, which predicts the concentration at a downwind point based on a three-dimensional Gaussian, or statistically normal, distribution (Figure 10.8). This creates a scenario where the plume moves following the wind and spreads in the vertical, along-wind, and crosswind directions. If we do not consider processes such as deposition and chemical transformations in the atmosphere, which act to change the total mass of the airborne plume, the mass will remain unchanged during its transport.

These models have the benefit that they can be run quickly and with minimal meteorological inputs and tend to provide conservative results, which is to say that they will tend to overpredict rather than underpredict the downwind concentration. This is important for cases where Gaussian models are used to provide rapid predictions for situations where time is of the essence, such as accidental releases and public emergency response actions. By providing conservatism, these models can help ensure that protective actions are taken to prevent accidental exposure of human health to hazardous situations.

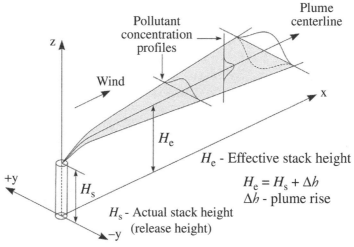

Figure 10.8 Schematic description of a Gaussian plume, showing the Gaussian distribution of concentration in the horizontal and vertical directions as the plume moves downwind as well as the potential change in effective stack height due to a forced or naturally buoyant plume (Leelossy et al., 2014).

The key pieces of required information to run a Gaussian model are the source of the release, atmospheric stability, wind speed, and downwind locations where predictions of airborne concentration are desired. While they require only minimal meteorological inputs (a single location is sufficient to generate a prediction), this does act to limit the spatial extent to which the model can be applied. A general rule of thumb is that Gaussian models are only reliable for predictions of 30–40 km distance. Beyond that distance, the meteorological conditions being used by the model will tend to become unrepresentative of the actual meteorological conditions further away, thus introducing greater uncertainties in the accuracy of the predictions.

A weakness of the Gaussian plume model is that it also becomes less reliable in conditions where wind speed is low and variable. In these conditions, the model assumes that the plume will move along in the direction of the wind measurement, when in fact the wind direction may be changing from one measurement to the next or exhibiting more variability in space. This can lead to errors where the actual plume is moving in directions not predicted by the model. The Gaussian model also is less reliable in nighttime stable conditions where topography can create local variations from the mean wind flow. For example, river valleys will often have wind conditions that flow downhill to lower elevations. This may not be represented by the meteorological measurement and would need to be considered by the person running the model to identify these difficult-to-predict situations.

For a very short or instantaneous release, the Gaussian plume equation can be given as:

$$\chi(x,y,z) = \frac{Q}{(2\pi)^{3/2}\sigma_x\sigma_y\sigma_z}\left[\exp\frac{-y^2}{2\sigma_y^2}\right]\left[\exp\frac{-(z+H)^2}{2\sigma_z^2} + \exp\left(\frac{-(z-H)_2}{2\sigma_z^2}\right)\right]$$

$$\quad\quad\quad\quad\quad\quad\quad\quad\quad\quad\quad\text{I}\quad\quad\quad\quad\quad\quad\text{II}\quad\quad\quad\quad\quad\text{III}\quad\quad\quad\quad\quad\quad\text{IV}$$

where the key components that predict the downwind plume concentration at a particular point in space $\chi(x,y,z)$ is determined by Q, the rate of release at the plume source, σ_x, σ_y, and σ_z, the dispersion coefficients in the along-wind, cross-wind and vertical directions, respectively, and H, the effective stack height. The dispersion coefficients of the plume width in the vertical and horizontal directions can be difficult to determine exactly. These values are often determined by the estimated atmospheric stability class and based on decades of field work that describe the characteristic dimensions of the plume based on a range of atmospheric conditions. The goal of these coefficients is to describe the width and height of the plume as a function of distance traveled. As the distance traveled increases, the plume is expected to grow with size, so the coefficients will grow as well.

Term I in the previous equation describes the centerline concentration along the plume. The centerline represents the concentrations along a horizontal line straight along the mean direction of transport at the same level as the source (or the effective source height in the case that the plume has natural or forced vertical motion upon release). Term II describes the change in concentration in the lateral direction (perpendicular to the horizontal) of the plume as you move away from the centerline. The concentration in a Gaussian plume will decrease as it moves farther from the center of the plume. Term III is similar to Term II but describes the decrease in the vertical direction moving away from the centerline.

The final term in this equation is theoretically similar to Term III but is necessary to predict the reflection of the plume off the surface. If the Gaussian curve is allowed to extend underground, we begin artificially losing airborne material in our predictive model. To ensure we don't have artificial plume loss, we model a second source that is mirrored underground. When we combine the concentration prediction from both the actual aboveground source and the hypothetical belowground source, we get an answer that includes all the released material and acts, essentially, as the plume curve reflecting off the earth's surface. Term IV captures this "reflection" dynamic and ensures that we do not have artificial loss of concentration in our predictions.

In the event that the release is ongoing for a period of time, we must modify the original Gaussian equation to account for time-dependent accumulation during the length of the release. That equation is given as:

$$\chi(x, y, z) = \frac{Q}{2\pi\sigma_y\sigma_z U} \left[\exp \frac{-y^2}{2\sigma_y^2} \right] \left[\exp \frac{-(z + H)^2}{2\sigma_z^2} + \exp \left(\frac{-(z - H)^2}{2\sigma_z^2} \right) \right]$$

where U is the wind speed, and the primary difference is that we are now focused only on the cross-wind and vertical spread of the plume and ignore the along-wind diffusion of the plume. We ignore the along-wind component because, as the release is ongoing, we are essentially simulating a series of instantaneous plumes that are overlapping each other. This effectively creates a fully integrated along-wind plume where the diffusion of each individual plume acts to cancel out the diffusion of others, and we no longer need to consider the horizontal diffusion in the along-wind direction.

If we have more than a single meteorological measurement that we can use, such as time-evolving wind fields, these can be included in the Gaussian model by predicting the plume in small jumps, where we assume a constant meteorology for a short time interval and predict the changes to the plume characteristics during that time. We then do a series of short time-interval predictions, assessing the changes that the plume would undergo during those times to assess a time-dependent evolution of the airborne plume. In the event that weather data is available from multiple sources, it is possible to extend the relevancy of the Gaussian model to greater distances.

10.5.3 Eulerian models

For more complex representations of atmospheric dispersion predictions, we require models that have increasing complexity, usually consisting of a meteorological model and an atmospheric transport model that may be run separately or concurrently. The more complex nature of the models necessitates increased computer resources are needed to complete the model, but most atmospheric transport predictions can be completed on the order of minutes (if meteorological model data is already available such as from the operational models run by the National Weather Service) to hours (if meteorological modeling must be run to drive the atmospheric transport model).

The more complex modeling has the benefit that the evolution of the plume can be monitored over larger spatial and temporal ranges. These models can be run in two or three dimensions, providing flexibility for how complex of a simulation the user requires by producing a wider range of output variables which allows for examination of more complex processes that may be occurring transport such as chemical transformations, deposition to the ground surface, or rainout if the plume passes through precipitation in the atmosphere.

The Eulerian method of predicting atmospheric dispersion adopts a gridded structure and predicts the concentration of the plume within each model grid cell and the transfer of the plume between adjacent grid cells using the following general equation:

$$\frac{\partial c}{\partial t} = -\mathbf{V} \cdot \nabla c + D\nabla^2 c + S_{\text{source}} - S_{\text{deposition}}$$

where the term on the lefthand side of the equation represents the change in concentration in a single grid cell with time, and on the righthand side of the equation, the first term describes the rates of plume advection in three dimensions, the second term describes diffusion of the plume by turbulent and molecular diffusion, the third term describes the addition of materials from sources, and the final term describes the loss of material due to removal either by deposition or chemical transformation (Figure 10.9). While assessing the mean movement of the plume between grid cells generally follows the mean wind speed, determining an appropriate method to predict the turbulent transport and diffusion of the plume between grid cells is more difficult. A variety of solutions have been proposed ranging from simpler box models, which assume that diffusion is negligible relative to the advective component of plume transport, to complex turbulence closure models, which seek to resolve or parameterize the very small-scale motions of the atmosphere based on predicted quantities like turbulence kinetic energy in the meteorological model.

The benefits of Eulerian techniques are they tend to be moderately computationally intensive relative to Gaussian and Lagrangian techniques but may still incorporate

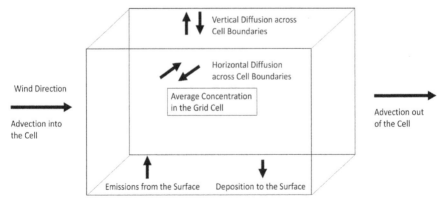

Figure 10.9 Graphical depiction of a simple Eulerian model. All of the processes identified by *arrows* are taken from model estimates or parameterized into a single calculation of the average concentration of the plume within the grid cell.

elements of particle or chemical changes that can take place in the atmosphere. These changes are generally driven by humidity and temperature conditions, which can change substantially either horizontally or vertically. By providing a three-dimensional field of these variables to the dispersion model, it allows for these processes to be predicted as the model is being run. However, the parameters of the Eulerian model relative to the size of the grid cells must be chosen carefully to ensure that the movement of the plume between grid cells is handled realistically. It can be easy to generate inaccurate predictions by choosing model parameters that lead to the plume either skipping grid cells because it is being advected too quickly or artificially hindering the growth of the plume by not allowing it to sufficiently spread between grid cells.

10.5.4 Lagrangian models

Like Eulerian models, Lagrangian models also rely on inputs from meteorological models to provide a two- or three-dimensional atmosphere for plume prediction. While the Eulerian method relies on a set grid and predicts the plume concentration of each grid cell as a scalar quantity, Lagrangian models follow the individual particles that are being simulated (Figure 10.10). By simulating each particle, we can characterize the pseudo-random turbulent motions on the individual particles as well as the specific changes to particle characteristics that may be influenced by atmospheric conditions. This creates a model framework that can rapidly become computationally expensive due to the need to simulate thousands to millions of particles to adequately resolve the properties of the plume, but it also creates a framework where it is easy to assess statistical properties of the plume because we know the properties of each individual particle that we simulate.

The Lagrangian dispersion equation is given by

$$X_{x,y,z}(t + \Delta t) = X_{x,y,z}(t) + \left[\overline{U}_{u,v,w}(t) + U'_{u,v,w}(t) \right] \Delta t$$

where the position of the particle in each of the three dimensions at the next model timestep $X_{x,y,z}(t + \Delta t)$ is calculated based on the particles' current positions at the current timestep $X_{x,y,z}(t)$, the mean wind speed in each of the three dimensions $\overline{U}_{u,v,w}(t)$, the turbulent portion of the wind $U'_{u,v,w}(t)$, and the length of the timestep Δt. The fluctuating portion of the wind is often prescribed using a Markov chain, which relates the current fluctuation of the wind speed to the previous value and the height of the particle. The theory behind this is that particles higher in the atmosphere are subject to larger turbulent circulations, which will be more consistent from one timestep to the next, while particles near the surface will be controlled by smaller turbulent eddies, which exhibit motions that can be characterized as more random fluctuations.

In Lagrangian models, we treat each particle as a separate entity. This means that we determine the specific meteorological environment surrounding each particle separately. Often, this is done by interpolating the values from the nearest grid points of the meteorological model to the specific location of each particle. In this way, every particle's movement is independent of the other particles.

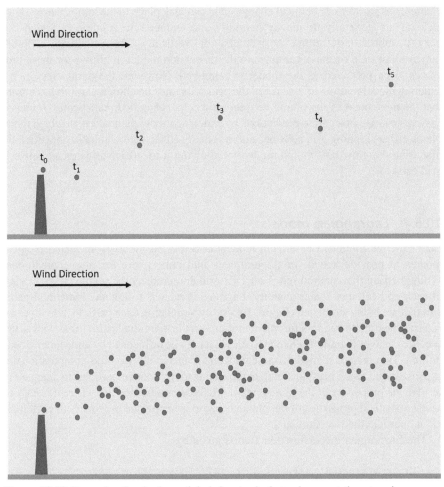

Figure 10.10 Example of a single particle being tracked at subsequent timesteps in a Lagrangian model (top) and a resulting plume when multiple particles are being tracked, leading to a comprehensive plume whose characteristics can be calculated and tracked (bottom).

To estimate the small turbulent motions that we cannot resolve in the meteorological model, we rely on parameterizations based on other quantities such as turbulence kinetic energy, which are often treated in a way that they are pseudo-random. While turbulence in the atmosphere can often be described as essentially a random distribution, particles in the atmosphere have enough mass that they maintain a certain amount of inertia, with larger particles having larger inertia. Thus, a particle that is moving upward will not suddenly turn around and start moving downward between one timestep and the next. A common practice in accounting for inertia is to implement a first-order Markov chain, which determines the turbulent component of the particle's movement as a function of its previous turbulent component and a new random component. Weights are prescribed to these two values that depend on parameters such as the particle's position, mass, and the model timestep. The components of the new turbulent component of motion will be weighted more heavily to the previous turbulent motion (the inertial portion) for heavier particles and more toward the random component for lighter particles that exhibit less inertia. Combined, we generate a motion for the Lagrangian particle that maintains the turbulent diffusion we observe while also accounting for the inertia of the particle.

10.6   Plume depletion processes

10.6.1   Dry deposition

In the case of gaseous releases in the atmosphere, the atmospheric dispersion modeling methods described above, which seek to address diffusion of the plume in the atmosphere, may be generally sufficient to describe dispersion in the atmosphere. For plumes comprised of solid particles or liquid droplets, we must also consider the gravitational effects on the plume.

Gravitational settling describes how quickly the plume constituents are pulled toward the surface by gravity. The rate of settling is determined by the size and shape of the particle. A distribution of gravitational settling speeds shows that the gravitational settling speed increases approximately logarithmically with increasing particle diameter for particles that have a diameter larger than a micron (Figure 10.11). Particles in this size range are large enough that the gravitational settling effects are generally stronger than the turbulent vertical motions that can act to lift particles in the atmosphere. Particles that are smaller than a micron are influenced more by the

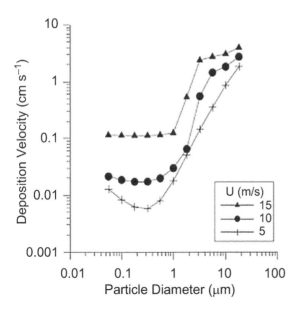

Figure 10.11 Example of the dependence of deposition velocity on the size of the particle and the wind speed. Higher wind speeds create higher deposition velocity through the generation of mechanical turbulence, which can help drive particles toward the ground (Chen et al., 2015).

small turbulent motions of the atmosphere, which can overcome the very small gravitational settling term. A visible example of this we see almost every day is clouds, which are composed of very small water droplets that can be held aloft until they reach a size where they can no longer be held up by the upward motions of the atmosphere and fall to the surface as rain.

The rate of gravitational settling for particles larger than a micron is often described by Stokes' law, which relates the diameter of a spherical particle with its gravitational settling speed and is given by

$$V_g = \frac{d^2 g \rho}{18 \mu}$$

where d represents the effective diameter of the particle, g is gravitational acceleration, ρ is the density of the particle, and μ is the dynamic viscosity of air. The effective diameter of a particle will equal the diameter of the particle if the particle is spherical.

If the particles we are interested in are not spherical, the effective diameter is generally determined so that the particle can be simulated by a representative spherical particle in numerical models. Irregular particles that have different cross-sectional areas when viewed from different sides will also produce differing amounts of air resistance. When a larger cross-sectional area is perpendicular to the vertical force of gravity, additional air resistance is generated by the downward motion of the particle and the gravitational settling term is decreased. It is impractical to try and model the specific orientation of each particle we release, so we instead identify a spherical particle that represents the average properties of the irregularly shaped particle we want to model. The diameter of the representative spherical particle is the effective diameter of the irregularly shaped particle, and the gravitational settling speed of the spherical particle represents the average gravitational settling of the irregularly shaped particle that we would expect to observe. Often, experimental results in controlled settings are required to develop a good estimate of the appropriate effective diameter of a particle.

When the particles in a plume are close to the surface, they can be removed through interactions with the surface such as impaction against the ground or water surface or against obstacles such as vegetation or buildings above the surface. When a particle impacts the ground and is removed from the airborne plume, it is considered deposited and no longer contributes to the airborne concentration calculated by a model. Because the surface characteristics are often too complicated to be reliably depicted within a model, the process is simulated by calculating the flux of particles into the surface, which is given by:

$$V_d = \frac{F}{c}$$

where F is the flux, which is typically defined as the mass moving vertically downward through a $1\,\text{m}^2$ area, V_d is the deposition velocity (which is equal to the gravitational settling described above for large particles), and c is the concentration of the plume near the surface. By determining the mass settling on to the surface, we can know how much mass the plume has lost.

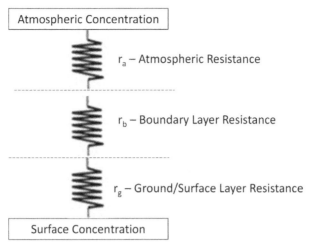

Figure 10.12 Depiction of deposition velocity as a series of resistances. Atmospheric resistance refers to the resistance to deposition generated by the wind, boundary layer resistance refers to resistance to deposition generated by small motions along and near the surface, and ground/surface layer resistance refers to resistance to deposition based on the surfaces' ability to collect airborne material.

For gaseous plumes, dry deposition can be more complicated because these plumes are not driven by gravitational settling and depletion of the plume comes through other processes such as being taken up by vegetation or undergoing chemical transformations in the atmosphere in addition to impacting the ground as the plume diffuses. In these cases, deposition velocity is typically calculated by defining resistances to gaseous uptake or diffusion into the surface for different processes similar to identifying electrical resistances for a circuit (Figure 10.12). These resistances are calculated by experimentally derived relationships and are primarily dependent on wind speed. Generally, higher wind speeds will generate greater amounts of turbulence near the surface of the atmosphere, which will increase the amount of deposition that occurs. When addressing deposition related to uptake by vegetation, then deposition velocities can change based on whether it is day or night as a result of the stomatal opening or closing that occurs on a diurnal cycle and is further influenced by environmental conditions. Thus, dry deposition is not necessarily a constant value and can be varied as a result of the type of surface (e.g., deserts vs plains vs forests) and atmospheric conditions.

10.6.2 Chemical transformations or particle modification

Chemical transformations in the atmosphere can act to deplete the primary constituent of the plume by creating secondary or tertiary chemical compounds as a result of interactions with sunlight or other chemical compounds in the atmosphere. These relationships will be specific to the constituent of interest and may be further complicated by environmental conditions such as atmospheric humidity and temperature. These transformations can also lead to create other chemical compounds that may be of interest.

Often, chemical transformations are addressed in applications of smog or pollution. Common chemical reactions that take place in the atmosphere involve chemicals released from anthropogenic sources mixing with light, NO_x compounds, or ozone in the atmosphere. Another common source of chemical compounds is from vegetation, which emit carbon-based compounds that will mix with other chemicals in the atmosphere to create aerosols or other small particles that can eventually act to promote cloud formation.

In other instances, we may be interested in studying how particles in the atmosphere change their shape or other properties as they are dispersed in the atmosphere. In these cases, we may be interested in knowing how mass or shape of the particle changes, such as with pollen grains or other organic-based particles that may contain moisture but will slowly lose it to the atmosphere if humidity is low. In these cases, the changes can be important because the physical changes can alter how the particles move in the environment. If a particle is becoming lighter or changes shape, then its settling and deposition velocity will be affected. In the case that the deposition velocity is reduced, this will allow the particle to potentially travel farther.

10.6.3 Wet deposition processes

Wet deposition refers to the removal of plume material through rainfall. Rainfall acts to scavenge material by collecting it as the raindrops fall through the atmosphere and then depositing the material at the surface when the raindrops impact the ground. Other forms of wet deposition can include the deposition of fog droplets at the ground that may interact with a plume prior to impacting the surface, or the absorption of small particles or gases into cloud droplets.

Wet deposition is generally given by the equation

$$V_w = w_r p$$

where w_r is a washout ratio, which is dependent on the chemical species or particulate comprising the plume, and p is the precipitation intensity or rainfall rate. The term V_w is a deposition velocity, like V_d, except that V_w is directly attributable to precipitation, while V_d accounts for normal gravitational settling or enhanced settling by mechanical turbulence. The value for wet deposition generally exceeds that of dry settling, as rain is generally considered a good scavenger and able to collect most if not all of the plume material it comes into contact with. Deposition efficiencies of 75–100% have been observed and are primarily dependent on the strength of the rainfall.

10.7 Summary

Atmospheric transport can be broken down into three primary events: the plume's source, the movement of the plume in the atmosphere, and depletion of the plume either through deposition or chemical transformation. Each stage of the evolution of particles or gases in the atmosphere has its own complexities. For sources, we must be aware of the characteristics of what's being emitted and answer questions like "How fast is it being emitted?" and "Is it warmer or colder than the environment it's being released into?" so we identify an appropriate release height. During transport, we must assess the atmospheric stability and turbulence to predict the movement and lateral or vertical diffusion of the plume. This can also require an assessment of the appropriate model to use to analyze plume details based on how quickly an analysis is needed and how much detail we require. Finally, the depletion of the plume requires that we understand our surface characteristics and how they are impacting turbulence and plume constituents near the surface. The rate at which plume material is removed is a key factor in determining what the downwind impacts of a plume may be and whether the plume is even detectable.

Understanding the fate of material in the atmosphere is necessary to understand what the impacts of an airborne plume may be on the environment or human health, whether from an industrial source or from natural sources such as fires or volcanos. Several tools and modeling techniques exist to help predict and analyze atmospheric dispersion processes, but to understand the movement of particles in the atmosphere we need to understand how primarily how plumes are influenced by the diurnal cycles of the PBL and how they interact with the surface.

References

Chen, L.-W.A., Tai, A.Y.-C., Wang, X., Holms, H., Chow, J., Watson, J., 2015. Evaluating atmospheric deposition of particulate matter in the Lake Tahoe basin. US Forest Service Report https://www.fs.fed.us/psw/partnerships/tahoescience/documents/p094_LakeTahoeDeposition_FinalReport.pdf.

Gifford, F.A., 1976. Turbulent diffusion-typing schemes: a review. Nucl. Saf. 17, 68–86.

Leelossy, A., Molnar Jr., F., Izsak, F., Havasi, A., Lagzi, I., Meszaros, R., 2014. Dispersion modeling of air pollutants in the atmosphere: a review. Cent. Eur. J. Geosci. 6, 257–278.

Pasquill, F., 1961. The estimation of the dispersion of windborne material. Meteorol. Mag. 90, 33–49.

Pasquill, F., 1974. Atmospheric Diffusion. John Wiley Publishing, New York. 222 pp.

Povinec, P.P., Gera, M., Holy, K., Hirose, K., Lujaniene, G., Nakano, M., Plastino, W., Sykora, I., Bartok, J., Gazak, M., 2013. Dispersion of Fukushima radionuclides in the global atmosphere and the ocean. Appl. Radiat. Isot. 81, 383–392.

Richardson, L.F., 1926. Atmospheric diffusion shown on a distance-neighbour graph. Proc. R. Soc. A 110, 709–737.

Roberts, O.F.T., 1923. The theoretical scattering of smoke in a turbulent atmosphere. Proc. R. Soc. A 104, 640–654.

Taylor, G.I., 1921. Diffusion by continuous movements. Proc. Lond. Math. Soc. 20, 196–211.

US Environmental Protection Agency, 2000. Meteorological monitoring guidance for regulatory modeling applications. EPA-454/R-99-005, 171pp.

Work it! Turning knowledge into power

Peter K. Hall, Jr
Avangrid Renewables, Portland, OR, United States

11.1 Introduction

Windmills have existed for millennia, converting the kinetic energy of moving air into mechanical energy for work such as crushing grains and pumping water (Ackermann and Söder, 2002). Utilizing wind to create electrical energy, via a wind turbine, was first developed toward the end of the 19th century, but it was not until the late 20th century that wind power became a viable source for utility-scale electricity (Manwell et al., 2009, chapter 1). The early utility-scale wind turbines produced tens of kilowatts, were less than 50 m (164 ft) tall, and had blade lengths of less than 30 m (98 ft; Hoen et al., 2018; Manwell et al., 2009, chapter 1). Today, wind turbines can produce more than 2.5 MW (2500 kW) of electrical power, are three to five times taller than their 20th-century cousins (Manwell et al., 2009, chapter 1), and capture energy from wind over a 10,000 to 20,000 m^2 area (Wiser et al., 2020; approximately 107,000–215,000 ft.^{2} or over three football fields!). Electric power from turbines is now a major part of the electric grid, with more than 120 GW (120 million kilowatts) of installed capacity (American Clean Power (ACP), 2020), and generates more than 8% of the electricity in the United States (www.eia.gov/energyexplained/wind/elec tricity-generation-from-wind.php).

Wind turbines continue to get larger in size and generating capacity or output (Wiser et al., 2020; Hansen and Hansen, 2007). While operational machines currently function in the lowest 10–200 m (98–656 ft.) of the atmosphere, soon wind turbines could reach over 250 m (more than 800 ft!) tall (www.seattletimes.com/business/ ges-monster-wind-turbine-is-upending-an-industry/). Coupling larger wind turbines with more and more wind farms on the grid means weather events have a greater significance to people. Not only can weather impact the electrical grid (e.g., damaging power lines, determining load forecasts based on temperatures), but weather can also drive supply on the grid. A consequence of these points is a need to better understand how wind turbines and wind farms interact with their fuel source: the wind.

The height of wind turbines places these machines in the surface and Ekman layers of the atmospheric boundary layer. Thus, the characteristics and daily evolution of the boundary layer are of great importance to the production of energy from the wind. The remainder of this chapter introduces you to electrical power and the power grid, describes the key parts of a wind turbine, and explains how those machines convert wind into electrical power. Once you are familiar with wind turbines, concepts relating to atmosphere–turbine interactions will be presented.

For this chapter, the term utility-scale wind farm will refer to a facility with a generating capacity of 20 MW or greater per hour. Additionally, the potential output of utility-scale wind turbines will be at least 1.5 MW/h.

11.1.1 You have the power

Energy comes in many forms (e.g., mechanical, electrical, thermal) and is either transferred between objects and systems to create work or is stored for potential use (Halliday et al., 1997a, chapters 7 and 8). Recall that work (in joules) is force over a distance, and power is work or energy over time. Thus, power is in the units of joules per second (J/s) or often expressed as a watt (W). In electricity, force is the coulomb (ampere per second), and when displaced as voltage, the result is electrical work (Halliday et al., 1997b, chapter 27). Electrical work also is measured in watts and can be easily calculated by multiplying voltage (V) by amperes (A). Electrical power is expressed in kilowatts (kW; 10^3 W), megawatts (MW; 10^6 W), and gigawatts (GW; 10^9 W). Production and consumption of electricity is often measured in kilowatt hours (kWh). For example, if you use an electrical device that consumes 500 W when active, and use that device for 1 h, then the total consumption is 0.5 kWh.

Kinetic energy (KE) refers to transfers of work through motion (Halliday et al., 1997a, chapter 7). Wind turbines convert the KE in the wind to electrical power. An equation that describes the amount of KE in the wind is (Manwell et al., 2009, chapter 1):

$$E_{wind} = \frac{1}{2}\rho A U^3, \tag{11.1}$$

where E_{wind} is the total kinetic energy (W), ρ (rho) is air density (kg/m^3), A is the rotor area (m^2), in this case a blade-swept area of a turbine, and U is the wind speed (m/s). Thus, wind energy is directly dependent on air density and the size of the blade-swept area, and increases to the cube of the wind speed (U). It is important to understand that Eq. (11.1) does not mean you can extract all the energy of the wind for electrical power. First, wind turbines converting all the energy from the wind would result in no airflow downwind of the wind turbine – this does not occur in reality (e.g., Crespo et al., 1999; Frandsen et al., 2006). Second, there is an upper limit to how much KE from the wind can be converted to power by the wind turbine, described by Betz (2013) to be 59.3%. Third, there is the matter of efficiencies of conversion. The wind moves the turbine blades, which spins a drive shaft, and coverts the energy to electrical power. How well a wind turbine converts KE to electrical energy depends on mechanical and electrical efficiencies and rotor power coefficients (Manwell et al., 2009, chapter 3). These deductions to the overall KE vary across wind turbine models (and is outside the scope of this chapter). The efficiency terms vary based on the turbine design and can be related to wind speed (Wagner et al., 2011). Manwell et al. (2009, chapter 2) estimate that up to 45% of the KE in the wind can be utilized for power production. For example, at an air density of $1.225 \, \text{kg/m}^3$, a rotor area of $8000 \, \text{m}^2$, and wind speed of 10 m/s, the KE of the wind is 4.9 MW. Applying the 45% utilization limit results in 2.2 MW of potential electrical power. Finally, it is important to understand that turbine models have a maximum power output, called rated output or rated capacity (Chiras, 2017, chapter 1). For example, a machine with a capacity of 2 MW might reach peak power at 15 m/s (33.5 mph), but that is the highest electrical power output the turbine can produce. Wind speed increases above that speed do not contribute to greater electrical output.

To understand how much power is produced by a utility-scale wind farm, first consider how much power is consumed by the average US household. The US Energy Information Administration (EIA) calculated that the average residential customer uses nearly 11,000 kWh of electricity per year (Energy Information Administration, 2020). This translates to more than 900 kWh per month or more than 1 kW per h (recall 1 kWh is a kW sustained for one hour). The next consideration when figuring how much power a wind farm produces is the capacity of the wind farm itself. If the wind farm under consideration has wind turbines that can produce up to 4 MW (4000 kW) of electrical power, and you have 25 wind turbines in the wind farm, the peak output is 100 MW (100,000 kW) each hour. Dividing the 100,000 kWh of maximum power output by the average hourly consumption of 1 kWh per residential customer, results in 100,000 homes being powered by the wind farm every hour. But wait! The hourly consumption of electricity and the hourly production of power from wind farms are not constant nor always matched (Bakke, 2016, chapter 1). The trend in daily power consumption is beyond the scope of this chapter, but consider when you use electricity more and less. In reality, wind turbines are not always at full production and consumer consumption of electricity is not always synchronized with wind electricity production. If on average the wind farm is producing at 36% of its maximum capacity (Wiser et al., 2021), the power going to homes is 36 MWh (36,000 kWh). Therefore, this example wind farm could power 36,000 homes.

11.1.2 Get on the grid

What is meant by "the grid"? The grid is a term used to describe the electric power system (Bakke, 2016; Schewe, 2007). That system transfers electricity from generating sources to consumers. Figure 11.1 illustrates what comprises the grid. The grid is the combination of generation sources, transmission lines, and distribution. Generation sources can include turbine generators fueled by coal, natural gas, oil, nuclear fission reactors, water, wind, and concentrating solar. Generation also can come from the photovoltaic modules that comprise typical solar energy facilities (no turbines needed). These generation sources are typically great distances away from the consumers. Thus, transmission lines are built to transport high-voltage electricity from multiple sources to areas where the electricity is needed. Electricity from any of the sources can travel on transmission lines. Electricity is then distributed in local areas using substations to step down or decrease the voltage, and finally arrives to the consumer.

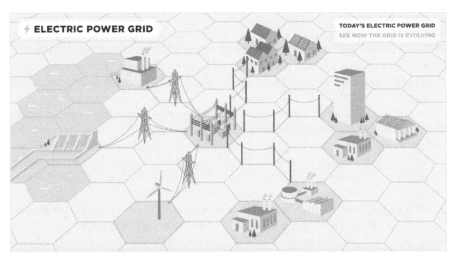

Figure 11.1 The electric power grid illustrating points of generation (e.g., wind farms, coal fired power plants), transmission from the generation points to distribution, and the distribution to homes and industry.
From the EPA.

In the United States and Canada, grid standards are set by the North American Electric Reliability Corporation (NERC; www.nerc.com/AboutNERC). There are four major interconnection areas in the United States and Canada, where electricity could originate from one area and be consumed in another part of that same interconnection (Energy Information Administration, 2016). Within these interconnections are regional transmission organizations (RTOs) and independent system operators (ISOs), although some interconnections also function as the transmission operators (Energy Information Administration, 2016; Brown and Sedano, 2004). These RTOs and ISOs further govern the electricity flowing on transmission lines and set rules for connecting new power sources (via interconnection agreement) to the grid. Balancing authorities within these networks help maintain grid stability (Energy Information Administration, 2016).

With proper approvals, you can put your utility-scale electricity on the grid (Bakke, 2016, chapter 4; Brown and Sedano, 2004). However, you are not guaranteed a set price for your production. A wind farm developer and/or operator can decide to sell electricity on the open market, known as a merchant power plant (Badissy et al., 2020). Prices for electricity can change each hour, and to sell electricity from your facility, you must be willing to take the market price. If you are in an area with a lot of wind farms, and a strong cold front impacts the area, there could be a lot of electricity generation. This surplus of generated electricity could lead to decreased price per megawatt produced (Mills et al., 2019).

To minimize price fluctuations, owners of wind farms can enter into a power purchase agreement (PPA; Badissy et al., 2020). PPAs are created between the seller of electricity and an off-taker, typically a utility (Badissy et al., 2020). A PPA sets a price for the electricity the off-taker agrees to pay, while the generator may guarantee production of a certain amount of power at that price (Badissy et al., 2020). Contracts can vary; however, a rough example would be an agreement of 10 years to buy 1000 MW hours from a generator for 50 dollars per megawatt. If the generator produced more electricity than the agreement, the off-taker is not required to buy the excess, and the extra electricity exists on the grid to be used elsewhere at market prices. If the generator cannot produce that agreed-to amount of electricity, the generator must buy power from other sources to cover the gap in production. This could lead to a net loss of revenue for the generator during those times.

A major challenge with wind power generation is wind power is not easily dispatchable. Dispatchable power is a generating source that can be turned up when demand is higher and turned down when demand decreases (www.visionofearth. org/industry/electricity-grid-key-terms-and-definitions/). For example, a natural gas power plant can increase output when needed, if it is not already at full capacity. Some facilities can be held in reserve to smooth out unexpected changes (e.g., everyone comes home early to catch the big game on TV). You cannot tell the wind to change speed, so a wind farm alone cannot increase output on demand in most cases. An operator can curtail production (i.e., reduce output) of the wind farm to alleviate congestion on the transmission lines (i.e., too much power). In this case, the operator can send remote commands to wind turbines, instructing the machines to slow or shut down completely. When the transmission congestion is gone, you can bring turbines back online to increase production.

11.1.3 Find the wind

Wind and weather originate from the variable heating of the earth's surface via the sun (see Chapter 1). Differences in air density, which is a function of pressure and temperature, must be balanced. The balance can be in the form of thermal, dynamic, or radiative processes. This chapter will not reiterate the many processes involved with the weather (for synoptic scale explanations, please see Wallace and Hobbs, 1977). Just recall that wind is the movement of air from areas of high pressure to low pressure, and that the Coriolis force and frictional force have their say in the magnitude and direction of wind flow on differing spatial and temporal scales (Williams, 1997, chapter 3).

Global wind patterns aside, where do you start looking for wind? You can examine climatological patterns in wind speed and direction, but those are typically based on surface winds (approximately 10 m above ground level), or upper air soundings (hundreds to thousands of meters above a potential wind farm), with few observations over large bodies of water. You can use climatological data and/or numerical weather prediction outputs to generate wind maps and identify areas of interest for wind farm development. Organizations such as the National Renewable Energy Laboratory (NREL) create wind maps for the United States with a spatial resolution of 2 km (Figure 11.2; www.nrel.gov/gis/wind.html). Maps like the one in Figure 11.2 allow a wind farm developer to focus on areas of interest.

When synoptic conditions are weak, mesoscale processes and local terrain features can aid in increasing wind speeds (Fernando et al., 2019). If the predominant

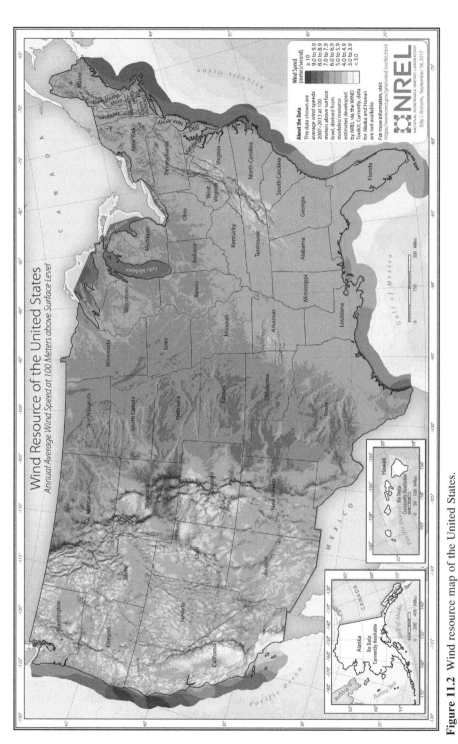

Figure 11.2 Wind resource map of the United States. From the National Renewable Energy Laboratory.

large-scale flow is parallel to a valley, that terrain feature can channel the winds (Fernando et al., 2019). Even broad and somewhat shallow valleys can help to channel and accelerate wind speeds (Emeis, 2013, chapter 4). Escarpments, which are long, steep slopes that usually separate two plateaus, offer a favorable area to place turbines. The turbines can be placed near the upper edge of the escarpment to capture enhanced flow, and avoid flow separation that can occur in steep, two-sided ridges (Barthelmie et al., 2016). Other flow patterns created by orographic features include mountain/valley winds, which can have effects a few hundred meters to a few hundred kilometers away and mountain-plain flows that are similar to land-sea breezes (Emeis, 2013, chapter 4). The mountain-plain flows take multiple hours to develop and can be relevant up to 100 km away from the foothills (e.g., Wilczak and coauthors, 2019). The air over the mountains is heated more than the plains and a broad circulation is created. Katabatic winds are the reverse of this process in that the cooler air in the mountains becomes less dense and a broad drainage flow develops (Emeis, 2013, chapter 4). Speaking of land-sea breezes, an area does not need terrain to generate mesoscale wind flow. A mesoscale process that can influence where you look for favorable wind farm locations is areas with frequent low-level jets (LLJ). The Great Plains of the United States are a favorable region for utilizing the nocturnal LLJ for wind power (Banta et al., 2002). In this region, Banta et al. (2002) measured more than half of the maxima of the nocturnal LLJ occurred below 140 m. Therefore, knowing synoptic and mesoscale patterns of a region, as well as the orography, are crucial to the selection of a potential wind farm.

There are many additional considerations once you have areas targeted for development of wind farms. You need to consider how far away your project is from transmission lines. Building additional transmission lines from your project to an interconnection point means more materials and greater cost. You need to consider how much land is available and landowner participation. Terrain, soil types, vegetation, and local government restrictions all must be studied and factored into a wind farm layout. Market factors are another important consideration when deciding where to build a wind farm. For example, states could have favorable renewable portfolio standards (RPS) increasing the need for power from wind sources (www.nrel.gov/state-local-tribal/basics-portfolio-standards.html). Additionally, a utility company may need to add renewable capacity. Those cases can lead a developer to pursue opportunities within those areas. These factors are examined by a team of engineers and analysts, including energy resource meteorologists.

11.2   Anatomy of a turbine

There are many technologies theorized and developed to extract power from the wind. Manwell et al. (2009, chapter 1) summarized and catalogued those myriad ideas. This chapter focuses on the predominant wind turbine type: a three-blade, horizontal axis machine facing into the wind (upwind facing; Chiras, 2017, chapter 3). A horizontal axis wind turbine (HAWT) is composed of four main sections: rotor, nacelle, tower, and base (Manwell et al., 2009, chapter 1).

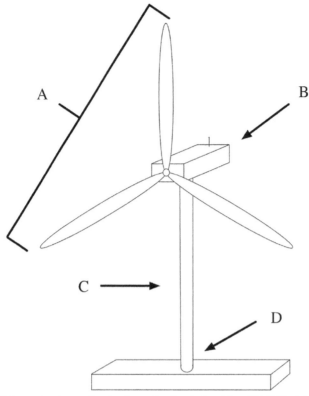

Figure 11.3 Schematic of a modern horizontal axis wind turbine (HAWT). *A* is the rotor, *B* is the nacelle, *C* is the tower, and *D* is the base.

Figure 11.3 depicts a modern HAWT. The rotor (Figure 11.3A) is made up of the blades and the hub (the rotor plane can be called the blade-swept area). The nacelle (Figure 11.3B) houses the drive shaft, gearbox or direct-drive mechanism, generator, yaw motors, cooling system, hydraulics, and control systems. The nacelle has a two-dimensional wind sensor (sonic anemometer or cup anemometer and wind vane) affixed to the outside that informs the control system of any changes in wind direction or speed so the turbine can respond accordingly. The nacelle mounts to the top of the tower (Figure 11.3C), and the nacelle-rotor combination can pivot (also known as yaw) to any horizontal direction. Besides supporting the rotor and nacelle, the tower (Figure 11.3C) contains the electric cables and access ladders. Some towers are divided into sections, so a wind turbine technician does not have to climb a 250-ft. ladder all at once. The base of the wind turbine (Figure 11.3D) has an access hatch for the tower, internal system control access, and the reinforced concrete foundation. The turbine has a transformer in the nacelle or at the tower base to step up or increase voltage. The electricity is sent via wires to a substation and finally added to the grid.

So how does it all work? Manwell et al. (2009, chapters 3–5) provides a detailed explanation of turbine blade aerodynamics and power generation. Air flows over the blades, creating lift. The lift starts the blades to move radially, spinning the hub and attached drive shaft (Figure 11.3A). The drive shaft turns and within the nacelle (Figure 11.3B) the generator section spins magnets in spools of copper. As a result, a flow of electricity is created. Taking a deeper dive, the yaw motors, at the nacelle–tower interface, turn the rotor and nacelle into the wind. The dominant wind turbine technology is pitch-controlled (Hansen and Hansen, 2007), the control system can change the angle of the blades relative to the hub (also known as feather) to maximize lift at different wind speeds. Blades can be feathered to reduce and even limit lift, thus stopping the rotation. All of the systems are controlled automatically by onboard computer; however, manual override is possible. The various wind turbine manufacturers have their own specific designs and configurations, but they all operate the same way.

Wind turbine models are typically described with three identifying characteristics: hub height, rotor diameter, and generator capacity. The hub height refers to the distance from the ground to the hub at the center of the rotor area. Wiser et al. (2020) found that the average hub height for onshore wind turbines in the United States increased from 60 m (197 ft.) in 1998 to 90 m (295 ft.) in 2019. Furthermore, the range of hub heights for turbines installed in 2019 was 80–112 m (262–367 ft.; https://emp. lbl.gov/wind-power-technology-trends). The rotor diameter is linked to the blade length. A machine with a 100-m rotor diameter has blades that are nearly 50 m (164 ft.) in length. For scale, that is just under half the length of a U.S. football field. The distance from the center of the hub to the tip of the blade will then be 50 m. The average rotor diameters of US wind turbines have steadily increased from 2009 to 2019, going from approximately 80 m (262 ft.) to 120 m (394 ft.; Wiser et al., 2020). Since 1998, the range of rotor diameters has been 40–136 m (131–446 ft.; https://emp.lbl.gov/wind-power-technology-trends). While not generally part of turbine descriptions, the tip height generally is the hub height plus one-half the rotor diameter. If you take the recent average hub height and rotor diameter, 90 and 120 m, respectively, you find the tip height of this example is nearly 150 m (492 ft.). Finally, the generator capacity describes the maximum output from the wind turbine and can range from 0.5 to more than 3.5 MW (Wiser et al., 2020).

When planning for a future wind farm, project developers typically acquire power curves from various wind turbine manufacturers. Power curves are the nominal power production at different wind speeds and are calculated and reported by the manufacturers based on different rotor, generator, and hub-height combinations. The power curve is the theoretical output for a given model, but actual output can vary (Lee et al., 2020). Figure 11.4 depicts a generic power curve for consideration. In the low end of the power curve, 2–6 m/s, there is not a lot of change in power production with change in wind speed, but the power production values are not large either. The cut-in speed, when the blades start moving in response to wind flow (Manwell et al., 2009, chapter 1), is 3 m/s for this example. When wind speeds increase from 6 to 9 m/s, there is approximately a fourfold increase in power production. This narrow band of wind speed fluctuation results in significant changes in power production. At 12 m/s the machine is fully rated (Manwell et al., 2009, chapter 1), at full power, and will limit production to 2 MW of power, reducing harmful loads. A speed change from 12 to

Power Production based on Wind Speed

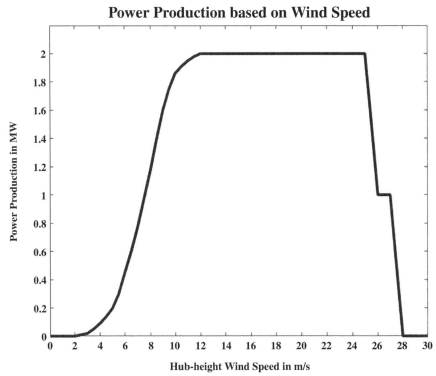

Figure 11.4 Idealized power curve for 2-MW wind turbine. Power production in MW is related to different wind speed bins.

25 m/s does not impact the power production, but the control system of the turbine continues to adjust the blades so as to not catch as much air and compensate for those higher speeds. In this example (Figure 11.4), there is a feature called a derate. At 26 m/s the control system feathers the blades to reduce lift to protect the turbine, and as a result the power is reduced. Not all models have a derating range, many will just cut out or feather the blades enough to slow and stop the rotation of the hub and drive shaft. From the example in Figure 11.4, if the wind speed rises to 28 m/s or greater, the turbine will go into a shutdown mode for safety until wind speeds decrease to a certain level after a certain amount of time.

Often a wind farm developer will have multiple machines in mind and consider different scenarios with the different machines. As a project progresses, the developer will work with a selected turbine manufacturer to pick a favorable machine for the conditions at the project. Blade length is important, you can catch more wind with longer blades, but the blades are more susceptible to turbulence in the atmosphere. Increasing the hub-height can expose your machine to faster winds, but it will require larger foundations and taller towers to support the machines. Those added materials increase the cost to build the wind turbine structures. Selecting the optimal turbine type, balancing rotor diameter, hub height, generator size, and project cost, is crucial for making a competitive wind project.

Selection of the rotor diameter and blade configuration will impact the thrust coefficient, C_T. Thrust coefficient is dimensionless and relates the momentum change of the wind through the rotor area compared to a dynamic response of the rotor. The thrust coefficient is expressed as (Manwell et al., 2009, chapter 3):

$$C_T = \frac{T}{\frac{1}{2}\rho A U^2},$$
(11.2)

where T is equal to the change of momentum through the rotor area. The denominator of Eq. (11.2) is the culmination of the dynamic force. For pitch-regulated wind turbines, the thrust coefficient changes in response to the entry wind flow (Manwell et al., 2009, chapter 3). Furthermore, the thrust coefficient is not constant (Barthelmie et al., 2010). Thrust coefficient is highest at low wind speeds to capture as much wind as possible. Once the wind speeds increase into the near-rated and rated zones of the power curve, the turbines use a smaller portion of the available kinetic energy resulting in a smaller percentage change in momentum. Thus, the thrust coefficient decreases.

Wind speed and air density are important to the power output of a wind turbine (Eq. 11.1). Another important atmospheric condition for wind turbines is turbulence. Large changes in wind speed and direction over short time periods mean the machine has to keep adjusting the pitch of the blades and/or yaw the turbine into the instantaneous wind direction. This wears out components and can lead to reduced power production because the turbine keeps trying to find the wind (Herbert-Acero et al., 2014). Furthermore, strong wind shear with height across the rotor plain can increase stress on turbine blades. Thus, turbulence intensity (TI) is measured and considered for proposed wind turbine locations. Wind turbine TI uses the standard deviation of the horizontal wind speed (σ) divided by the average wind speed (u) over the same period of time (Stull, 1988).

$$TI = \frac{\sigma}{u},$$
(11.3)

Both wind farm developers and turbine manufacturers can examine the TI per wind speed bin (e.g., 8–9 m/s, 9–10 m/s, etc.) at a given project to identify how the observed turbulence compares to industry-standard curves (e.g., International Electrotechnical Commission, 2019). Turbine configurations are classified into wind turbine classes based on average wind speeds, gusts, and TIs (International Electrotechnical Commission, 2019). These classes help the project developer and turbine manufacture select a machine that will hold up to the typical conditions encountered at the wind farm location. Figure 11.5 shows the plots of reference TI from the IEC standard (two of the classes), as well as a depiction of an example TI (what is measured at a proposed wind farm location). The graph shows the decrease in turbulence as wind speed increases. The solid curve depicts the average on-site conditions per wind speed bin. This curve never crosses any of the TI class thresholds (Class C is the dashed line and Class A+ is the dotted line), thus the location can qualify as a Class C site, the gentlest of conditions.

Wind speed is not constant with height over the rotor diameter/blade swept area. Wagner et al. (2011) addresses that wind speed variability over the blade-swept area with the proposed use of a rotor equivalent wind speed (REWS). REWS accounts for

Turbulence Intensity over Wind Speed Bin

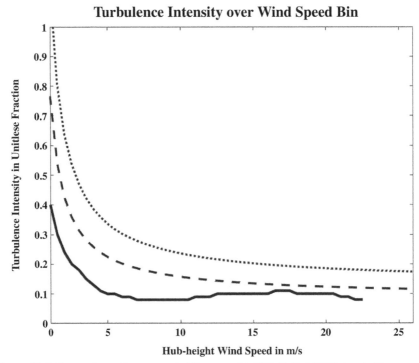

Figure 11.5 Comparison of an example TI curve with IEC 6140-1:2019 threshold curves.

the wind speed over the entire rotor area experienced by that turbine, as opposed to a point-estimate of wind speed at hub height. The equation modified from Wagner et al. (2011):

$$U_{REWS} = \left(\frac{1}{A} \sum_{i=1}^{n} U_i^3 A_i \right)^{1/3},\tag{11.4}$$

where A is the total blade swept area of the turbine (in m^2), and A_i is a part of the total area with an associated wind speed U_i (m/s). The easiest way to measure REWS is with ground-based, active remote sensing. That way you can measure wind speed at 10- or 20-m slices, determine the area of rotor for each slice, dependent on model and rotor diameter, and scale the sum based on the total area. Finally, the cube root of the area-weighted wind speed is taken to get your units back to m/s. REWS can be used when formulating power curves and to gauge turbine efficiency (Wagner et al., 2011).

11.3 Down on the farm

The previous section focused on an individual turbine. Now it is time to consider the structure of a wind farm. Wind turbines are arranged in rows also known as strings and are laid out in ways to maximize efficiency while also reducing construction costs.

Wind turbines could be grouped close together to reduce the length of electrical cables used and services roads installed (potentially reducing costs), but that setup could greatly impact power output from downwind wind turbines (Meyers and Meneveau, 2012). Conversely, the machines could be spread far apart to limit impacts from one to the other, but that plan can increase materials used to support the wind farm. Thus, there is a balance that must be achieved when designing and constructing a wind farm (Kusiak and Song, 2010). Additionally, there is the question of how much land is needed. Land requirements are highly variable and often depend on terrain, landowner participation, distance away from roads and structures (also known as set-back requirements), and other regulations.

A way to normalize distances when we talk about turbines and wind farms is through the use of the turbine rotor diameter, D. For example, if your turbines have a 100-m rotor diameter, then a project with $10D$ spacing has turbines with 1000 m of space between them. If turbulence from a turbine is measured $20D$ away, then the turbulence was measured 2,000 m away. Often, spacing is dependent on typical wind direction (Stevens et al., 2014). If wind flow is fairly constant from one direction, you can place turbines closer together on each string or row that is perpendicular to that flow.

Power lines from turbines typically run underground to a collection point. Multiple collection points feed into the wind farm's substation, where power is stepped up to transmission voltage. At this point the electricity enters the grid. The maximum power the wind farm can produce is called the nameplate capacity. This number is simply the number of turbines times the maximum power created by the model of turbine. For example, a wind farm with 50 machines, each rated at 2 MW, has a nameplate capacity of 100 MW. Developers can figure how many machines they can fit in an area, select the size of the generator, and determine the nameplate capacity for interconnection with the grid. On the other hand, the developer may have an option for a 50-MW interconnection window and may decide to install 25 2-MW machines or 10 5-MW machines.

Figure 11.6 presents an example wind farm layout. There are 25 locations denoted, with each row (sometimes called a string) containing 5 machines. Within each row, the spanwise (S_y) spacing is $5D$. If the rotor diameter for these example wind turbines is 100 m, that makes S_y equal to 500 m. There are five rows, each spaced $15D$ in the

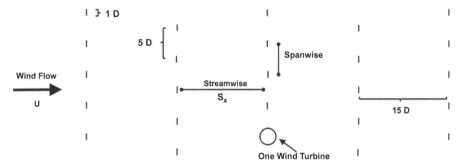

Figure 11.6 A plan view of an example wind farm layout. There are 25 wind turbines denoted and the wind flow is from left to right. The row-to-row distance is called streamwise (S_x) spacing, while spanwise (S_y) spacing denotes the distance between machines within a row.

streamwise (S_x) direction. Again using 100 m for D, S_x equals 1,500 m (nearly one mile). The overall size of this example wind farm is 2,000 by 6,000 m, or over 1 mile spanwise by nearly 4 miles streamwise. Stevens (2016) modeled the energy output from a wind farm with turbines inline and turbines with offset alignment (offset alignment depicted in Figure 11.6). The author found offset alignment had higher power output. Meyers and Meneveau (2012) recommended the 15D streamwise spacing to mitigate wind speed losses due to energy extraction. Mehta et al. (2014) described zones downwind of a turbine and noted that turbines should not be closer than 2–4D spanwise. Winds from consistent directions throughout the year could allow a designer to use a smaller spanwise spacing because the likelihood of winds going along a row are low.

There are some other considerations for wind farms. You want a site that is windy but not too windy. A 50 m/s wind has a lot of energy in it, but can be hazardous to many manufactured structures, including wind turbines. Weather conditions that are too hot or too cold in temperature can lead to machinery inefficiencies, for example derating due to high temperatures. Ice accumulation impacts the turbines as well. As ice accumulates on the blades it can reduce lift and stress the blades (Froese, 2017). Air temperature is important for the machines at the nacelle level. Electrical and mechanical performance can be impacted by extreme temperatures. There are even instances, depending on turbine type, where the machine may derate in response to air temperature. Understanding the air density at a site allows you to select the proper power curve(s) to use when modeling power production (recall Eq. 11.1).

11.3.1   Turbine–atmosphere interactions

Wind turbines impact the free stream (i.e., ambient) atmosphere upwind (ahead) of the turbine, at the turbine, and downwind (behind) of each wind turbine (Porté-Agel et al., 2020). The structure of the ambient boundary layer (e.g., stable, unstable) affects the magnitude of those impacts (e.g., Abkar and Porté-Agel, 2015; Sescu and Meneveau, 2015). These turbine–atmosphere interactions are important to understand because they affect the power production of the wind farm. Other effects are being discovered but do not necessary impact power production. This section discusses the interplays of wind turbines, wind farms, and the atmosphere. An examination of flow around one wind turbine is introduced, followed by flow in and around a full wind farm.

11.3.2   One turbine

First, consider the ambient wind flow approaching a single wind turbine. The wind turbine blades, tower, and nacelle are a barrier to that airflow. In response to this barrier, the wind speed decreases slightly before interacting with all parts of the turbine (Figure 11.7). This zone is known as the induction region (Porté-Agel et al., 2020) and can extend from the rotor plane to 3D upwind of the turbine (Medici et al., 2011). The induction region depends on the turbine's rotor diameter and a "rotor induction factor" (Porté-Agel et al., 2020) as well as the thrust coefficient of the turbine (Troldborg and Meyer Forsting, 2017). At 1D upwind, the wind velocity deficit (reduction from

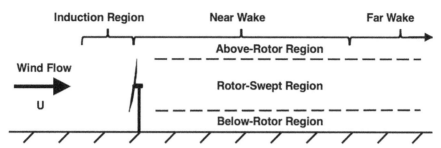

Figure 11.7 Vertical and horizontal regions of the atmosphere around a wind turbine. The induction region has been measured to extend 1.5*D* (Simley et al., 2016) and modeled up to 3*D* upstream (Medici et al., 2011). The near-wake region behind a turbine can range from 2 to 4*D* downwind (Vermeer et al., 2003). Beyond that is the far wake region, which can extend many rotor diameters away from the turbine (Hirth and Schroeder, 2013; Platis et al., 2018). The vertical zones were adapted from Frandsen (1992), Calaf et al. (2010), and Newman et al. (2013).

ambient) has been measured at 1–3% (Simley et al., 2016). Aitken et al. (2014) noted, "velocity deficits are greatest when the wind speed is below rated because of variations in the turbine thrust coefficient."

Next, air comes in contact with the blades of the turbine in the rotor plane. A strong pressure difference is created as the air flows over the blade. This causes lift on the blade and a rotation of the rotor. Air is then shed from the blades, resulting in the generation of vortices. Vortices form at the tip of the blades, called tip vortices, and the root of the blades, called root vortices (Mehta et al., 2014). These vortices cause a significant decrease in wind speed, increase in wind shear, and increase in turbulence. Because of the rotation of the blades, the tip vortices form a vortex tube and can be imagined as a cork-screw shape with the axis centered on the hub of the turbine (Troldborg et al., 2010). The vortex tubes expand toward the centerline of the axis and also outward until there is a disk of shear. Further downwind of the turbine, the wake breaks down due to entrainment of the surrounding wind field.

Wind turbine wake can be divided into two main parts: near wake and far wake (Crespo et al., 1999; Vermeer et al., 2003). Those zones are immediately behind and downwind of the turbine. The near-wake area is a highly turbulent zone and can extend downwind 2–4*D* (e.g., for a 100-m rotor diameter, this means 200–400 m downwind). Crespo et al. (1999) notes the near-wake region is dominated by turbulent diffusion of momentum. As the vortex tubes break down, the flow becomes fully turbulent and reaches maturity. This transition marks the start of the far-wake zone. Far-wake areas start 2–4*D* away from a turbine and can extend nearly 2 km for onshore wind turbines (e.g., Hirth and Schroeder, 2013) to more than 40 km for offshore turbines (e.g., Platis et al., 2018), depending on wind speed and atmospheric conditions. Thrust coefficient for the turbine and TKE produced by the rotor are the major factors in the far-wake characteristics (Crespo et al., 1999). Regions and their transitions are listed in ranges because the distance is dependent on ambient conditions. For example, if the predominant wind is strong, vortex tubes will not stay formed as long, as those tubes will be transformed to full turbulent flow more quickly (Troldborg et al., 2010).

Distinguishing between the two wake regions is important when modeling the wake from a turbine and for spacing turbines in a wind farm. Processes in the near wake are very complex and require computational fluid dynamics (CFD) modeling to resolve the life cycle of tip vortices (Mehta et al., 2014). However, in the far-wake region, large eddy simulation (LES) can resolve the broader features of the wake at a lower computational cost (Mehta et al., 2014). For turbine spacing, the designer of the wind farm should avoid placing machines within the near-wake zones. The turbulent stresses can reduce the life of the turbine in this region, and you have a significant loss in wind speed, which leads to lower power production (Frandsen, 2007). Placing turbines outside of far-wake regions is not feasible, but knowing the kind of reduction in wind speed can allow you to better predict energy production.

Velocity deficits and turbulence intensities within wakes have been modeled and measured. Up to a 60% velocity deficit (wake wind speed compared to free-stream wind speed) can occur in the near wake of the wind turbine (e.g., Hirth and Schroeder, 2013; Aitken and Lundquist, 2014). The velocity deficit is roughly Gaussian (think bell curve), with the peak velocity deficit centered on the axis of the hub height. The overall shape of the velocity deficit, and the wake as a whole, is not symmetrical in the vertical sense. The wake it is limited by the ground below and is influenced by ambient flow aloft (Mehta et al., 2014). The free-stream wind speed is only $1D$ above the turbine hub height (Barthelmie et al., 2003), showing just how vertically limited the wake can be. Horizontally, the wake is more symmetric but can meander (e.g., Hirth and Schroeder, 2013) and start to interact with wakes from other turbines. Studies by Crespo et al. (1999) and more recently Porté-Agel et al. (2020) show the maximum turbulence intensity is found at the tip height of the turbine.

Aitken and Lundquist (2014) noted that the ambient wind speed can impact the magnitude of the wind speed deficit. The velocity deficit depends on the thrust coefficient of the machine, and that thrust coefficient is related to the entry wind speed. For pitch-regulated turbines, the thrust coefficient is not constant. On the higher end of the power curve, approaching rated wind speeds, the thrust coefficient is at its lowest point – the turbine does not have to catch as much air to generate lift. Thus, the wind speed deficit in the wake is not as large. Conversely, in the low end of the power curve, the turbine adjusts the pitch of the blades to catch more air and is at its maximum thrust coefficient. As a result, the velocity deficit is highest at these lower wind speeds.

Wakes do not persist forever, nor do they spread infinitely. Downstream of the turbine the slower wind speeds in the wake interact with the ambient flow. This interaction speeds up the wind velocity in the wake, but also slows some of the ambient flow. As such, the wake zone widens and the deficit in wind velocity decreases. The rate of wake expansion is slow, about 5 degrees of spread in the horizontal (Barthelmie et al., 2010). How soon the conditions recover depends on atmospheric stability, ambient turbulence intensity, turbine size (both rotor diameter and hub height), turbine

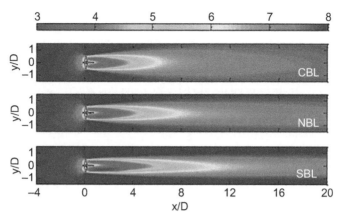

Figure 11.8 Wake modeled during different atmospheric stability conditions, top down. Cooler colors depict areas of velocity deficit due to the extraction of energy from the wind by the wind turbine. Warmer colors show where the wind flow is returning to the free-stream condition of 8 m/s. The scale is based on the rotor diameter both stream and span wise. The wake dissipation/velocity recovery occurs over a shorter downwind distance in convective cases (CBL) than in stable cases (SBL). NBL denotes neutral boundary layer.
From Abkar and Porté-Agel (2015).

spacing, turbine thrust coefficient, surface roughness, and terrain. Figure 11.8 illustrates what the wake looks like from above a model wind turbine.

The velocity deficit and the spread of wake can be approximated. Modifying wake equations explained by Frandsen et al. (2006), Aitken et al. (2014) presented simplified formulas for calculating the expansion of the wake region. This equation only considers the rotor diameter and distance away from the turbine.

$$w(x) = 1.3x^{0.33}, \tag{11.5}$$

where x is the downwind distance away from the wind turbine. The term x is expressed relative to the rotor diameter. For example, consider a 100-m rotor diameter turbine ($D = 100$ m). At x equals 1, the resulting wake width is $1.3D$. Multiply 100 (your D in meters) by 1.3 to convert back to 130 m of wake width. At $5D$ (500 m) downwind, the width becomes $2.2D$ (220 m), and at $10D$ (1 km) the width of the wake is approximately $2.8D$ (280 m). Aitken et al. (2014) also provided a simplified velocity deficit equation based on multiple past works,

$$VD = 56x^{-0.57}, \tag{11.6}$$

where VD is the velocity deficit in percent, and x is again expressed as the rotor diameters. The same 100-m diameter wind turbine is expected to create a 56% drop in wind speed in the immediate wake $1D$ or 100 m away. The velocity recovers (wake decays) to approximately 15% of the free-stream flow at $x = 10$ ($10D$), which in this case is 1 km away.

11.3.3   Full wind farm

Once you start considering multiple wind turbines, the impacts to the boundary layer become more complicated. The wind turbines react to the atmosphere and the turbines push back on the atmosphere. Bleeg et al. (2018) described the effect of "wind-farm-scale blockage." As mentioned previously, the induction zone of one turbine causes a slight decrease in wind speed before reaching a wind turbine. The culmination of many rows and columns of turbines creates a blockage on the atmosphere up to 10D in front of (upwind) of the turbine rows (Bleeg et al., 2018). If the turbine rotor diameter is 100 m, this means a blockage effect 1 km upwind of the wind farm. Bleeg et al. (2018) noted that while the reduction from free-stream wind speed for one turbine could be around 1.5%, the culmination of a wind farm can have more than a 3% decrease in wind speed before reaching turbines; a significant amount when considering energy production. Thus, wind turbines in a farm never see the free-stream wind flow. Segalini and Dahlberg (2020) noted the magnitude of the slowdown depends on the configuration of the wind farm, such as row spacing, turbine thrust coefficient, and turbine rotor diameter. Sescu and Meneveau (2015) discovered through modeling that in unstable (for example daytime) scenarios, the blockage effect is lower than during stable conditions.

Wakes from adjacent machines interact downwind and depend on the entry wind flow into the wind farm. Only the leading edge of a wind farm will experience (near) free-stream flow. All other turbines after the leading edge will be in a waked area. Interestingly, the wake effects do not compound when progressing from row to row in a wind farm. Early experiments summarized by Crespo et al. (1999) examined wakes when turbines were aligned so that each turbine was behind and downwind of each other. In this setup the first line of turbines in a wind farm produced expected power based on the ambient wind and the power curve of the machine. The second row of turbines experiences a significant loss in wind velocity from being in the wake of the first row and thus less power production. However, each row afterwards does not experience additional losses. In support of those findings, Frandsen et al. (2009) showed the wind speed at each row after the leading row experienced similar wind speeds and power production in non-complex terrain. Frandsen et al. (2006) noted a wind farm cannot have a situation where there is zero wind speed on the last line of turbines. Transferring of momentum, via entrainment, from higher altitudes or horizontally from the edges of wind farm replenishes some of the lost velocity. The entry flow does matter and will impact the magnitude of wind speed (and power) reduction. There are some reductions that reach an asymptote of reduction downwind (Barthelmie et al., 2010). The ratio between hub-height wind speed and free stream at the same height formulated by Emeis (2010) depends on the hub height of the turbines, the roughness of the surface, the Monin–Obukhov length, and an extraction coefficient c_t (not to be confused with thrust coefficient, C_T). The extraction ratio depends on the turbine characteristics, the size of the turbine (rotor diameter and height), the spacing of the turbines, and the drag imparted by the machine. Thus, the momentum extraction can be expressed as

$$c_t = A\left(\frac{N}{A_f}\right)Dc_i,$$ (11.7)

where N/A_f is the number of turbines over a given area, A is the rotor area, D is the rotor diameter, and c_i is the drag coefficient of a single turbine. Thus, an increase in rotor diameter and associated rotor area as well as more turbines per fixed areas lead to a larger extraction coefficient, even if turbine drag does not change (Emeis presented a c_i of $0.04\,\mathrm{m}^{-1}$).

Emeis (2013) calculated the best recovery of wind speed, or replenish rate (Crespo et al., 1999), and by extension power, occurs when the ambient atmosphere is unstable and the surface roughness around the wind farm is high (e.g., $Z_o = 1\,\mathrm{m}$). Further, recovery of ambient wind speed for the next farm downwind is greatest in unstable conditions and rough surface. Conversely, a stable and smooth surface, like one that is found in the marine boundary layer, has longer recovery distances between strings or from one wind farm to the next. Thus, the wake dissipates across shorter distances for land-based turbines vs their sea-based counterparts. Abkar and Porté-Agel (2015) modeled that convective atmospheric regimes lead to quicker wake recovery due to more efficient entrainment from above. The positive buoyancy, enhancement of TKE, and vertical momentum transport in convective situations work to mix out wind speed deficits. However, stable conditions diminish the wake recovery rate (Abkar and Porté-Agel, 2015). Allaerts and Meyers (2015) discovered the overall capping inversion of the ABL, both its height and strength also impact power production. When the strength of the inversion increased from 2.5 to $10\,\mathrm{K}$, the power output of a modeled wind farm decreased 13% (Allaerts and Meyers, 2015).

Wakes from wind turbines can be detected via satellite (e.g., Christiansen and Hasager, 2005), aircraft (e.g., Platis et al., 2018), radar (e.g., Hirth and Schroeder, 2013), lidar (e.g., Aitken and Lundquist, 2014), and sodar (e.g., Barthelmie et al., 2003). Many of the studies are limited to domain size (e.g., radar beam increasing with height) and challenges with building algorithms to automatically detect wakes (e.g., Aitken et al., 2014). In the absence of extensive meteorological measurements, wakes can be detected in the individual production from each wind turbine. The windward edge of turbines could have higher production than the trailing rows of turbines. Comparing the expected production based on wind entering the wind farm to the actual power production can show the reduction in production within the wind farm. Knowing the magnitude and expanse of those production deficits is important to wind farm design and forecasting. Measuring wake is challenging, and often requires the addition of computer models to estimate (or explain) the wake in a given wind farm.

11.4 The wind turbine atmospheric boundary layer

The concept of the wind turbine atmospheric boundary layer (WTABL) was first defined by Calaf et al. (2010). This work discussed the possibility of a wind farm becoming large enough (infinite) that the internal boundary layer created by the turbines' interaction with the ambient flow becomes its own fully developed boundary layer. The concept of an "infinite wind farm" also means that the addition of more rows of turbines does not affect the overall flow characteristics (Calaf et al., 2010). A wind farm could become an infinite wind farm when the length of the wind farm is 10 to 20 times as long as the ambient ABL is deep. In the example wind farm from Figure 11.6, which had a length of 6000 m, it could develop infinite wind farm flow when the boundary layer height is 300–600 m above ground level. Sescu and Meneveau (2015) demonstrated that the mere presence of wind turbines leads to an increase in the boundary layer height. Further, Abkar and Porté-Agel (2015) noted, "It is evident that the presence of the wind turbines has significant effects on the growth of the boundary layer."

Figure 11.9 illustrates two cases of the WTABL (Wu and Porté-Agel, 2017). Porté-Agel et al. (2020) further describe the regions in Figure 11.9. The induction zone is made up of all the induction regions of the turbines and contains the blockage area. Some ambient flow can be deflected above and around the wind farm. The entrance and development region forms at and above the wind farm. At this point it is an internal boundary layer as the wakes interact with the ambient flow. The fully developed region is the point where the "entire boundary-layer flow if fully adjusted to the wind farm" (Porté-Agel et al., 2020). At this point the flow is mostly homogeneous and the power extraction from the airflow is balanced by turbulent vertical transport of KE from above. At this point the wind farm is "infinite." The wind farm wake region has velocity deficits that will continue to mix out away from the wind farm. Porté-Agel et al. (2020) notes that there could be a 2% wind speed deficit even 5–20 km downwind. In strong inversion situations, an exit region can develop between the fully

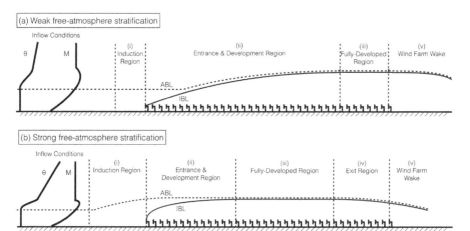

Figure 11.9 Flow regions in an infinite wind farm. The IBL created by the wind farm develops to a point where it meets the ABL. Potential temperature and mean velocity (M) for the ambient environment are displayed. Two cases are displayed: (A) a weak and (B) strong stratification of the free atmosphere (Wu and Porté-Agel, 2017).

developed and wind farm wake regions. In the exit region, there could be an acceleration of wind speeds that could increase production on the last rows of turbines.

The way wakes impact the boundary layer can be examined in three different zones. Frandsen et al. (2006) initially proposed and Meyers and Meneveau (2012) enhanced the model of three layers of the wind turbine ABL. The wake mixing layer is roughly the rotor area, extending downwind. Interactions in this layer already were presented earlier. Below the wake mixing layer is the inner layer of the wind turbine ABL. This layer is important to intra-farm interactions and impacts on the surface layer. Above the turbine is the outer layer. This layer interacts with the free atmosphere by way of an internal boundary layer. In convective (unstable) atmospheres the boundary layer can be 16% higher over a wind farm than if a wind farm is not present (Lu and Porté-Agel, 2015).

Wind turbines can be considered as "distributed roughness elements" (Crespo et al., 1999). Calaf et al. (2010) notes that a wind farm that is longer than the ABL is tall is large enough to develop its own boundary layer. Because the turbines extract power and reduce wind speed, the machines can be a drag on the overall airflow. Fitch et al. (2012) notes that wind farms can be modeled as momentum sinks on ambient flow. Further, the machines are converting kinetic energy to electrical energy, thus wind farms are TKE sources. Abkar and Porté-Agel (2015) modeled TKE production from a wind turbine. They detected the production of TKE was at the top and bottom of the rotor area, but the production of TKE was largest at the top of the rotor.

Takle et al. (2019) offered some insight into the ways wind turbines impact the boundary layer in a non-complex wind farm. They divided the boundary layer into three sections: the below-rotor level (BRL), rotor-swept level (RSL), and above-rotor layer (ARL). Their results were divided into the standard stability classes of neutral, unstable, and stable. The study further split results into measurements behind a single turbine and bulk wakes (conditions that would be seen downwind of a wind farm). Figure 11.10 illustrates the key points from their work. There were no differences in humidity from outside or inside the wind farm below the blade-swept area, however humidity was increased in the blade-swept layer in the wind farm during stable conditions. Additionally, temperature increased below and in the blade-swept area. The authors noted the presence of the wind farm suppressed surface cooling as opposed to heating the surface. Rajewski et al. (2013) was the first to examine this with a testbed of measurements and an operational wind farm. They measured little-to-no cooling of the surface temperature during the day, and a small ($<2\,°C$) increase in temperature, when compared to surrounding locations. Greater TKE was measured at night, due to higher wind speeds. No significant influence on sensible heat flux during the was observed. The authors of this study also found a slight acceleration of near-surface winds speeds under the wake of the turbines. The wind farm parameterization (Dudhia, 2011) of the WRF model (Skamarock et al., 2008) was able to create an acceleration of surface wind speed around 11% when modeling an offshore wind farm (Fitch et al., 2012).

Rajewski et al. (2014) were able to detect changes in fluxes of momentum, sensible heat, and carbon dioxide above a crop canopy. Downwind of turbines, where the wake

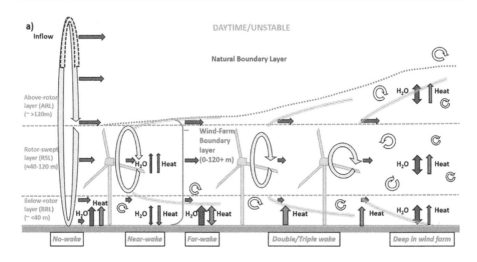

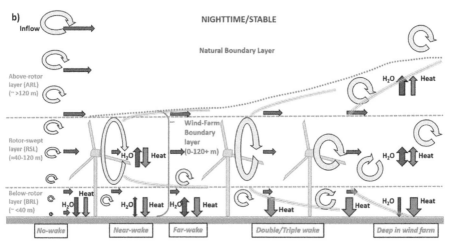

Figure 11.10 Conceptualization of wind farm modification of surface and boundary layer microclimate for (A) daytime/unstable conditions and (B) nighttime/stable conditions based on available field measurements, wind tunnel measurements, and numerical simulations. *Wider arrows* denote larger values in heat (*red*) and water vapor (*blue*) fluxes, and *narrower arrows* denote smaller values in fluxes. *Double-headed arrows* denote fluxes that could be either direction, zero, or unknown. Scales of turbulence are denoted by *light blue swirls*, with the top portion of the largest daytime eddy having a *dashed line* to infer boundary layer depth of several hundreds of meters or a few kilometers above the wind farm boundary layer.
From Takle et al. (2019).

intersects with the ground, the authors found moisture and carbon dioxide fluxes increase by a factor of 5. Carbon dioxide venting and sensible heating toward the surface were detected. In a later study by Rajewski et al. (2016), the authors noted "Air in the near wake at night is cooled, but air in the far wake and double wake is warmed; air temperature during the day is little affected by the turbines." Furthermore, Baidya Roy et al. (2004) modeled that a wind farm could warm temperatures near the surface and cool temperatures above hub height. While this effect was present in overnight (stable) conditions, the warming and cooling zone were negligible during the day (Baidya Roy et al., 2004).

The impacts of both individual turbines and large windfarms on the boundary layer can cause challenges for numerical modeling. Based on WRF model simulation, Tomaszewski et al. (2018) showed WRF wind farm parameterization can simulate wake impacts on the atmosphere and those wake impacts are more pronounced in stable conditions. More specifically, simulated wakes were modeled in times when sensible heat was negative, indicating cooling. Modeled with WRF and verified with satellite observations, Xia et al. (2017) showed surface warming in an operational wind farm region but cooling downwind of the wind farm. Xia et al. (2018) continued this work and examined the mechanisms that contribute to those heating and cooling trends, concluding, "[wind farm] induced change is [sensible heat] flux is the dominant factor for the simulated temperature changes at the surface." Further, the surface warming was driven by the TKE effect, while a momentum sink effect was responsible for surface cooling. Changes in the vertical temperature were noted in the modeled scheme. Finally, temperature advection is important for the surrounding, non-wind farm areas (Xia et al., 2018).

References

Abkar, M., Porté-Agel, F., 2015. Influence of atmospheric stability on wind-turbine wakes: a large-eddy simulation study. Phys. Fluids 27, 035104.

Ackermann, T., Söder, L., 2002. An overview of wind energy-status 2002. Renew. Sust. Energ. Rev. 6 (1), 67–128.

Aitken, M.L., Lundquist, J.K., 2014. Utility-scale wind turbine wake characteristics using nacelle-based long-range scanning lidar. J. Atmos. Ocean. Technol. 31 (7), 1529–1539.

Aitken, M.L., Banta, R.M., Pichugina, Y.L., Lundquist, J.K., 2014. Quantifying wind turbine wake characteristics from scanning remote sensor data. J. Atmos. Ocean. Technol. 31 (4), 765–787.

Allaerts, D., Meyers, J., 2015. Large eddy simulation of a large wind-turbine array in a conventionally neutral atmospheric boundary layer. Phys. Fluids 27, 065108.

American Clean Power (ACP), 2020. ACP Market Report: Fourth Quarter 2020. 27 pp. https://cleanpower.org/wp-content/uploads/2021/02/ACP_MarketReport_4Q2020.pdf. (Accessed 13 June 2021).

Badissy, M., Evans, A., Ewelukwa, N., Govender, J., Ketchum, R.T., Loraoui, M., Lienbenberg, C., Nagarajan, S., Ndahumba, G.T., Pavry, J., Vajeth, O., 2020. Understanding Power Purchase Agreements, second ed. U.S. Commercial Law Development Program, U.S. Department of Commerce. https://cldp.doc.gov/programs/cldp-in-action/understanding-power-purchase-agreements-second-edition.

Baidya Roy, S., Pacala, S.W., Walko, R.L., 2004. Can large wind farms affect local meteorology? J. Geophys. Res. 109. https://doi.org/10.1029/2004JD004763.

Bakke, G., 2016. The Grid: The Fraying Wires Between Americans and Our Energy Future. Bloomsbury, New York.

Banta, R.M., Newsom, R.K., Lundquist, J.K., Pichugina, Y.L., Coulter, R.L., Mahrt, L., 2002. Nocturnal low-level jet characteristics over Kansas during CASES-99. Bound.-Layer Meteorol. 105, 221–252.

Barthelmie, R.J., Folkerts, L., Ormel, F.T., Sanderhoff, P., Eecen, P.J., Stobbe, O., Nielsen, N. M., 2003. Offshore wind turbine wakes measured by sodar. J. Atmos. Ocean. Technol. 20 (4), 466–477.

Barthelmie, R.J., Pryor, S.C., Frandsen, S.T., Hansen, K.S., Schepers, J.G., Rados, K., Schlez, W., Neubert, A., Jensen, L.E., Neckelmann, S., 2010. Quantifying the impact of wind turbine wakes on power output at offshore wind farms. J. Atmos. Ocean. Technol. 27 (8), 1302–1317.

Barthelmie, R.J., Wang, H., Doubrawa, P., Giroux, G., Pryor, S.C., 2016. Effects of an escarpment on flow parameters of relevance to wind turbines. Wind Energy 19, 2271–2286.

Betz, A., 2013. The maximum of the theoretically possible exploitation of wind by means of a wind motor. Wind Eng. 37 (4), 441–446. https://doi.org/10.1260/0309-524X.37.4.441.

Bleeg, J., Purcell, M., Ruisi, R., Traiger, E., 2018. Wind farm blockage and the consequences of neglecting its impact on energy production. Energies 11 (6), 1609.

Brown, M.H., Sedano, R.P., 2004. Electricity Transmission: A Primer. National Council on Electricity Policy, ISBN: 1-58024-352-5. 77 pp.

Calaf, M., Meneveau, C., Meyers, J., 2010. Large eddy simulation study of fully developed wind-turbine array boundary layers. Phys. Fluids 22, 015110.

Chiras, D., 2017. Power from the Wind: Achieving Energy Independence, a Practical Guide. New Society Publishers, Gabriola Island, BC.

Christiansen, M.B., Hasager, C.B., 2005. Wake effects of large offshore wind farms identified from satellite SAR. Remote Sens. Environ. 98 (2), 251–268.

Crespo, A., Hernández, J., Frandsen, S., 1999. Survey of modelling methods for wind turbine wakes and wind farms. Wind Energy 2 (1), 1–24.

Dudhia, J., 2011. WRF version 3.3. New features and updates. In: Proceedings of the 12th Annual WRF Users' Workshop, Boulder, CO, NCAR, 1.1.

Emeis, S., 2010. A simple analytical wind park model considering atmospheric stability. Wind Energy 13 (5), 459–469.

Emeis, S., 2013. Wind Energy Meteorology. Springer, Heidelberg.

Energy Information Administration, 2016. U.S. Electric System Is Made up of Interconnections and Balancing Authorities. https://www.eia.gov/todayinenergy/detail.php?id=27152.

Energy Information Administration, 2020. How Much Electricity Does an American Home Use? https://www.eia.gov/tools/faqs/faq.php?id=97&t=3, United States Government, Accessed 22 September 2020.

Fernando, H.J.S., Mann, J., Palma, L.M.L.M., Lundquist, J.K., Barthelmie, R.J., Belo-Pereira, M., Brown, W.O.J., Chow, F.K., Gerz, T., Hocut, C.M., Klein, P.M., Leo, L.S., Matos, J.C., Oncley, S.P., Pryor, S.C., Bariteau, L., Bell, T.M., Bodini, N., Carney, M.B., Courtney, M. S., Creegan, E.D., Dimitrova, R., Gomes, S., Hagen, M., Hyde, J.O., Kigle, S., Krishnamurthy, R., Lopes, J.C., Mazzaro, L., Neher, J.M.T., Menke, R., Murphy, P., Oswald, L., Otarola-Bustos, S., Pattantyus, A.K., Veiga Rodrigues, C., Schady, A., Sirin, N., Spuler, S., Svensson, E., Tomaszewski, J., Turner, D.D., van Veen, L., Vasilijevic, N., Vassallo, D., Voss, S., Wildmann, N., Wang, Y., 2019. The Perdigão: peering into microscale details of mountain winds. Bull. Am. Meteorol. Soc. 100 (5), 799–819. https://doi.org/10.1175/BAMS-D-17-0227.1.

Fitch, A.C., Olson, J.B., Lundquist, J.K., Dudhia, J., Gupta, A.K., Michalakes, J., Barstad, I., 2012. Local and mesoscale impacts of wind farm as parameterized in a mesoscale NWP model. Mon. Weather Rev. 140 (9), 3017–3038.

Frandsen, S., 1992. On the wind speed reduction in the centerof large clusters of wind turbines. J. Wind Eng. Ind. Aerodyn. 39, 251–265.

Frandsen, S.T., 2007. Turbulence and Turbulence-Generated Structural Loading in Wind Turbine Clusters. Risø National Laboratory. Risø-R-1188(EN).

Frandsen, S., Barthelmie, R., Pryor, S., Rathmann, O., Larsen, S., Højstrup, J., Thøgersen, M., 2006. Analytical modelling of wind speed deficit in large offshore wind farms. Wind Energy 9 (1), 39–53.

Frandsen, S.T., Jørgensen, H.E., Barthelmie, R., Rathmann, O., Badger, J., Hansen, K., Ott, S., Rethore, P., Larsen, S.E., Jensen, L.E., 2009. The making of a second-generation wind farm efficiency model complex. Wind Energy 12 (5), 445–458.

Froese, M., 2017. Cracking the Icing Problem on Turbine Blades. Windpower Engineering & Development https://www.windpowerengineering.com/cracking-icing-problem-turbine-blades/. (Accessed 7 March 2022).

Halliday, D., Resnick, R., Walker, J., 1997a. Fundamentals of Physics Part 1, fifth ed. John Wiley & Sons, Inc., New York.

Halliday, D., Resnick, R., Walker, J., 1997b. Fundamentals of Physics Part 3, fifth ed. John Wiley & Sons, Inc., New York.

Hansen, A.D., Hansen, L.H., 2007. Wind turbine concept market penetration over 10 years (1995-2004). Wind Energy 10 (1), 81–97.

Herbert-Acero, J.F., Probst, O., Réthoré, P.E., Larsen, G.C., Castillo-Villar, K.K., 2014. A review of methodological approaches for the design and optimization of wind farms. Energies 7, 6930–7016. https://doi.org/10.3390/en7116930.

Hirth, B.D., Schroeder, J.L., 2013. Documenting wind speed and power deficits behind a utility-scale wind turbine. J. Appl. Meteorol. Climatol. 52 (1), 39–46.

Hoen, B.D., Diffendorfer, J.E., Rand, J.T., Kramer, L.A., Garrity, C.P., Hunt, H.E., 2018. United States Wind Turbine Database (ver. 4.0, April 2021): U.S. Geological Survey. American Clean Power (ACP) Association (formerly American Wind Energy Association), and Lawrence Berkeley National Laboratory data release. https://doi.org/10.5066/F7TX3DN0.

International Electrotechnical Commission (Ed.), 2019. Wind Energy Generation Systems – Part 1. IEC 61400-1:2019, fourth ed. International Electrotechnical Commission.

Kusiak, A., Song, Z., 2010. Design of wind farm layout for maximum wind energy capture. Renew. Energy 35 (3), 685–694.

Lee, J.C.Y., Stuart, P., Clifton, A., Fields, M.J., Perr-Sauer, J., Williams, L., Cameron, L., Geer, T., Housley, P., 2020. The power curve working group's assessment of wind turbine power performance prediction methods. Wind Energy Sci. 5, 199–223.

Lu, H., Porté-Agel, F., 2015. On the impacts of wind farms on a convective atmospheric boundary layer. Bound. Lay. Meteorol. 157 (1), 81–96.

Manwell, J.F., McGowan, J.G., Rogers, A.L., 2009. Wind Energy Explained, second ed. Wiley, Chippenham.

Medici, D., Ivanell, S., Dahlberg, J.-Å., Alfredsson, P.H., 2011. The upstream flow of a wind turbine: blockage effect. Wind Energy 14 (5), 691–697.

Mehta, D., van Zuijlen, A.H., Koren, B., Holierhoek, J.G., Bijl, H., 2014. Large eddy simulation of wind farm aerodynamics: a review. J. Wind Eng. Ind. Aerodyn. 133, 1–17.

Meyers, J., Meneveau, C., 2012. Optimal turbine spacing in fully developed wind farm boundary layers. Wind Energy 15 (2), 305–317.

Mills, A.D., Millstein, D., Wiser, R., Seel, J., Carvallo, J.P., Jeong, S., Gorman, W., 2019. Impact of Wind, Solar, and Other Factors on Wholesale Power Prices. Lawrence Berkeley National Laboratory https://emp.lbl.gov/publications/impact-wind-solar-and-other-factors.

Newman, Meneveau, Castillo, 2013. Streamwise development of the wind turbine boundary layer over a model wind turbine array. Phys. Fluids 25, 085108.

Platis, A., Siedersleben, S.K., Bange, J., Lampert, A., Bärfuss, K., Hankers, R., Cañadillas, B., Foreman, R., Schulz-Stellenfleth, J., Djath, B., Neuman, T., Emeis, S., 2018. First in situ evidence of wakes in the far field behind offshore wind farms. Sci. Rep. 8 (1), 1–10.

Porté-Agel, F., Bastankhah, M., Shamsoddin, S., 2020. Wind-turbine wind-farm flows: a review. Bound. Lay. Meteorol. 174 (1), 1–59. https://doi.org/10.1007/s10546-019-00473-0.

Rajewski, D., Takle, E., Lundquist, J., Oncley, S., Prueger, J., Horst, T., Rhodes, M., Pfeiffer, R., Hatfield, J., Spoth, K., Doorenbos, R., 2013. Crop wind energy experiment (CWEX): observations of surface-layer, boundary layer, and mesoscale interactions with a wind farm. Bull. Am. Meteorol. Soc. 94 (5), 655–672. https://doi.org/10.1175/BAMS-D-11-00240.1.

Rajewski, D.A., Takle, E.S., Lundquist, J.K., Prueger, J.H., Pfeiffer, R.L., Hatfield, J.L., Spoth, K.K., Doorenbos, R.K., 2014. Changes in fluxes of heat, H_2O, and CO_2 caused by a large wind farm. Agric. For. Meteorol. 194, 175–187.

Rajewski, D.A., Takle, E.S., Prueger, J.H., Doorenbos, R.K., 2016. Toward understanding the physical link between turbines and microclimate impacts from in situ measurements in a large wind farm. J. Geophys. Res. Atmos. 121 (22), 13392–13414.

Schewe, P.F., 2007. The Grid: A Journey Through the Heart of our Electrified World. Joseph Henry Press, Wasington, DC.

Segalini, A., Dahlberg, J.-Å., 2020. Blockage effects in wind farms. Wind Energy 23 (2), 120–128.

Sescu, A., Meneveau, C., 2015. Large-eddy simulation and single-column modeling of thermally stratified wind turbine arrays for fully developed, stationary atmospheric conditions. J. Atmos. Ocean. Technol. 32 (6), 1144–1162.

Simley, E., Angelou, N., Mikkelsen, T., Sjöholm, M., Mann, J., Pao, L.Y., 2016. Characterization of wind velocities in the upstream induction zone of a wind turbine using scanning continuous-wave lidars. J. Renew. Sustain. Energy 8, 013301.

Skamarock, W.C., et al., 2008. A Description of the Advanced Research WFR Version 3. NCAR Technical Note, NCAR/TN-475+STR. 125 pp.

Stevens, R.J.A.M., 2016. Dependence of optimal wind turbine spacing on wind farm length. Wind Energy 19 (4), 651–663.

Stevens, R.J.A.M., Gayme, D.F., Meneveau, C., 2014. Large eddy simulation studies of the effects of alignment and wind farm length. J. Renew. Sust. Energy 6, 023105.

Stull, R., 1988. Chapter 2: Some mathematical & conceptual tools: Part 1. Statistics. An Introduction to Boundary Layer Meteorology. Kluwer Academic Publishers.

Takle, E.S., Rajewski, D.A., Purdy, S.L., 2019. The Iowa atmospheric observatory: revealing the unique boundary layer characteristics of a wind farm. Earth Interact. 23 (2), 1–27.

Tomaszewski, J.M., Lundquist, J.K., Kaffine, D.T., Duvivier, K.K., Wilden, C., 2018. Quantifying wake impacts on downwind wind farms using the WRF wind farm parameterization. In: Ninth Conference on Weather, Climate, and the New Energy Economy 7–11 January 2018, Austin, TX. American Meteorological Society. 5.3.

Troldborg, N., Meyer Forsting, A.R., 2017. A simple model of the wind turbine induction zone derived from numerical simulations. Wind Energy 20, 2011–2020. https://doi.org/10.1002/we.2137.

Troldborg, N., Sørensen, J.N., Mikkelsen, R., 2010. Numerical simulations of wake characteristics of a wind turbine in uniform inflow. Wind Energy 13 (1), 86–99.

Vermeer, L.J., Sørensen, J.N., Crespo, A., 2003. Wind turbine wake aerodynamics. Prog. Aerosp. Sci. 39, 467–510.

Wagner, R., Courtney, M., Gottschall, J., Lindelöw-Marsden, P., 2011. Accounting for the speed shear in wind turbine power performance measurements. Wind Energy 14 (8), 993–1004.

Wallace, J.M., Hobbs, P.V., 1977. Atmospheric Science: An Introductory Survey. Academic Press, San Diego. 467 pp.

Wilczak, J.M., coauthors, 2019. The second wind forecast improvement project (WFIP2): observational field campaign. Bull. Am. Meteorol. Soc. 100 (9), 1701–1723.

Williams, J., 1997. The Weather Book. USA Today, second ed. Random House, New York. 227 pp.

Wiser, R., Bolinger, M., Hoen, B., Millstein, D., Rand, J., Barbose, G., Darghouth, N., Gorman, W., Jeong, S., Mills, A., Paulos, B., 2020. Wind Energy Technology Data Update: 2020 Edition. Lawrence Berkeley National Laboratory. 87 pp.

Wiser, R., Bolinger, M., Hoen, B., Millstein, D., Rand, J., Barbose, G., Darghouth, N., Gorman, W., Jeong, S., Mills, A., Paulos, B., 2021. Land-Based Wind Market Report: 2021 Edition. Lawrence Berkeley National Laboratory. 72 pp.

Wu, K.L., Porté-Agel, F., 2017. Flow adjustment inside and around large finite-size wind farms. Energies 10, 2164.

Xia, G., Cervarich, M.C., Roy, S.B., Zhou, L., Minder, J.R., Jimenez, P.A., Freedman, J.M., 2017. Simulating the impacts of real-world wind farms on land surface temperatures using the WRF model: validation with observations. Mon. Weather Rev. 145 (12), 4813–4836.

Xia, G., Zhou, L., Minder, J.R., Jimenez, P.A., Fovell, R.G., 2018. Simulating impacts of real-world wind farms on land surface temperature using the WRF model: physical mechanisms. In: Ninth Conference on Weather, Climate, and the New Energy Economy 7–11 January 2018, Austin, TX. American Meteorological Society. 5.4.

Further reading

Monteiro, C., Heko, H., Bessa, R., Miranda, V., Botterud, A., Wang, J., Conzelmann, G., 2009. A Quick Guide to Wind Power Forecasting: State-of-the-Art 2009. Argonne National Laboratory. ANL/DIS-10-2.

Petersen, E.L., Mortensen, N.G., Landbert, L., Højstrup, J., Frank, H.P., 1998. Wind power meteorology. Part I: climate and turbulence. Wind Energy 1 (1), 2–22.

Petersen, E.L., Mortensen, N.G., Landbert, L., Højstrup, J., Frank, H.P., 1998. Wind power meteorology. Part II: siting and Models. Wind Energy 1 (1), 55–72.

Smith, R.B., 2010. Gravity wave effects on wind farm efficiency. Wind Energy 13 (5), 449–458.

Troen, I., Petersen, E.L., 1989. European Wind Atlas. Risø National Laboratory, Roskilde.

Wharton, S., Lundquist, J.K., 2012. Assessing atmospheric stability and its impacts on rotor-disk wind characteristics at an onshore wind farm. Wind Energy 15 (4), 525–546.

The times they are a changing: How boundary layer processes cause feedbacks and rectifiers that affect climate change and earth system modeling

12

Russell K. Monson
Department of Ecology and Evolutionary Biology, University of Colorado, Boulder, CO, United States

12.1 Introduction

Computer models are inexact representations of the real world, and it's inevitable that the logic and equations underlying models will evolve as knowledge increases and scientists pursue improved accuracy. Over the past 2 decades, the forms of Earth System Models (ESMs), the primary tools for climate prediction, have regularly undergone modification and transition, progressing from a focus on climate change attribution to one of future adaptation and ecosystem management. As a result, ESMs have been honed to add new biophysical processes that extend the prognostic time horizon and include feedbacks and linkages among processes (Kawamiya et al., 2020). One challenge in the design and implementation of upgraded ESMs is the reconciliation of predictions across a progressively broader set of temporal scales (Meehl et al., 2014, 2021). ESMs were originally designed to predict climate change at scales of years-to-a few decades, but their prognostic targets are now extending across a century or more (Brune and Baehr, 2020). As model parameters are required to operate across multiple timescales, and the extent of observational constraint and speed of computation remain limited, modelers will continue to rely on time-averaging to produce composite parameters and drivers, as well as a continued focus on relatively slow biogeochemical processes. These tendencies will facilitate model tractability but are likely to exacerbate an already limited capability to address prediction uncertainties.

One remedy to this state of affairs is to improve representation of the higher frequency processes that underlie the slower phases in the current generation of ESMs. Such an emphasis would not only create opportunities for a more accurate representation of physics and geochemistry, but it would also provide a means for the increased engagement of system feedbacks. The potential impediments to this type of model improvement are both empirical and theoretical. Many of the faster processes within the real Earth system have only recently been revealed. This is a result of new

Conceptual Boundary Layer Meteorology. https://doi.org/10.1016/B978-0-12-817092-2.00002-3

observational approaches and instrumentation that have been improved to operate at higher sampling frequencies, and improved methods for integrating data and models. In terms of theory, one ongoing challenge has been the mathematical description and incorporation of non-linearities among process interactions. Many of the most relevant non-linear relations are due to processes that are sensitive to daily or seasonal solar cycles. Solar-driven cycles cause oscillations and covariances in process variables, but given the complexities of ESMs, these patterns are often linearized through time-dependent averaging and therefore mute in computational interactions.

In this chapter, I consider the faster timescales of selected Earth system processes and their potential to cause uncertainty in the climate-change predictions produced by ESMs. I focus on processes associated with the atmospheric boundary layer (ABL), most of which operate at the scale of 30 min or less. There are many examples of ABL processes that I could have discussed within the scope of climate prediction. I have chosen to focus on two as case studies: (1) ocean–atmosphere thermal coupling and its influence on the formation and persistence of stratocumulus cloud banks and (2) flux-gradient atmospheric CO_2 coupling with regard to the identification of global sources and sinks in the terrestrial carbon cycle. In both cases, I take a mechanistic approach to describing the "fast" processes and their relevance to non-linear linkages between model inputs and outputs. I then present a few examples of how some of the processes that cross diurnal-to-seasonal solar cycles produce rectifier effects in the real world and how those effects cause errors in ESM predictions. However, prior to presenting these more advanced topics, and in an effort to ensure that all readers start with the same background, I use the initial section of the chapter to consider general aspects of recent climate change and the importance of global cloud banks and greenhouse gas accumulation as climate-forcing variables.

12.2 Climate change in the Anthropocene

The earth's climate is, and always has been, dynamic. In the "climate change" debate of the past few decades, there are so-called skeptics or deniers who use this fact to discount projections of "unnatural" or anthropogenically forced climate change. However, there are many indicators that show the most recent warming is far from natural. While it is not possible to finely calibrate rates of climate warming from distant history, it is clear that the rate of warming since the 19th-century advent of the Industrial Revolution has been significant (IPCC, 2021), and since the mid-1970s it has been faster than at any time in the past 2000 years (Pages 2k Consortium, 2019). Furthermore, unlike some past "natural" warming events (e.g., the Medieval Warm Anomaly), the current phase is not limited in regional extent – it is, instead, globally coherent (Neukom et al., 2019). In fact, the warming since the beginning of the Industrial Revolution has catalyzed discussion of a new geologic epoch, separate from the Holocene – what is often called the Anthropocene (Steffen et al., 2011).

Based on analyses from physics and geochemistry, the primary cause of Anthropocene warming is clear – it is due to anthropogenic radiatively active (greenhouse) gases, principally CO_2, CH_4, O_3, N_2O, and chlorofluorcarbons (CFCs). These greenhouse gases have long lifetimes, allowing them to accumulate over decades in the ABL. The Sixth Report from the Intergovernmental Panel on Climate Change (IPCC), which was released in 2021, concluded that human activities have unequivocally increased atmospheric greenhouse gas concentrations, which are the greatest factor causing $\sim 1.1\,°C$ (within a statistically acceptable range of 0.8–1.3 °C) of mean global warming since 1900 (IPCC, 2021). The recent warming has been partially moderated by increases in atmospheric aerosol particles, which scatter incoming solar radiation and have a cooling effect on the earth's surface (Liao et al., 2004). Increases in aerosol concentrations since the Industrial Revolution are also linked to human activities, primarily through the pollution caused by fossil fuel combustion. However, the cooling influence of anthropogenic aerosols only offsets about 30% of the warming due to greenhouse gases (IPCC, 2013). Increased aerosol loading in the atmosphere has also been linked to increases in the formation and persistence of low-level cloudbanks and associated increases in planetary albedo (Christensen et al., 2020). These mechanisms controlling aerosol–cloud relations have not been included in most current-generation ESMs.

The total anthropogenic flux of CO_2 to the atmosphere, due to fossil fuel use in 2019, was ~ 37 billion metric tons (Friedlingstein et al., 2019). It is difficult to place the scale of this impact into its proper perspective given that it represents only 4–5% the annual global gross exchanges of CO_2 due to photosynthesis and respiration. One might ask: How can a "trickle" of CO_2 into the lower atmosphere, due to human activities, exert such large climate effects given the "immense" size of the natural fluxes? The answer lies in the fact that the global carbon budget had been in balance throughout the Holocene epoch, since the end of the last global glaciation. That is, the annual additions of CO_2 due to global respiration were approximately balanced by the annual removals due to photosynthesis. As a result, the CO_2 concentration of the atmosphere remained relatively stable (Figure 12.1). The balanced nature of the pre-industrial, natural carbon cycle provides the clarity required to conclude that anthropogenic activities are the primary reason for observed anomalies to the long-term mean CO_2 concentration.

The degree to which the lower atmosphere and surface temperatures change in response to a forced physical or geochemical perturbation, such as an imbalance in the sources and sinks of a greenhouse gas, is called climate sensitivity. Climate sensitivity is formally defined as the change in global mean surface temperature per unit change in a radiative forcing. Typically, it is calculated according to a standardized metric referenced to CO_2; for example, climate sensitivity is the warming caused by the increased concentration of any chosen trace gas, relative to that caused by a doubling in CO_2 concentration. The units of the sensitivity parameter are °C (or K). Climate sensitivity is determined through computer modeling, and the value of the derived sensitivity multiplier is dependent on how surface-atmosphere feedbacks are treated in the model. Two of the most important feedbacks are those due to (1) the temperature sensitivity of the radiated energy from atmospheric trace gases and surface mass according to the Stefan–Boltzmann law and (2) the influence of tropospheric water vapor concentration on the formation of clouds.

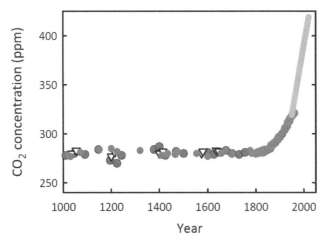

Figure 12.1 Atmospheric CO_2 concentration obtained from ice cores (prior to the 20th century) or direct observations (since the late 20th century). The near-constant CO_2 concentration during most of the past millennium reflects balanced CO_2 fluxes into and out of the atmosphere, on average. The increase in CO_2 concentration over the past century is due to an imbalance in CO_2 fluxes, with more emitted to the earth's atmosphere than extracted from it. The different symbols represent different datasets.
Redrawn from the Intergovernmental Panel on Climate Change (IPCC), 2001, Policy Makers Report, Working Group 1, p. 6.

The Stefan–Boltzmann law relates radiant flux density from a blackbody mass to the fourth power of its temperature (i.e., $R = \sigma T^4$, where σ is the Stefan–Boltzmann constant and equals $5.673 \times 10^{-8} \, \text{J s}^{-1} \, \text{m}^{-2} \, \text{K}^{-4}$). For objects other than blackbodies, the Stefan–Boltzmann law is modified according to the mass' radiative emittance (ε_L; the fractional emission from a non-blackbody relative to a blackbody). The modified Stefan–Boltzmann law is represented as $R = \varepsilon_L \sigma T^4$. The modified Stefan–Boltzmann law is used to estimate greybody emission (with $0 < \varepsilon_L < 1$). As an example, applying the Stefan–Boltzmann law to the earth's surface (assumed to be at an average temperature of 288 K), we can predict a blackbody emission of $390 \, \text{W m}^{-2}$. This example is highly simplified as the earth is in fact a greybody emitter, with variable long-wave emissivities (between 0.6 and $\sim$1). Nonetheless, the exercise provides an example of the thermodynamic linkage between body surface temperature and the loss of radiant energy (see Chapter 1 for a more discussion of radiation laws).

From the Stefan–Boltzmann law we predict that the loss of radiant energy from a warm body will create a negative feedback on surface temperature – the warmer the body, the more energy radiated, which works to cool the body. This radiative feedback would also cause a decrease in climate sensitivity. A warmer atmosphere, due to greenhouse gases and their trapping of upwelling longwave radiation from the earth's surface, will radiate more energy outward to space than a cooler atmosphere will. This represents a decreasing sensitivity of atmospheric warming to increasing atmospheric temperature, which is, in turn, transferred as a negative feedback to increases in surface temperature. The increase in radiative emission of a warmed body described by the Stefan–Boltzmann law is derived from a more fundamental thermodynamic relation known as Planck's Radiation Distribution law, and so the feedback that results from the temperature–radiation relation is often referred to as Planck's feedback.

A second feedback is caused by the formation of clouds at the top of the ABL. Clouds are known to have a large impact on the global energy budget, but the uncertainties in quantifying that impact are huge and continue to challenge global climate modelers (Arakawa, 1975; Medeiros et al., 2008; Vial et al., 2019). Clouds potentially cause both positive and negative feedbacks that influence climate sensitivity. A negative feedback occurs in the case of low-altitude cumulus and stratocumulus clouds, as these clouds tend to be thick and act like near-earth reflectors, thereby reducing the solar photon flux to the earth's surface. Because of their low altitude, cumulus clouds exist at temperatures similar to the earth's surface and therefore emit longwave radiation upward, from the cloud tops, at about the same rate as it is absorbed from the surface. This, combined with their high albedo, renders cumulus and stratocumulus clouds as net contributors to surface cooling and thus components of a negative feedback on climate sensitivity (Figure 12.2A). A positive feedback occurs in the case of high-altitude cirrus clouds, which are thin and "cold," being composed mostly of ice crystals. The cold nature of cirrus clouds means that at their top surface longwave radiation is emitted into space at lower rates than that absorbed from earth-surface emissions. This is balanced by a downward net flux of radiant energy. The reflection of shortwave solar radiation (albedo) from the tops of cirrus clouds has a weaker influence on surface warming than the net downward re-radiation of thermal energy. This renders cirrus clouds as net contributors to surface warming and thus components of a

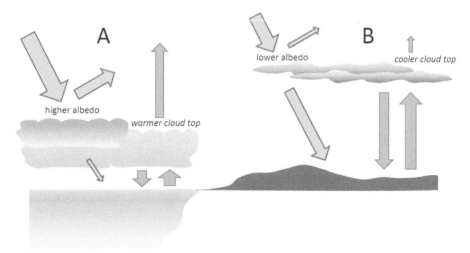

Figure 12.2 (A) Low-altitude stratocumulus cirrus clouds have relatively high albedo from cloud top which, along with relatively high upward rates of long-wave radiation, result in a net cooling effect of the earth's surface. (B) High-altitude cirrus clouds have lower cloud top albedos and lower long-wave radiation rates from cloud tops, relative to bottoms, which result in a net warming of the earth's surface.

positive feedback on climate sensitivity (Figure 12.2B). The complex influences of cloud feedbacks represent the largest uncertainty in models that quantify climate sensitivity (Cess et al., 1990; Bony and Dufresne, 2005; Webb et al., 2006).

12.3   Marine clouds and climate change

Some of the most important changes to the earth's energy budget are due to moisture transport between the oceans and atmosphere. Atmospheric water vapor is a greenhouse gas, although of non-anthropogenic origin. The downward re-emission of longwave radiation by water vapor contributes to temperature increases in the lower troposphere and on the earth's surface. The diurnal and seasonal patterns by which water vapor is evaporated from the ocean surface, transported in the atmosphere, and condensed to form cloud droplets largely controls the influence of atmospheric water vapor on planetary warming. In the absence of cyclones, the MABL is typically less than 1 km in depth (Zeng et al., 2004), and it does not reach the extreme heights that can be observed during deep convection above terrestrial, equatorial regions. Shallow marine cumulus and stratocumulus clouds, particularly in trade wind regions, cover vast stretches of the near-equatorial MABL, and because of their high albedo and influence on tropical convection, exert important controls over climate sensitivity (Neggers et al., 2007; Medeiros et al., 2008). Globally, stratocumulus clouds (which form between the surface and the height equivalent of 700 hPa atmospheric pressure) have been estimated to cover, on average, 23% of the marine surface, and to be responsible for most of the observed seasonal and interannual variability in mean global albedo (Hartmann et al., 1992; Wood, 2012). Through a simple representation of the global energy budget, it can be shown that a 10% error in our estimation of mean stratocumulus albedo, even considering its limited global extent, is carried through to an 8% error in modeled estimates of planetary albedo (Cahalan et al., 1994). The energy exchanges forced by marine clouds provide some of the best examples of important climate-associated processes that operate at short timescales and introduce significant uncertainties into the current generation ESMs.

12.3.1   Clouds in the equatorial MABL and the shifting nature of atmospheric albedo

The broad swaths of marine stratocumulus clouds that cover tropical trade wind regions are most commonly formed from thinner, stratus clouds that originate at the top of the MABL and are thickened through downward convection driven by longwave radiative emission and associated cooling at cloud tops. These processes interact within marine upwelling regions along the coasts of subtropical oceans. Cold water that is brought to the surface from deep ocean pools meets air that is warmed by the descending branch of deep atmospheric circulation cells, creating a stably stratified, inverted MABL (cool air below warm air). Stratus clouds form in the upper layers of the MABL thermal gradient (Klein, 1997; Stevens, 2005). As the incipient stratus layer thickens and moistens, increased radiative cooling at the cloud top creates

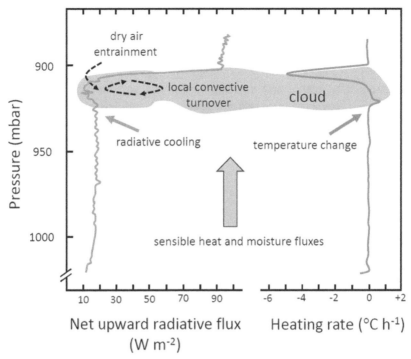

Figure 12.3 Vertical profiles of radiative infrared cooling flux (*orange*) and temperature change (*green*) for a nocturnal stratocumulus cloud layer at equatorial latitudes. Upward sensible heat and moisture fluxes from the ocean surface provide the buoyancy and water vapor required for cloud formation, and radiative cooling from the cloud top provides downward entrainment and mixing of dry air from aloft to feed cloud thickening.
Redrawn from Slingo et al. (1982) and Wood (2012).

a thin cold layer that sinks and initiates localized convective turnover (Figure 12.3). The convective motions within the developing cloud bank drive cycles of lifting and sinking, coincident with cycles of condensation and entrainment of drier air from aloft, all of which continues to thicken the bank to eventual stratocumulus status (Stevens, 2005; Wood, 2012).

One important MABL process that is facilitated through short timescale changes in sea-surface temperature (SST) and driven in part by low-level stratocumuli coverage is the Madden–Julian Oscillation (MJO) (Flatau et al., 1997; Bernie et al., 2005; Feng et al., 2015). The MJO influences wind and precipitation events within the entire equatorial troposphere, as it circles the globe in a 1–2-month cycle. However, its effects are most obvious in the equatorial regions of the Indian and Western Pacific Oceans, where it interacts with Indo-Pacific Warm Pool (De Deckker, 2016). The observed mean SST anomaly during the MJO is on the order of 0.5–1.0 °C, similar in scale to other equatorial Pacific oscillations, such as El Niño Southern Oscillation (ENSO). The MJO is an organized atmospheric envelope that forms in late March to early May with alternating phases of suppressed/enhanced rainfall that shift eastward at the rate of 4–8 m s^{-1}. Thus, the MJO is a traveling atmospheric oscillation, with dynamics that play out at the timescale of days to weeks, unlike those associated with the stationary, seasonal tendencies of ENSO. At any one moment in time, approximately half of the equatorial Earth exists in the suppressed phase with the other half in the active phase (Figure 12.4). As the MJO envelope moves eastward, it controls local alternations between cloud formation and dissipation (Riley et al., 2011). Prediction of the MJO has been attempted in some ESM analyses, though uncertainties remain high (Meehl et al., 2021). In fact, our understanding of the causes of the oscillation, with the level of detail that would enable representation in models, remains rudimentary (Zhang et al., 2020). This is one type of "fast" climate process that has the potential to exert large influences on the MABL and global climate but is still inadequately reconciled in ESM projections.

Stratocumuli albedos can be variable depending on solar zenith angle and liquid water path (LWP; the mass of liquid water between two points in the atmosphere), which defines cloud optical thickness. Optical thickness is influenced by water droplet concentration and size. Water droplet concentration, in turn, is dependent on aerosol particle concentration, which can vary from 50 particles cm^{-3} in aerosol-depleted atmospheres, such as above the remote ocean, to more than 500 particles cm^{-3} in aerosol-repleted atmospheres, such as those above continental shelves (Martin et al., 1994). An increase in aerosol particle concentration increases the density of

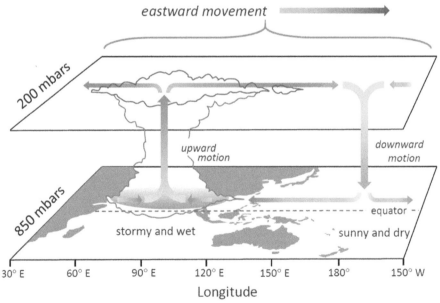

Figure 12.4 Progression of the Madden-Julian Oscillation (MJO) showing alternating phases of enhanced and suppressed stratocumulus regimes that continually move eastward. Climate features like the MJO represent shorter time-scale dynamics with longer time-scale energy-budget ramifications that are challenging to represent in ESMs.
Redrawn from Climate.gov by Fiona Martin (see Gottschalck, 2014).

cloud condensation nuclei (CCN), which causes a related increase in droplet concentration and reduction in droplet size. Both the increase in droplet number *and* decrease in droplet size are important factors determining the photon-scattering potential of stratocumuli cloud banks. The collision and coalescence of cloud droplets at critically high densities and sizes, and with longer residence times, causes liquid droplet formation and cloud drizzle (Feingold et al., 1996). Drizzle is a destabilizing influence on stratocumuli depth and longevity and thus a suppressant to regional albedo (Wood, 2012). As droplet size decreases, drizzle formation is reduced and cloud optical depth is increased (Pincus and Baker, 1994), causing changes in both the spectral nature of the scattered photon stream and within-cloud droplet circulation, both of which lead to an increase in the scattering of photons (Lu and Seinfeld, 2006).

In the remote oceans, atmospheric aerosols are derived from surface sea-salt sprays, produced by bubble-bursting and wind shear as well as the emission of organic compounds, such as dimethylsulfide (DMS), from phytoplankton (Andreae et al., 2003; Pierce and Adams, 2006). Many of the aerosol particles that drive near-coastal cloud banks originate in terrestrial ecosystems and are advected to the coastal atmosphere by onshore winds (de Leeuw et al., 2014). In trade wind regions, coastal atmospheric circulation mixes dust from terrestrial ecosystems with cool marine air. Some of the most extensive stratocumuli banks occur off the western coast of South America, where regional wind patterns are influenced by the Andes Mountains, delivering aerosol dusts to the atmosphere above areas of coastal upwelling. The Chilean cloud banks can extend up to 2000 km out into the Pacific Ocean, arching from central Chile northward toward the equator. Dust blown from the Saharan Desert has been observed thousands of kilometers away from its source, in the marine aerosol load of the Caribbean Sea (Carlson and Prospero, 1972). Transported dust has a twofold effect on marine stratoculumi. First, it exerts a direct forcing on the regional radiative balance by reflecting sunlight, which cools the earth's surface and warms the upper troposphere. This forcing inhibits cloud formation and forces those cloud banks that form to rest at higher altitudes. Second, it exerts an indirect influence through an increase in CCN concentration and promotion of cloud droplet formation. In the case of long-distance dust transport from the Sahara to the North Atlantic trade wind region, the direct effect has been shown to outweigh the indirect effect (Gutleben et al., 2019).

Pollution from industrial activity and the fuel exhaust from ships also have the potential to contribute CCNs to the MABL, promote cloud droplet formation, and increase solar scattering. This is generally referred to as the Twomey Effect (Twomey, 1974). Numerous studies have focused on the influences of ship-track pollution on cloud brightening, with most showing evidence of a significant Twomey Effect (e.g., Hobbs et al., 2000; Berner et al., 2015; Possner et al., 2018; although see Toll et al., 2019). Strategies have been proposed to take advantage of the Twomey Effect to reduce the global solar input and cool the earth's surface (e.g., Latham et al., 2008). In one proposal, aerosol precursors, such as those in sea spray, would be dispersed into the atmosphere beneath stratocumulus cloud banks as a means of forcing new particle entrainment, enhancing water droplet concentration and reducing water droplet radius, thus increasing albedo (Salter et al., 2008).

12.3.2   Clouds in the Arctic MABL and the dynamic state of sea ice albedo

Cloud–albedo relations differ in several ways for the case of Arctic cloud banks compared to those at equatorial latitudes. The vertical structure of the Arctic MABL can be complex, including frequent temperature inversions and episodic separation among layers of the ABL (Shupe et al., 2013; Brooks et al., 2017). An understanding of the unique nature of Arctic stratocumuli banks is important because they largely control the rate of Arctic Sea ice melt and its accompanying influences on regional sea-surface albedo (Curry et al., 1993; Uttal et al., 2002; Serreze and Barry, 2011). In the late Arctic summer, as sea-ice content reaches a minimum, a surface mixed layer (SML) forms within the lowest part of the troposphere, frequently with a cloud mixed layer (CML) above. The SML is cooled by the cold ocean/ice surface, which suppresses convective exchange and limits the overall depth of the MABL. The SML most often exists at near-neutral stability, with local mixing sustained through shear-induced turbulence. The overtopping CML is seldom influenced by bottom-up processes, such as surface radiative forcing and associated convection, as occurs at equatorial latitudes. Rather, it is driven by local turbulence created by cloud-top radiative cooling. The CML includes a persistent stratocumulus layer composed of a suspended mixture of liquid water and ice (McFarquhar et al., 2011; Morrison et al., 2012). There are summer periods when the SBL and the lowest layers of the CML are connected through upward, shear-driven mixing. The continuity of this process, however, is frequently disrupted by formation of a thin temperature inversion that develops above the sea surface – disconnecting turbulent exchange between the SML and CML (Shupe et al., 2013; Brooks et al., 2017).

The coupled and decoupled interactions between the SML and CML are clearly evident in past studies that have been conducted in the central Arctic Ocean (Figure 12.5). Observations of vertical profiles in atmospheric turbulence have been made using a combination of atmosphere- and surface-borne instruments and characterized according to the gradient Richardson number (Brooks et al., 2017). The Richardson number (Ri) is a dimensionless ratio commonly used to distinguish domains and transitions among dynamically unstable and stable conditions within the atmospheric surface layer. A critical Ri (Ri_c) is generally identified at $\sim$0.25, below which flow is turbulent and the atmosphere is dynamically unstable. The Ri at which turbulent flow fully transitions to the near-laminar flows of stable atmospheres (Ri_T) is generally identified at $\sim$1.0. In the data presented in Figure 12.5, the Ri profile of the uncoupled Arctic atmosphere shows frequent flows with greater stability that are initiated at $\sim$100 m and continue to develop up to $\sim$250 m, before trending back toward greater turbulence (Figure 12.5A). Relatively high convectively driven turbulence in the CML is clearly seen above $\sim$400 m, while more modest shear-driven turbulence is seen in the SML below $\sim$100 m. The Ri profile of the coupled Arctic atmosphere shows turbulent continuity between the SML and CML, with significant stability only appearing above the CML at 700–800 m (Figure 12.5B). During the 20-day, late-summer observation period depicted in Figure 12.5, the MABL existed in the uncoupled state $\sim$75% of the time. The presence of the inversion, and its relatively high frequency of formation, have important ramifications for understanding how

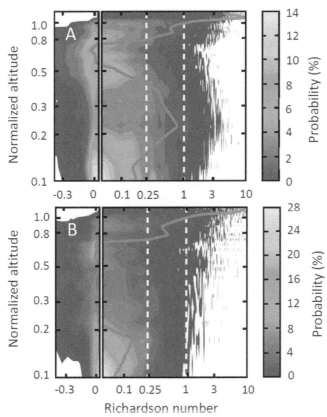

Figure 12.5 Normalized vertical profiles and probabilities (%) for the Richardson number (Ri) for all observations during the Arctic Summer Cloud-Ocean Study (ASCOS) in the summer of 2008. Time periods have been binned for those when the atmospheric surface mixed layer (SML) and cloud mixed layer (CML) were (A) decoupled or (B) coupled.
Redrawn from Brooks et al. (2017).

the Arctic stratocumulus layer is sustained. One of the primary questions concerns the source of CML water vapor in the absence of an SML connection. The relation of Arctic stratocumuli to surface evaporation must be fundamentally different than that for equatorial stratocumuli.

One observed state of the Arctic marine atmosphere that likely contributes to humidification of the CML in both coupled and uncoupled conditions is the frequent occurrence of specific humidity inversions (SHIs). The typical mid-to-equatorial atmosphere exhibits a vertical decrease in specific humidity with increasing altitude (Wood, 2012). In the Arctic region, SHIs, which exhibit altitude-dependent increases in humidity, have been observed during past research campaigns (Uttal et al., 2002; Tjernström et al., 2014). The summertime delivery of moisture to atmospheric layers above the CML is attributed to the advective transport of warmer, moist air from lower latitudes (Naakka et al., 2018; Sotiropoulou et al., 2018). The arrival of moist air above the CML provides a source for cloud humidification through cloud-top radiative cooling and downward turbulent entrainment (Solomon et al., 2014; Egerer et al., 2021). The progressive removal of moisture from cooled, lower portions of advected air masses, due to condensation and precipitation as they interact with cold, ice-covered surfaces, provides the vertical shape of the SHI.

12.4   Rectifier effects and their influence on ABL processes

In both ocean and terrestrial components of the earth system, interactions among processes produce rectifier-type covariances that complicate time-dependent averaging. Conventionally, a rectifier is an electronic circuit that converts an alternating current to a direct current; thereby, producing the unidirectional, unfluctuating flow of electrons required for many electronic applications. Rectifiers function as an electronic filter, permitting current to flow only under specified conditions. The imposition of those conditions causes an output flow that is not linearly proportional to the forced input flow. In the presence of rectification, the flow of electrons can only be accurately related to the original forcing function if the rectifier conditions are known and accommodated in the mathematical model that relates output to input.

Atmospheric and ocean rectifiers are similar to electronic rectifiers in that they are composed of process interactions that condition the directional flows of energy and mass and, as a result, obscure proportionality in the relation between cause and effect. In the case of the earth system, rectification occurs through the process of averaging – converting oscillating fluxes of mass or energy (e.g., between sources and sinks) into a single mean. The value of the mean cannot accurately reflect the contributions of its component fluxes and their dependencies on climate drivers without explicit resolution of the oscillations. The general nature of non-linear dependencies in the earth system, with appropriate analogies to electronic systems, is reviewed in Rial et al. (2004). The most well-studied Earth system rectifiers occur in the presence of diurnal or seasonal variations in solar energy inputs. Alternations between high and low solar energy states cause covariances between, for example: (1) patterns of sea-surface energy exchange and near-surface ocean mixing and (2) rates of CO_2 source (respiration) or sink (photosynthesis) exchange and convection in the lower atmosphere. In both cases, non-linearities between energy or mass fluxes and climate forcing elements must be reconciled before climate means can be used as accurate forcing drivers in models. In the following sections, I will discuss the issue of rectification within the context of two case studies: (1) cloud dynamics, especially in the equatorial trade wind regions and (2) determination of global patterns of CO_2 source and sink distribution through inverse modeling of the interhemispheric atmospheric CO_2 gradient.

12.4.1   Rectification of ocean–atmosphere coupling in the equatorial regions

The formation and persistence of shallow cumulus cloud banks in the equatorial MABL is dependent on processes, such as surface-atmosphere convection and turbulent mixing at cloud top, both of which occur at short timescales and vary according to solar cycles (Johnson et al., 1999; Vial et al., 2017). Due to high uncertainties in the mathematical representation of these processes, they are generally not resolved in ESM grids (Nuijens et al., 2015). Rather, cloudiness in most ESMs is determined by prescribed parameters representing mean climate states. The recognition and resolution of uncertainties in this approach have been slow to occur. However, several recent studies have revealed that avoidance of the short timescale processes has the potential to cause significant errors in model projections (e.g., Danabasoglu et al., 2006; Ham et al., 2010; Klingaman and Woolnough, 2014; Seo et al., 2014; Ruppert, 2016).

One area of focus has concerned the effect of SST fluctuations on the formation and propagation of large-scale climate oscillations, such as those associated with the MJO (Inness and Slingo, 2003). Recall that the active and suppressed phases of the MJO occur across scales of days to weeks. The eastward progression of the MJO, if represented at all in ESMs, is driven by relatively slow processes parameterized by the daily mean climate. Bernie et al. (2005) used a one-dimensional ocean mixed layer model to show that failure to account for diurnal (hourly) SST fluctuations resulted in a significant underestimation of projected intra-seasonal SST gradients, relative to the daily-mean approach. The modeled effect was caused by non-linear dynamics involving the diurnal covariance of near-surface thermal dynamics and ocean convection. During the day, heat accumulates near the surface, creating a positive (upward) buoyancy force with vertical stratification below the surface and relatively shallow mixing. At night, the surface cools due to radiative heat loss and sensible heat exchange, creating a negative (sinking) buoyancy force in the surface water that results in increased vertical mixing. Thus, at night, accumulated surface heat is distributed to greater depths. It is the covariance between heat accumulation at shallower depth during the day and heat loss to greater depths at the night that creates a non-linear relation between energy-budget forcings and water temperature (input vs output), and thus drives diurnal cycles in SST.

The suppressed phase of the MJO is associated with high rates of daytime solar heating and light winds, which enhance diurnal differences in SST. The accumulation of heat across successive days amplifies diurnal differences even further as the suppressed phase progresses. The active phase of the MJO is characterized by thick clouds and higher winds, which mute the accumulation of near-surface heat and cause deeper daytime ocean mixing. Thus, the influence of the diurnal cycle on SST dynamics exhibits positive covariance with the intra-seasonal gradient in SST – strong diurnal effects in the suppressed phase and weak diurnal effects in the active phase. This is the basis for rectification of the intra-seasonal SST gradient by the faster processes underlying the diurnal SST cycle.

A cloud-resolving model analysis of the Indo-Pacific (tropical) warm pool revealed the influences of the SST diurnal cycle on atmospheric processes (Ruppert and Johnson, 2016; Ruppert, 2016; Figure 12.6). Model results revealed that local suppressed and active diurnal phases are involved in shallow cumulus formation. Starting from an initial morning state of subsidence and convective suppression, the lower atmosphere will progressively warm and dry, causing existing clouds to thin. This is shown as a late-morning and early-afternoon period of relative stability in the

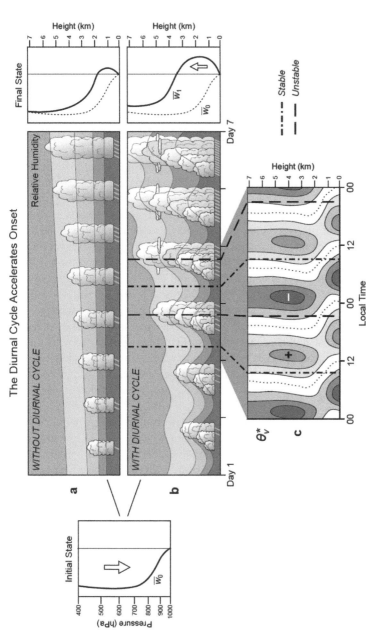

Figure 12.6 The model-simulated influence of a diurnal cycle on the progressive formation of low-altitude stratocumuli clouds above the Indo-Pacific warm pool for a 1-week period. The covariance between diurnal cycles in humidity and static stability yields greater divergence in mean vertical wind speed profiles from time 0 ($\overline{w_0}$) to 1 week ($\overline{w_1}$), and stronger daily-mean moist convection and diabatic heating for the case with a diurnal cycle. Stronger shallow heating for the case of a diurnal cycle, supports greater rising motion and column moistening, thereby accelerating the onset of deep convection, as shown in the lower figure for the diurnal-anomaly, vertical profile of virtual potential temperature (θ_v^*). From Ruppert (2016).

alignment of Figure 12.6B and C. As solar heating continues, local SST will reach a maximum in the late afternoon, accompanied by increased convective exchange and a shift to dynamic instability. In essence, the late afternoon shift in stability covaries with the period of highest atmospheric humidity, amplifying the potential for deep moisture transport. The increases in surface-to-atmosphere fluxes will lead to a late afternoon cloud moistening phase, which drives deeper cumulus development and drizzle. Evening cloud-top cooling would contribute to this process and extend cumulus buildup through the night. In the absence of a diurnal diabatic cycle, cloud development above the warm pool is considerably weaker with little rainfall (Figure 12.6A). Potential rectification of the diurnal SST cycle into the processes governing daily cloud dynamics also produces the potential amplification of cumulus development across timescales of days to weeks (Figure 12.6B). As the diurnal SST cycle is repeated across successive days, accumulated heat is progressively transferred downward to the near-surface, ocean mixed layer (Schneider and Müller, 1990; Weller and Anderson, 1996). As that heat cycles back and forth vertically during successive buoyancy cycles, it will complement solar heating at the surface. This progressive amplification of SST will, in turn, potentially amplify the late-afternoon cloud moistening phase.

In summary, cyclic coupling between SST and cloud processes in trade wind regions, especially at the diurnal scale, appear to have roles in causing some of the most important non-linearities and associated rectifier effects that connect the fast and slow scales of equatorial climate dynamics. Future breakthroughs and modifications in model logic, parameter estimation schemes, and time-step resolution will have to be made to adopt rectifier mechanisms in ESMs and accurately predict their influences on ocean-driven climate processes.

12.4.2 Rectification, inverse modeling, and the global carbon budget

Surface-to-atmosphere CO_2 fluxes are coupled to atmospheric CO_2 concentrations through the theoretical conditions of thermodynamic continuity and the laws of mass balance (Monson and Baldocchi, 2014). Briefly, continuity ensures that any divergence in mass flux between two connected reservoirs must be balanced by a proportional divergence in the mass concentrations composing the two reservoirs. Conditioned by this constraint, continuous monitoring of atmospheric CO_2 concentrations can be used to derive time-dependent divergences in surface CO_2 fluxes.

There are currently more than 150 global monitoring sites dedicated to collecting long-term records on tropospheric greenhouse gas concentrations (World Data Centre for Greenhouse Gases, https://gaw.kishou.go.jp/about_wdcgg/wdcgg). The longest running CO_2 record is from the Mauna Loa Observatory (MLO), which has produced an iconic trend of increasing northern hemisphere, tropospheric CO_2 concentration from 1958 to present (known as the Keeling Curve; Keeling, 1998). The MLO and other global CO_2 monitoring stations have been carefully located to avoid local (boundary layer) influences, and instead record CO_2 dynamics aloft in the well-mixed free troposphere. Since 1980, scientists at the US National Oceanic and Atmospheric Administration (NOAA) have complemented observations at fixed monitoring stations with ship sampling, providing mobile-platform ocean transects. The NOAA CO_2 sampling database provides evidence of an interhemispheric gradient in mean annual concentration (from the South to North Poles for the years 1981–87) of approximately +3 ppm (Figure 12.7; Tans et al., 1990). The gradient exists because interhemispheric mixing of tropospheric air is slow (1–1.5 years; Heimann and Keeling, 1986), relative to seasonal and annual dynamics in CO_2 source and sink activity. The observed gradient reflects quantitative insight into hemispheric-scale interactions between surface CO_2 fluxes and atmospheric transport. The key to extracting that insight is to deploy models to determine the most parsimonious combination of sources, sinks, and transport vectors that explain the CO_2 gradient.

Interhemispheric CO_2 Gradient   Seasonal Atmospheric Rectifier

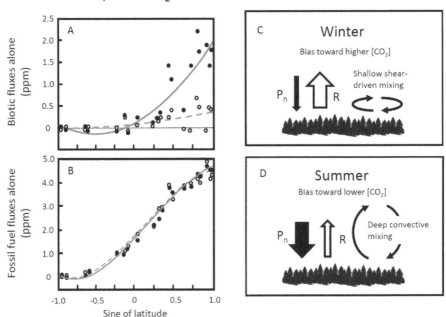

Figure 12.7 (A) The predicted interhemispheric (CO_2) gradient due to biotic CO_2 exchange (photosynthesis and respiration) determined from the Colorado State University General Circulation Model (GCM; *solid circles*), which includes an atmospheric rectifier effect, vs the Goddard Institute for Space Studies GCM (*open circles*), which does not include a rectifier effect. (B) The predicted (CO_2) gradient due to anthropogenic fossil fuel CO_2 fluxes determined by both models. Both (A) and (B) are rupp. (C) Rectifier effects illustrated as the covariance between biotic CO_2 exchanges and during shallow vertical mixing in winter when respiration (R) is higher than photosynthesis (P_n). (D) Rectifier effects illustrated for the case of deeper vertical mixing during summer.
Panel A: Redrawn from Denning et al. (1995).

Using a 3D tracer transport model and some a priori assumptions about source and sink strengths, Tans et al. (1990) partitioned the interhemispheric gradient into those fractions due to underlying processes. Approximately 0.25 ppm of the pole-to-pole difference in CO_2 was due to atmospheric transport patterns alone. The emission of CO_2 from fossil fuel combustion, which is considerably higher in the Northern Hemisphere, was predicted to contribute between 3.8 and 4.6 ppm to the total gradient, depending on assumptions about atmospheric chemistry. Tropical deforestation (a low-latitude net CO_2 source) together with terrestrial ecosystem CO_2 exchange was predicted to contribute another 0.6 ppm to the gradient. The ocean CO_2 sink, which is higher in the Southern Hemisphere compared to the Northern Hemisphere, was predicted to increase the gradient by an additional 0.7–2.3 ppm, depending on assumptions about ocean processes. Thus, considering all best-guess contributions, and building from the bottom up, the total pole-to-pole change in the interhemispheric CO_2 gradient was predicted to be between 5 and 8 ppm. This analysis revealed a gap of 2–6 ppm between the observed (see Figure 12.7) and predicted gradients. By deduction, these researchers concluded that a relatively large and unrecognized terrestrial CO_2 sink in the Northern Hemisphere is required to close the gap. The analytical approach used by Tans et al. (1990), whereby inferences about different combinations of sources and sinks are tested against an observed CO_2 gradient, is referred to as "inverse modeling." In inverse modeling, *causes* (sources and sinks) are inferred from *effects* (changes in CO_2 concentration), which is the opposite of forward modeling. Global inverse modeling is most often deployed with a 3D global transport model and applied to the annually averaged CO_2 gradient.

In its original deployment, the interhemispheric CO_2 gradient was most useful as a two-dimensional analytical tool. Three-dimensional perspectives entered the analysis only as a means to generate modeled transport vectors. Vertical gradients in CO_2 were not part of the model constraint. Because of this deficiency, there were no means available to assess covariances between surface fluxes and vertical mixing, which could have influenced estimated surface fluxes. This does not mean that the issue was unrecognized. For example, Tans et al. (1990, p. 1434) wrote: "The covariance of seasonal transport and seasonal CO_2 sources and sinks may lead to annually averaged concentration differences between different sites, both in the model and in the atmosphere, even in the absence of net annual sources: If transport is less vigorous during the season when a surface region is a source rather than when it is a sink, a positive net annual concentration anomaly will result." However, at the time of the Tans et al. (1990) analysis, there was limited availability of tracer observations and models that could be used to evaluate local or regional convective mixing processes. In the absence of observations, prescribed monthly mean convection statistics were used to condition atmospheric transport models.

In a later study, aimed at evaluating sub-grid covariance between surface fluxes and atmospheric convection, Denning et al. (1995) used a transport model that included hourly dynamics in vertical transport and boundary layer depth. They used all other model components and data constraints from the original Tans et al. (1990) study and repeated their analysis. Similar to the results of the original study, the contribution of fossil fuel combustion to the gradient was determined to be ~4.4 ppm

(Figure 12.7A), indicating that inclusion of the vertical mixing scheme had little effect on the anthropogenic component. However, the seasonal correlation between ecosystem (biotic) CO_2 exchange and vertical mixing contributed an additional $\sim$1.5–2 ppm to the gradient, which had been missing from the original analysis (Figure 12.7B). The larger gap between the modeled and observed gradients required a re-evaluation of global sources and sinks and specifically led to the proposal of an even larger CO_2 sink in Northern Hemisphere ecosystems and a larger CO_2 source in tropical ecosystems than had been previously proposed (Denning et al., 1999). The seasonal correlation between surface CO_2 exchange and vertical mixing depth occurs because in terrestrial ecosystems: (1) both photosynthesis (net sink) and strong convective mixing occur during the summer and (2) respiration (net source) and weak vertical mixing occur during the winter. Thus, seasonality in vertical mixing correlated with CO_2 exchange, even in the presence of balanced annual fluxes, will cause a positive anomaly in near-surface mean CO_2 concentrations. This is the terrestrial atmospheric rectifier effect (Figure 12.7C and D). Knowledge about the rectifier has significantly influenced our understanding of global flux-gradient relationships (Denning et al., 1999).

The terrestrial rectifier has been observed directly through diurnal and seasonal studies of ABL height and turbulent mixing using atmospheric tracers such as ^{222}Rn (Williams et al., 2011; Pal et al., 2015). Evidence has also been obtained from remotely sensed observations of column CO_2 concentrations in which mid-troposphere CO_2 concentrations diverge to levels lower or higher than those during the winter and summer, respectively (Strow and Hannon, 2008). Rectifier-type correlations between ABL depth, turbulent mixing, and surface CO_2/H_2O exchange have also been studied effectively from a tall tower ($\sim$400 m height) in Wisconsin (Yi et al., 2004; Denning et al., 2008).

Stephens et al. (2007) extended the CO_2 gradient observations from the global 2D framework used in Tans et al. (1990) to include local vertical CO_2 gradients that extend beyond the boundary layer. With this approach they showed that the addition of a vertical mixing constraint on model predictions has a large influence on inferred distributions of sources and sinks and the variation is driven by differences in the transport models. Some models vent too much CO_2 aloft during the summer, producing weak vertical gradients. During the winter, most models provide relatively accurate predictions of vertical CO_2 gradients, with some models predicting slightly weaker gradients and others predicting slightly stronger gradients compared to observations. Those models that predict stronger gradients (i.e., trap higher amounts of CO_2 near the surface compared to observations) lead to predictions of larger extratropical sinks and tropical sources. The Stephens et al. (2007) study, and others (e.g., Gurney et al., 2004; Gaubert et al., 2019) that have occurred since the Tans et al. (1990) analysis, clearly establish that the representation of boundary layer venting in models matters to the accurate prediction of near-surface CO_2 gradients and the inferred distribution of sources and sinks.

Currently, the field of global carbon cycle modeling is in crucial need of clarity, especially with regard to the relationship of tropical vs extratropical sinks and the influences of climate change and atmospheric CO_2 fertilization on that relationship (see Schimel et al., 2015). The question remains as to why past inverse modeling studies have had such challenges identifying source and sink distributions when inferred from observed CO_2 gradients. There are obvious areas for improved clarity, including increased sampling of atmospheric CO_2 concentrations at tropical latitudes, longerterm records to capture the high levels of decadal-scale variability in tropical forest NPP, and even better resolution of the shorter timeframe 3D transport processes in atmospheric models.

12.5   Concluding statement

The earth's climate system changes relatively slowly. The atmospheric and oceanic fluxes that transfer mass and energy within the earth system operate along gradients embedded in large biogeochemical reservoirs. The climatic timescales used to describe adjustments in the contents of those reservoirs are defined by weeks to decades. From that perspective, it is difficult to imagine processes restricted to the atmospheric boundary layer or near-surface marine layers that are capable of participating in climate changes. These dynamic layers are defined according to processes that operate at the scale of minutes to hours. Yet, several recent analyses of atmospheric and marine processes at these shorter timescales have been shown to significantly affect climate model predictions. The faster processes have the potential to impose non-linearity in the functions relating flux scalars to climate drivers, which in turn have the potential to amplify or mute the linear predictions constrained by slower processes within the current generation of ESMs. In mathematical terms, Earth system processes are defined by a property called Jensen's Inequality whereby the predicted transformation of mean values using a convex (non-linear) function is not necessarily equal to the mean obtained after transformation. In other words, a modeled output using mean inputs does not equal the mean of model outputs using individual inputs (see Ruel and Ayres, 1999). The physics of mechanisms that are hidden in the faster covariances and associated non-linearities of the earth system appear to represent important determinant forcings in ESM functions.

In this chapter, I've used our understanding of ocean and atmosphere feedbacks and rectifier effects to illustrate the confounding errors that can appear when mean attributes are used in climate projections. Despite a general recognition of the potential for model errors, we remain challenged by a lack of information on the exact scope and magnitude of the problem. The evidence to date shows that processes at timescales shorter than those normally considered in ESMs do indeed matter to the accurate prediction of climate. However, observations at the shorter timescales, which are required to evaluate non-linear feedbacks and rectifiers, remain sparse. As more data becomes available at the shorter timescales, opportunities will arise to deploy new inverse modeling approaches, including direct tests of model structure and parameter estimation schemes. The sensitivity analyses that have been conducted to date have revealed a theoretical justification for the "times to change." The enactment of that change will require the development and integration of an even closer relationship between data and models.

References

Andreae, M.O., Andreae, T.W., Meyerdierks, D., Thiel, C., 2003. Marine sulfur cycling and the atmospheric aerosol over the springtime North Atlantic. Chemosphere 52, 1321–1343.

Arakawa, A., 1975. Modelling clouds and cloud processes for use in climate models. In: The Physical Basis of Climate and Climate Modelling, GARP Publication Series. vol. 16. World Meteorological Organization (WMO); International Council of Scientific Unions, pp. 181–197.

Berner, A.H., Bretherton, C.S., Wood, R., 2015. Large eddy simulation of ship tracks in the collapsed marine boundary layer: a case study from the Monterey area ship track experiment. Atmos. Chem. Phys. 15, 5851–5871.

Bernie, D., Woolnough, S., Slingo, J., Guilyardi, E., 2005. Modeling diurnal and intra-seasonal variability of the ocean mixed layer. J. Clim. 18, 1190–1202.

Bony, S., Dufresne, J.L., 2005. Marine boundary layer clouds at the heart of tropical cloud feedback uncertainties in climate models. Geophys. Res. Lett. 32. https://doi.org/10.1029/2005GL023851, L20806.

Brooks, I.M., Tjerstrom, M., Persson, P.O.G., Shupe, M.D., Atkinson, R.A., Canut, G., Birch, C.E., Mauritsen, T., Sedlar, J., Brooks, B.J., 2017. The turbulent structure of the Arctic summer boundary layer during the Arctic summer cloud-ocean study. J. Geophys. Res. Atmos. 122, 9685–9704.

Brune, S., Baehr, J., 2020. Preserving the coupled atmosphere–ocean feedback in initializations of decadal climate predictions. WIREs Clim. Change. https://doi.org/10.1002/wcc.637, e637.

Cahalan, R.F., Ridgway, W., Wiscombe, W.J., Bell, T.L., Snider, J.B., 1994. The albedo of fractal stratocumulus clouds. J. Atmos. Sci. 51, 2434–2455.

Carlson, T.N., Prospero, J.M., 1972. Vertical and areal distribution of Saharan dust over western equatorial North-Atlantic Ocean. J. Geophys. Res. 77, 5255–5265.

Cess, R.D., Potter, G.L., Blanchet, J.P., Boer, G.J., Delginio, A.D., Deque, M., et al., 1990. Intercomparison and interpretation of climate feedback processes in 19 atmospheric GCMs. J. Geophys. Res. 95, 16601–16615.

Christensen, M.W., Jones, W.K., Stier, P., 2020. Aerosols enhance cloud lifetimes and brightness along the stratus to cumulus transition. Proc. Natl. Acad. Sci. U. S. A. 117, 17591–17598.

Curry, J.A., Schramm, J.L., Ebert, E.E., 1993. Impact of clouds on the surface radiation balance of the Arctic Ocean. Meteorog. Atmos. Phys. 51, 197–217.

Danabasoglu, G., Large, W.G., Tribbia, J.J., Gent, P.R., Briegleb, B.P., McWilliams, J.C., 2006. Diurnal coupling in the tropical oceans of CCSM3. J. Clim. 19, 2347–2365.

De Deckker, P., 2016. The indo-Pacific warm pool: critical to world oceanography and world climate. Geosci. Lett. 3, 20. https://doi.org/10.1186/s40562-016-0054-3.

de Leeuw, G., Guieu, C., Arneth, A., Bellouin, N., Bopp, L., Boyd, P.W., et al., 2014. Ocean-atmosphere interactions of particles. In: Liss, P.S., Johnson, M.T. (Eds.), Ocean-Atmosphere Interactions of Gases and Particles. MT Book Series. Springer Earth System Sciences, pp. 171–246, https://doi.org/10.1007/978-3-642-25643-1_4.

Denning, A.S., Fung, I.Y., Randall, D., 1995. Latitudinal gradient of atmospheric CO_2 due to seasonal exchange with land biota. Nature 376, 240–243.

Denning, A.S., Takahashi, T., Friedlingstein, P., 1999. Can a strong atmospheric CO_2 rectifier effect be reconciled with a "reasonable" carbon budget? Tellus Ser. B Chem. Phys. Meteorol. 51, 249–253.

Denning, A.S., Zhang, N., Yi, C., Branson, M., Davis, K., Kleist, J., Bakwin, P., 2008. Evaluation of modeled atmospheric boundary layer depth at the WLEF tower. Agric. For. Meteorol. 148, 206–215.

Egerer, U., Ehrlich, A., Gottschalk, M., Griesche, H., Neggars, R.A.J., Siebert, H., Wendisch, M., 2021. Case study of a humidity layer above Arctic stratocumulus and potential turbulent coupling with the cloud top. Atmos. Chem. Phys. 21, 6347–6364.

Feingold, G., Cotton, W., Stevens, B., Frisch, A.S., 1996. The relationship between drop in-cloud residence time and drizzle production in numerically simulated stratocumulus clouds. J. Atmos. Sci. 53, 1108–1122.

Feng, Z., Hagos, S., Rowe, A.K., Burleyson, C.D., Martini, M.N., de Szoek, S.P., 2015. Mechanisms of convective cloud organization by cold pools over tropical warm ocean during the AMIE/DYNAMO field campaign. J. Adv. Model. Earth Syst. 7, 357–381.

Flatau, M., Flatau, P.J., Phoebus, P., Niiler, P.P., 1997. The feedback between equatorial convection and local radiative and evaporative processes: the implications for intra-seasonal oscillations. J. Atmos. Sci. 54, 2374–2385.

Friedlingstein, P., Jones, M.W., O'Sullivan, M., Andrew, R.M., Hauck, J., Peters, G.P., et al., 2019. Global carbon budget 2019. Earth Syst. Sci. Data 11, 1783–1838.

Gaubert, B., Stephens, B.B., Basu, S., Chevallier, F., Deng, F., Kort, E.A., Patra, P.K., Peters, W., Rodenbeck, C., Saeki, T., et al., 2019. Global atmospheric CO_2 inverse models converging on neutral tropical land exchange, but disagreeing on fossil fuel and atmospheric growth rate. Biogeosciences 16, 117–134.

Gottschalck, J., 2014. What is the MJO, and why do we care? Climate.gov (US National Oceanic and Atmospheric Administration) https://www.climate.gov/.

Gurney, K.R., Law, R.M., Denning, A.S., Rayner, P.J., Pak, B.C., Baker, D., et al., 2004. Transcom 3 inversion intercomparison: model mean results for the estimation of seasonal carbon sources and sinks. Glob. Biogeochem. Cycles 18. https://doi.org/10.1029/2003GB002111, GB1010.

Gutleben, M., Groß, S., Wirth, M., 2019. Cloud macro-physical properties in Saharan-dust-laden and dust-free North Atlantic trade wind regimes: a lidar case study. Atmos. Chem. Phys. 19, 10659–10673.

Ham, Y.G., Kug, J.S., Kang, I.S., Jin, F.F., Timmermann, A., 2010. Impact of diurnal atmosphere-ocean coupling on tropical climate simulations using a coupled GCM. Clim. Dyn. 34, 905–917.

Hartmann, D.L., Ockert-Bell, M.E., Michelsen, M.L., 1992. The effect of cloud type on earth's energy balance: global analysis. J. Clim. 5, 1281–1304.

Heimann, M., Keeling, C.D., 1986. Meridional eddy diffusion model of the transport of atmospheric carbon dioxide. Part 1, seasonal carbon cycle over the tropical Pacific Ocean. J. Geophys. Res. 91, 7765–7781.

Hobbs, P.V., Garrett, T.J., Ferek, R.J., Strader, S.R., Hegg, D.A., Frick, G.M., et al., 2000. Emissions from ships with respect to their effects on clouds. J. Atmos. Sci. 57, 2570–2590.

Inness, P.M., Slingo, J.M., 2003. Simulation of the Madden-Julian oscillation in a coupled general circulation model. Part I: comparison with observations and an atmosphere-only GCM. J. Clim. 16, 345–364.

IPCC, 2013. Detection and attribution of climate change: from global to regional. In: Stocker, T. F., Qin, G.-K., Plattner, M., et al. (Eds.), Climate Change 2013. The Physical Science Basis. Contribution of Working Group I to the Fifth Assessment Report of the Intergovernmental Panel on Climate Change. Cambridge University Press, Cambridge (86 pp., Chapter 10).

IPCC, 2021. Summary for policymakers. In: Masson-Delmotte, V., Zhai, P., Pirani, A., Connors, S.L., Péan, C., Berger, S., Caud, N., Chen, Y., Goldfarb, L., Gomis, M.I., Huang, M., Leitzell, K., Lonnoy, E., Matthews, J.B.R., Maycock, T.K., Waterfield, T., Yelekçi, O., Yu, R., Zhou, B. (Eds.), Climate Change 2021: The Physical Science Basis. Contribution of Working Group I to the Sixth Assessment Report of the Intergovernmental Panel on Climate Change. Cambridge University Press, Cambridge, United Kingdom and New York, NY, USA, pp. 3–32, https://doi.org/10.1017/9781009157896.001.

Johnson, R.H., Rickenbach, T.M., Rutledge, S.A., Ciesielski, P.E., Schubert, W.H., 1999. Trimodal characteristics of tropical convection. J. Clim. 12, 2397–2418.

Kawamiya, M., Hajima, T., Tachiiri, K., Watanabe, S., Yokohata, T., 2020. Two decades of earth system modeling with an emphasis on model for interdisciplinary research on climate (MIROC). Prog. Earth Planet Sci. 7, 64. https://doi.org/10.1186/s40645-020-00369-5.

Keeling, C.D., 1998. Rewards and penalties of monitoring the earth. Annu. Rev. Energy Environ. 23, 25–82.

Klein, S.A., 1997. Synoptic variability of low-cloud properties and meteorological parameters in the subtropical trade wind boundary layer. J. Clim. 10, 2018–2039.

Klingaman, N.P., Woolnough, S.J., 2014. The role of air-sea coupling in the simulation of the madden-Julian oscillation in the Hadley Centre model. Q. J. R. Meteorol. Soc. 140, 2272–2286.

Latham, J., Rasch, P., Chen, C.-C., Kettles, L., Gadian, A., Gettelman, A., Morrison, H., Bower, K., Choularton, T., 2008. Global temperature stabilization via controlled albedo enhancement of low-level maritime clouds. Philos. Trans. R. Soc. Lond. 366, 3969–3987.

Liao, H., Seinfeld, J.H., Adams, P.J., Mickley, L.J., 2004. Global radiative forcing of coupled tropospheric ozone and aerosols in a unified general circulation model. J. Geophys. Res. Atmos. 109, D16207.

Lu, M.-L., Seinfeld, J.H., 2006. Effect of aerosol number concentration on cloud droplet dispersion: a large-eddy simulation study and implications for aerosol indirect forcing. J. Geophys. Res. 111, D02207. https://doi.org/10.1029/2005JD006419.

Martin, G.M., Johnson, D.W., Spice, A., 1994. The measurement and parameterization of effective radius of droplets in warm stratocumulus clouds. J. Atmos. Sci. 51, 1823–1842.

McFarquhar, G.M., Ghan, S., Verlinde, J., Korolev, A., Strapp, J.W., Schmid, B., Tomlinson, J. M., Wolde, M., Brooks, S.D., Cziczo, D., et al., 2011. Indirect and semi-direct aerosol campaign (ISDAC): the impact of Arctic aerosols on clouds. Bull. Am. Meteorol. Soc. 92, 183–201.

Medeiros, B., Stevens, B., Held, I.M., Zhao, M., Williamson, D.L., Olson, J.G., Bretherton, C. S., 2008. Aquaplanets, climate sensitivity, and low clouds. J. Clim. 21, 4974–4991.

Meehl, G.A., Goddard, L., Boer, G., Burgman, R., Branstator, G., Cassou, C., et al., 2014. Decadal climate prediction: an update from the trenches. Bull. Am. Meteorol. Soc. 95, 243–267.

Meehl, G.A., Richter, J.H., Teng, H.Y., Capotondi, A., Cobb, K., Doblas-Reyes, F., Donat, M. G., England, M.H., Fyfe, J.C., Han, W.Q., et al., 2021. Initialized earth system prediction from subseasonal to decadal timescales. Nat. Rev. Earth Environ. 2, 340–357.

Monson, R.K., Baldocchi, D.D., 2014. Terrestrial Biosphere-Atmosphere Fluxes. Cambridge University Press, Cambridge. 487 pp., ISBN-13: 978-1107040656.

Morrison, H., de Boer, G., Feingold, G., Harrington, J., Shupe, M.D., Sulia, K., 2012. Resilience of persistent Arctic mixed-phase clouds. Nat. Geosci. 5, 11–17.

Naakka, T., Nygård, T., Vihma, T., 2018. Arctic humidity inversions: climatology and processes. J. Clim. 31, 3765–3787.

Neggers, R.A., Neelin, J.D., Stevens, B., 2007. Impact mechanisms of shallow cumulus convection on tropical climate dynamics. J. Clim. 20, 2623–2642.

Neukom, R., Stieger, N., Gomez-Navarro, J.J., Wang, J., Werner, J.P., 2019. No evidence for globally coherent warm and cold periods over the preindustrial common era. Nature 571, 550–554.

Nuijens, L., Medeiros, B., Sandu, I., Ahlgrimm, M., 2015. The behavior of trade-wind cloudiness in observations and models: the major cloud components and their variability. J. Adv. Model. Earth Syst. 7, 600–616.

Pages 2k Consortium, 2019. Consistent multi-decadal variability in global temperature reconstructions and simulations over the common era. Nat. Geosci. 12, 643–649.

Pal, S., Lopez, M., Schmidt, M., Ramonet, M., Gibert, F., Xueref-Remy, I., Ciais, P., 2015. Investigation of the atmospheric boundary layer depth variability and its impact on the ^{222}Rn concentration at a rural site in France. J. Geophys. Res. Atmos. 120, 623–643.

Pierce, J.R., Adams, P.J., 2006. Global evaluation of CCN formation by direct emission of sea salt and growth of ultrafine sea salt. J. Geophys. Res. 111. https://doi.org/10.1029/2005JD006186, D06203.

Pincus, R., Baker, M.B., 1994. Effect of precipitation on the albedo susceptibility of clouds in the marine boundary layer. Nature 372, 250–252.

Possner, A., Wang, H.L., Wood, R., Caldeira, K., Ackerman, T.P., 2018. The efficacy of aerosol-cloud radiative perturbations from near-surface emissions in deep open-cell stratocumuli. Atmos. Chem. Phys. 18, 17475–17488.

Rial, J.A., Pielke Sr., R.A., Beniston, M., Claussen, M., Canadell, J., Cox, P., et al., 2004. Non-linearities, feedbacks and critical thresholds within the Earth's climate system. Clim. Chang. 65, 11–38.

Riley, E.M., Mapes, B.E., Tulich, S.N., 2011. Clouds associated with the Madden-Julian oscillation: a new perspective from CloudSat. J. Atmos. Sci. 68, 3032–3051.

Ruel, J.J., Ayres, M.P., 1999. Jensen's inequality predicts effects of environmental variation. Trends Ecol. Evol. 14, 361–366.

Ruppert, J.H., 2016. Diurnal timescale feedbacks in the tropical cumulus regime. J. Adv. Model. Earth Syst. 8, 1483–1500.

Ruppert, J.H., Johnson, R.H., 2016. On the cumulus diurnal cycle over the tropical warm pool. J. Adv. Model. Earth Syst. 8, 669–690.

Salter, S., Sortino, G., Latham, J., 2008. Sea-going hardware for the cloud albedo method of reversing global warming. Philos. Trans. R. Soc. A366, 3989–4006.

Schimel, D., Stephens, B.B., Fisher, J.B., 2015. Effect of increasing CO_2 on the terrestrial carbon cycle. Proc. Natl. Acad. Sci. U. S. A. 112, 436–441.

Schneider, N., Müller, P., 1990. The meridional and seasonal structures of the mixed layer depth and its diurnal amplitude observed during the Hawaii-to-Tahiti shuttle experiment. J. Phys. Oceanogr. 20, 1395–1404.

Seo, H., Subramanian, A.C., Miller, A.J., Cavanaugh, N.R., 2014. Coupled impacts of the diurnal cycle of sea surface temperature on the Madden–Julian oscillation. J. Clim. 27, 8422–8443.

Serreze, M.C., Barry, R.G., 2011. Processes and impacts of Arctic amplification: a research synthesis. Glob. Planet. Chang. 77, 85–96.

Shupe, M.D., Persson, P.O.G., Brooks, I.M., Tjernström, M., Sedlar, J., Mauritsen, T., et al., 2013. Cloud and boundary layer interactions over the Arctic sea-ice in late summer. Atmos. Chem. Phys. 13, 9379–9400.

Slingo, A., Brown, R., Wrench, C.L., 1982. A field study of nocturnal stratocumulus; III. High resolution radiative and microphysical observations. Q. J. R. Meteorol. Soc. 108, 145–165. https://doi.org/10.1002/qj.49710845509.

Solomon, A., Shupe, M.D., Persson, O., Morrison, H., Yamaguchi, T., Caldwell, P.M., Boer, G. D., 2014. The sensitivity of springtime Arctic mixed-phase stratocumulus clouds to surface-layer and cloud-top inversion-layer moisture sources. J. Atmos. Sci. 71, 574–595.

Sotiropoulou, G., Tjernström, M., Savre, J., Ekman, A.M.L., Hartung, K., Sedlar, J., 2018. Large-eddy simulation of a warm-air advection episode in the summer Arctic. Q. J. R. Meteorol. Soc. 144, 2449–2462.

Steffen, W., Grinevald, J., Crutzen, J., McNeill, J., 2011. The Anthropocene: conceptual and historical perspectives. Philos. Trans. R. Soc. A369, 842–867.

Stephens, B.B., Gurney, K.R., Tans, P.P., Sweeney, C., Peters, W., Bruhwiler, L., et al., 2007. Weak northern and strong tropical land carbon uptake from vertical profiles of atmospheric CO_2. Science 316, 1732–1735.

Stevens, B., 2005. Atmospheric moist convection. Annu. Rev. Earth Planet. Sci. 33, 605–643.

Strow, L.L., Hannon, S.E., 2008. A 4-year zonal climatology of lower tropospheric CO_2 derived from ocean-only atmospheric infrared sounder observations. J. Geophys. Res. Atmos. 113. https://doi.org/10.1029/2007JD009713, D18302.

Tans, P.P., Fung, I.Y., Takahashi, T., 1990. Observational constraints on the global atmospheric CO_2 budget. Science 247, 1431–1438.

Tjernström, M., Leck, C., Birch, C.E., Bottenheim, J.W., Brooks, B.J., Brooks, I.M., Bäcklin, L., Chang, R.Y.-W., de Leeuw, G., Di Liberto, L., et al., 2014. The Arctic Summer Cloud Ocean Study (ASCOS): overview and experimental design. Atmos. Chem. Phys. 14, 2823–2869.

Toll, V., Christensen, M., Quaas, J., Bellouin, N., 2019. Weak average liquid-cloud-water response to anthropogenic aerosols. Nature 572, 51–55.

Twomey, S., 1974. Pollution and the planetary albedo. Atmos. Environ. 8, 1251–1256.

Uttal, T., Curry, J.A., McPhee, M.G., Perovich, D.K., Moritz, R.E., Maslanik, J.A., Guest, P.S., Stern, H.L., Moore, J.A., Turenne, R., Heiberg, A., Serreze, M.C., et al., 2002. Surface heat budget of the Arctic Ocean. Bull. Am. Meteorol. Soc. 83, 255–275.

Vial, J., Bony, S., Stevens, B., Vogel, R., 2017. Mechanisms and model diversity of trade-wind shallow cumulus cloud feedbacks: a review. Surv. Geophys. 38, 1331–1353.

Vial, J., Vogel, R., Bony, S., Stevens, B., Winkers, D.M., Cai, X., et al., 2019. A new look at the daily cycle of trade wind cumuli. J. Adv. Model. Earth Syst. 11, 3148–3166.

Webb, M.J., Senior, C., Sexton, D., Ingram, W., Williams, K., Ringer, M., et al., 2006. On the contribution of local feedback mechanisms to the range of climate sensitivity in two GCM ensembles. Clim. Dyn. 27, 17–38.

Weller, R.A., Anderson, S.P., 1996. Surface meteorology and air–sea fluxes in the western equatorial pacific warm pool during the TOGA coupled ocean atmosphere response experiment. J. Clim. 9, 1959–1990.

Williams, A.G., Zahorowski, W., Chambers, S., Griffiths, A., Hacker, J.M., Element, A., Werczynski, S., 2011. The vertical distribution of radon in clear and cloudy daytime terrestrial boundary layers. J. Atmos. Sci. 68, 155–174.

Wood, R., 2012. Stratocumulus clouds. Mon. Weather Rev. 140, 2373–2423.

Yi, C., Davis, K.J., Bakwin, P.S., Denning, A.S., Zhang, N., Desai, A., Lin, J.C., Gerbig, C., 2004. Observed covariance between ecosystem carbon exchange and atmospheric boundary layer dynamics at a site in northern Wisconsin. J. Geophys. Res. 109. https://doi.org/10.1029/2003JD004164, D08302.

Zeng, X., Brunke, M.A., Zhou, M., Fairall, C., Bond, N.A., Lenschow, D.H., 2004. Marine atmospheric boundary layer height over the eastern Pacific: data analysis and model evaluation. J. Clim. 17, 4159–4170.

Zhang, C., Adames, Á.F., Khouider, B., Wang, B., Yang, D., 2020. Four theories of the Madden-Julian oscillation. Rev. Geophys. 58. https://doi.org/10.1029/2019RG000685, e2019RG000685.

Index